A Voyage to South America

Describing at Large the Spanish Cities, Towns, Provinces, &c. on that Extensive Continent

VOLUME 1

ANTONIO DE ULLOA
EDITED AND TRANSLATED BY
JOHN ADAMS

CAMBRIDGE
UNIVERSITY PRESS

CAMBRIDGE UNIVERSITY PRESS

Cambridge, New York, Melbourne, Madrid, Cape Town,
Singapore, São Paolo, Delhi, Tokyo, Mexico City

Published in the United States of America by Cambridge University Press, New York

www.cambridge.org
Information on this title: www.cambridge.org/9781108031684

© in this compilation Cambridge University Press 2011

This edition first published 1806
This digitally printed version 2011

ISBN 978-1-108-03168-4 Paperback

CAMBRIDGE LIBRARY COLLECTION

Books of enduring scholarly value

Travel and Exploration

The history of travel writing dates back to the Bible, Caesar, the Vikings and the Crusaders, and its many themes include war, trade, science and recreation. Explorers from Columbus to Cook charted lands not previously visited by Western travellers, and were followed by merchants, missionaries, and colonists, who wrote accounts of their experiences. The development of steam power in the nineteenth century provided opportunities for increasing numbers of 'ordinary' people to travel further, more economically, and more safely, and resulted in great enthusiasm for travel writing among the reading public. Works included in this series range from first-hand descriptions of previously unrecorded places, to literary accounts of the strange habits of foreigners, to examples of the burgeoning numbers of guidebooks produced to satisfy the needs of a new kind of traveller - the tourist.

A Voyage to South America

Antonio de Ulloa (1716–95) was a Spanish scientist and mathematician. In 1734 he was asked by Philip V of Spain to join the French geodesic expedition to measure the circumference of the Earth at the equator, and accordingly in 1735 Ulloa and his fellow scientist Jorge Juan y Santacilia (1713–73) travelled to South America, staying until 1744. These two volumes contain the English translation of Ulloa's account of South America, first published in 1758. The work was very popular, producing five subsequent editions: this reissue is of the fourth edition of 1806. It provides valuable insights into the social, religious and economic institutions of colonial South America. Volume I contains detailed descriptions of the cities of Carthagena, Panama and Quito and their provinces, including historical, economic and geographical accounts of the cities, together with an ethnological discussion of indigenous people of Quito.

Cambridge University Press has long been a pioneer in the reissuing of out-of-print titles from its own backlist, producing digital reprints of books that are still sought after by scholars and students but could not be reprinted economically using traditional technology. The Cambridge Library Collection extends this activity to a wider range of books which are still of importance to researchers and professionals, either for the source material they contain, or as landmarks in the history of their academic discipline.

Drawing from the world-renowned collections in the Cambridge University Library, and guided by the advice of experts in each subject area, Cambridge University Press is using state-of-the-art scanning machines in its own Printing House to capture the content of each book selected for inclusion. The files are processed to give a consistently clear, crisp image, and the books finished to the high quality standard for which the Press is recognised around the world. The latest print-on-demand technology ensures that the books will remain available indefinitely, and that orders for single or multiple copies can quickly be supplied.

The Cambridge Library Collection will bring back to life books of enduring scholarly value (including out-of-copyright works originally issued by other publishers) across a wide range of disciplines in the humanities and social sciences and in science and technology.

Pl. III. for the Title Vol. I.

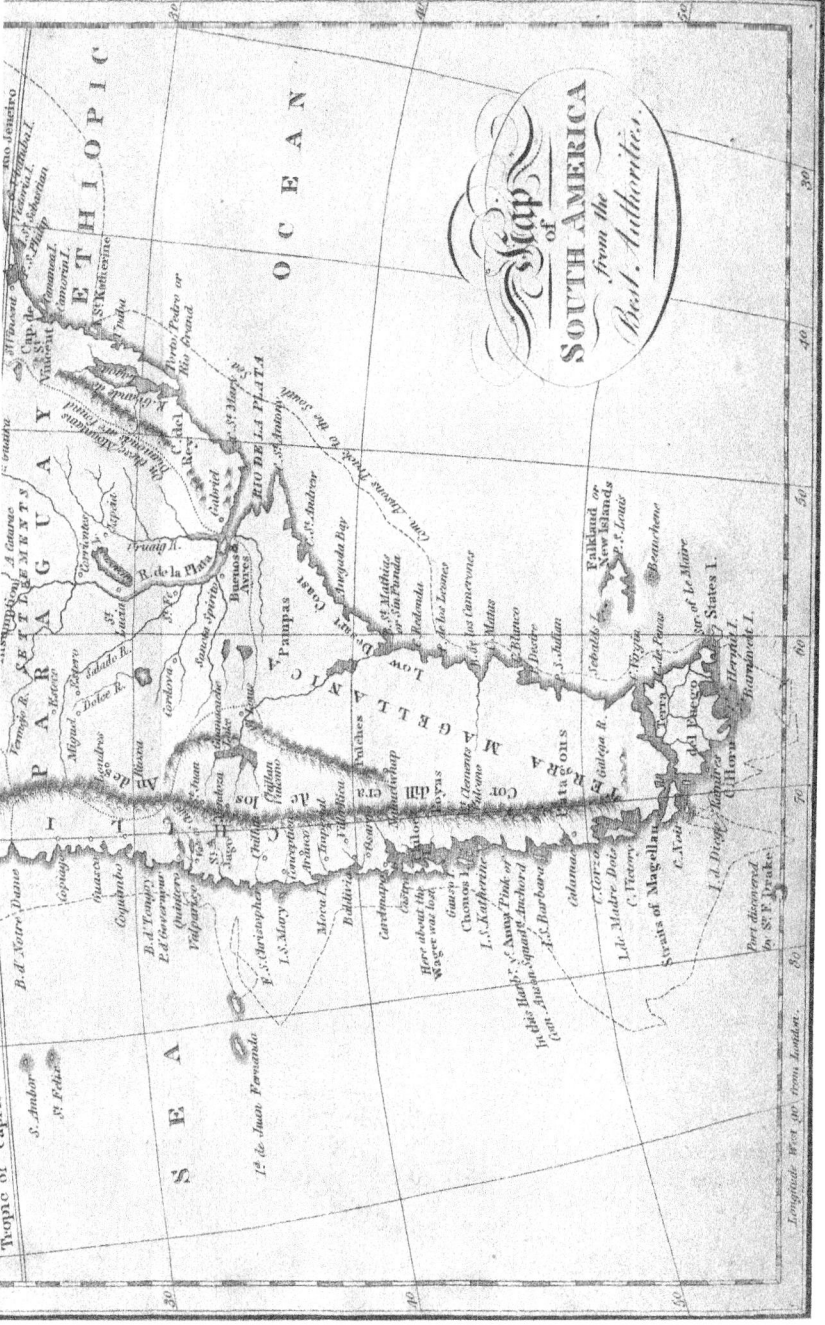

A

VOYAGE

TO

SOUTH AMERICA:

DESCRIBING AT LARGE

THE SPANISH CITIES, TOWNS, PROVINCES, &c.

ON THAT

EXTENSIVE CONTINENT:

UNDERTAKEN, BY COMMAND OF THE KING OF SPAIN,

BY

DON GEORGE JUAN,

AND

DON ANTONIO DE ULLOA,

CAPTAINS OF THE SPANISH NAVY,

FELLOWS OF THE ROYAL SOCIETY OF LONDON,
MEMBERS OF THE ROYAL ACADEMY AT PARIS, &c. &c.

TRANSLATED FROM THE ORIGINAL SPANISH;

WITH

NOTES AND OBSERVATIONS; AND AN ACCOUNT OF THE
BRAZILS.

By JOHN ADAMS, Esq. of Waltham Abbey;
Who resided several Years in those Parts.

THE FOURTH EDITION.
ILLUSTRATED WITH PLATES.

VOL. I.

LONDON:
PRINTED FOR JOHN STOCKDALE, PICCADILLY; R. FAULDER, BOND-
STREET; LONGMAN AND CO. PATER-NOSTER ROW; LACKINGTON
AND CO. FINSBURY SQUARE; AND J. HARDING, ST. JAMES'S STREET.

1806.

J. Brettell, Printer, Marshall-Street, Golden-Square.

TO

COMMODORE SIR HOME POPHAM, K<small>NT</small>.
&c. &c. &c.

S<small>IR</small>;

W<small>ITH</small> a Pride which, I venture to assert, I feel, only in common with every other British Subject, I take the liberty to dedicate to you this Republication.

I<small>T</small> contains, with other interesting matter, a slight sketch of Buenos-Ayres; that Territory which now is, and will most probably remain, from your Foresight, Ability, and personal Bravery, one of the richest Jewels in the United Crown.

I <small>AM</small> proud, Sir, in thus publicly offering my feeble tribute to your extraordinary Merits, to assure you of the respectful Esteem, high Regard, and sincere Friendship, with which

I have the honour to subscribe myself

Your ever obliged,

Obedient and humble Servant,

JOHN STOCKDALE.

London,
Sept. 26, 1806

a 2

PREFACE.

IT is certainly a very true, as well as trite observation, that knowledge is the food of the mind;
and if this be so, then certainly that ought to have
the preference, which is at once equally nutritive and
pleasant. On this account, books of voyages and travels have been in such general esteem, and at the same
time have been commended by persons of the greatest
sagacity, and in the highest reputation for superior
understanding. The pleasantness of this kind of
reading has attracted many, who had before no
relish for learning, and brought them by degrees to
enter upon severer enquiries, in order more effectually
to gratify that curiosity which this kind of study naturally excites. Men of higher abilities have turned
their thoughts on this subject, from the consideration
of its real utility. This induced the ingenious Hakluyt
to make that noble collection, which procured him the
patronage of queen Elizabeth's ablest minister. This
led the elder Thevenot, to enrich the French language
with a very copious collection of the same kind. And,
not to multiply examples, this made voyages and travels the favourite study of the judicious Locke, who
looked upon it as the best method of acquiring those
useful and practical lights, that serve most effectually
to strengthen and also to enlarge the human understanding.

IT

I⊤ is indeed true, that in respect to this, as well as other branches of science, there have been many productions, which for a time were applauded and admired, and which, notwithstanding, have served rather to mislead, than to instruct men's minds, by a display of specious falsehoods, highly acceptable to such as read merely for amusement. But these authors of marvellous, and very often incredible relations; of strange and surprizing adventures; these pompous describers of wonderful curiosities, which men of more penetration, but of sounder judgements, could never afterwards, though pursuing the same routes with their utmost diligence, discover; quickly lost that credit, which novelty alone gave them; and, being once exploded by sensible judges, gradually sunk, first into the contempt, and then into the oblivion, they deserved.

THESE books, however, are thus far useful, that they serve to give us a clearer idea of our wants, and a more just notion, than perhaps we could otherwise obtain, of the qualities requisite to render voyages and travels truly worthy of esteem. They demonstrate very fully, that, in the first place, it is of great consequence, to know the characters of the authors we peruse, that we may judge of the credit that is due to their reports; and this as well in point of abilities, as of veracity; for many writers impose on the world, not through any evil intention of deceiving others, but because they have been deceived themselves. They relate falsehoods, but they believe them: we cannot therefore justly accuse them of want of candour; the fault properly to be laid to their charge, is credulity. We are most in danger of suffering by those authors, who have either lived in, or passed through countries, that are rarely visited, and into which few are permitted to come. This protects their mistakes

         for

for a great length of time ; and we know that pre-
scription is a fortress in which error often holds
out a long siege. There cannot therefore be a more
acceptable tribute offered to the republic of letters,
than voyages or travels, composed by persons of established reputation for learning, and in equal esteem
for their probity. But the value of the present is
much enhanced, if these voyages or travels respect
countries little known, the coasts of which only,
perhaps, have been accidentally visited by seamen,
or harassed and plundered by privateers, persons often
of suspected faith, and almost always of very limited
capacities. Some few exceptions indeed there may be
to this general rule ; but even in regard to these, there
will be necessarily great defects ; and allowing them
their highest merit, they can only report truly the little they have seen : and what idea can we form of a
Turkey carpet, if we look only at the border, or, it
may be, at the selvage ?

THE authors, whose writings are now offered to the
public in an English dress, are men of the most respectable characters, men distinguished for their parts and
learning, and yet more for their candour and integri-
ty : men who did not travel through accident, but
by choice ; and this not barely their own, but approved by authority, and approved because they were
known to be equal to the task they undertook ; and
that task was, the examining every thing they went
to see, with all possible care and scrupulous attention,
in order to furnish the public with such lights as
might be entirely and safely depended on. This was
the design which they undertook : this design they
executed with the circumspection it deserved ; and
the punctuality with which they discharged it, has
procured them the just returns of favour from their
royal master, and the applause and approbation of
a 4   the

the best judges in their own and other countries.
These are circumstances that distinguish, in a very
singular degree, the following work; circumstances
that, no doubt,. will have their proper weight, and
which it would be entirely needless for us to enforce,
though it would have been inexcusable not to have
mentioned them.

THERE is however one other circumstance that deserves particular notice, which is, that, with respect to
the characters and abilities of these gentlemen, all does
not rest either upon their writings or foreign autho-
rities: they have been in this country; were seen and
known by those who were best able to judge of their
merit ; and, in consequence of that, are both of them
members of our royal society.   They continued here
some time,   conversed indifferently with all sorts of
people, and were unanimously allowed to have very
extensive views in respect to science ; great sagacity;
much application; were very assiduous and very accurate in their enquiries, as well as candid and commu-
nicative in relation to the discoveries and observations
which they had made in their travels.—Men of such talents, and such dispositions, must render themselves
agreeable every where; much more in a country of liberty, and where, without partiality, we may have leave
to say, the sciences are as deeply rooted, and flourish
in as high a degree, as in any other in Europe.   We
will add, that, from a knowledge of their merit and
candour, they not only received the greatest civilities,
but the most seasonable protection, to which, in some
measure, the world is indebted for this very performance, as the reader will learn in the perusal of it; accompanied with those marks of gratitude and respect,
which were due to their kind benefactors, more especially the late worthy president of the royal society,
whose memory is justly dear to all who had the honour
of

of being in the least acquainted with him*; and that
humane and polite patron of every useful branch of literature, Earl Stanhope; whose noble qualities reflect
honour on his titles, and who inherits the virtues of
his illustrious father, one of the bravest men, and
one of the most disinterested ministers, this nation
could ever boast.

AFTER doing justice to the authors, let us come to
the work itself. In pieces of this kind, there is, generally speaking, no part so tedious and unpleasant, at
least to the generality of readers, as what regards oc-
currences at sea; and yet these are allowed to have their
utility. In the following sheets, however, though
they are found pretty copiously, we shall see them
without those defects. If these writers mention the
variation of the compass, they explain the nature, enquire into the cause, and shew the uses that arise from
observing this phænomenon. In this manner, they
treat of calms, winds, currents, and other incidents,
in so succinct and scientific a method, as at the same
time to be very instructive, and not unentertaining.
In this respect, we may look upon their narratives as
a sort of practical introduction to the art of navigation,
which we not only read without disgust, but which,
when read with any tolerable attention, will enable us
to understand many passages in other writers of voyages, which we should otherwise pass over, as utterly
uninteresting and unintelligible. This observation,
the reader will find so fully verified, from his own
experience, that, I am confident, he will think it no
small recommendation to the book; and the more so,
because, though very necessary, and much wanted,

* Martin Folkes, Esq.; a gentleman not more conspicuous from
his extensive knowledge, than amiable for the politeness of his manners, and respectable for his excellent private character.

2   the

the difficulties attending it had hitherto, in a great measure, discouraged any such attempt.

THE geographical descriptions we have of the country about Carthagena, the isthmus of Darien, the Terra Firma, the countries of Peru and Chili, those watered by the vast river of the Amazons, and, in a word, of the greatest part of South America, are not only perfectly accurate, very methodical, and, in all respects, full, clear, and satisfactory; but also what we greatly wanted, and what we never had, at least in any comparison with what we now have, before this work appeared. These are countries that, from the time of their discovery, have maintained the reputation of being as pleasant, as fertile, and as valuable, as any upon the globe. But though we knew this in general, and, from the Spanish descriptions and histories, were not altogether unacquainted with many particulars relating to them; yet with respect to any distinct and precise delineation of their several provinces, their divisions and subdivisions, the distribution of mountains, rivers, plains, and other circumstances, with their relations to each other, and comparative values in all respects, they were things not barely unknown, but such as we could never expect to know, from the nature of the Spanish government, with any degree of certainty. But by the accident of these gentlemen going thither, with no other view than the improvement of knowledge, pursuing that view with the most lively zeal and assiduous application, and founding their reputation upon a plain and candid communication of all that knowledge, which, with so much pains and labour, they had acquired; we have now as clear, concise, and correct a representation of these extensive regions, as we can possibly desire: such a one, as will answer all the ends of information and instruction,
                                        enable

enable us to discover the errors and partialities in former accounts, and prevent our being amused or misled by any erroneous relations for the future; which are certainly circumstances of very great consequence.

THE natural history of these countries will be likewise found in the following sheets, in a manner no less perfect and pleasing. These gentlemen went about it in a proper method, and with the talents requisite to the complete accomplishment of their design. They saw things with their own eyes, they enquired carefully, but they took nothing on trust: on the contrary, they discovered, and they have disclosed, many errors of an old standing; exploded various common notions that were ill founded, and have left others in the state in which they ought to be left, as things not thoroughly proved, or absolutely disproved; but which are reserved for further examination. It is chiefly from the natural history, that we collect the value and importance of any country, because from thence we learn its produce of every kind. In these sheets we find the greatest care taken in this particular; all the riches of the mineral, vegetable, and animal kingdoms exhibited to our view, their places exactly assigned, their respective natures described, the methods of using, improving, and manufacturing them, pointed out; and, exclusive of a multitude of vulgar errors exposed, and mistaken notions refuted, an infinity of new, curious, and important remarks are made, all tending to explain and illustrate the respective subjects. Of these many instances might be given; but that would be to anticipate the reader's pleasure, and arrogate to ourselves the merits of the authors we celebrate.

IN

In respect to the civil history, the world in general was yet more in the dark, than as to the natural; knowing much less of the inhabitants than of the commodities of these countries; and in this respect, our authors have been as candid, as circumstantial, and as copious, as in the other. They not only acquaint us with the distribution and disposition of the Spanish governments; with the nature, extent, and subordination of those who preside in them; but have also given us a regular plan of their administration, and of the order and method in which justice is dispensed, and the civil policy maintained; the domestic œconomy of the Spaniards, their customs, manner of living, their way of treating the Indians, both subjects and savages, are stated with the same freedom and precision. In like manner they give us a succinct account of the Creoles, that is, such as are descended from the Spaniards, and have been longer or later settled in the Indies, with whatever is peculiar in respect to the genius, humour, virtues, and vices of these people; and more especially the points in which they differ from the native Spaniards. The state and condition of the Indians who live in subjection to the Spaniards, their tempers, employments, good and ill qualities, labours and diversions. The habitations of the free Indians, their customs, dress, manner of spending their lives, exercises, talents, religion, and method of preserving the remembrance of past transactions, as also the condition of the Negroes and Mulattoes, whether in the capacity of slaves, domestic servants, or in possession of their freedom, with whatever differences occur in the state of any of these people in different provinces.

But to the English reader, perhaps, nothing in the following pages will be more acceptable, as indeed
nothing

nothing seems to have been more carefully considered by the authors, than the commercial history of these countries. We find here, not only the principal commodities of every province distinctly enumerated; but we are also informed of the particular places where they grow, their different qualities and degrees in value, the method of collecting and curing most part of them, the manufactures of cotton, wool, and other materials, the produce of their mines and different kinds of metal, their potteries, and whatever else is the object of industry and skill: The manner of conveying them from one province to another, the great roads, the inland and coasting navigation, their commerce with Spain, their contraband trade, the manner of introducing, and the great consumption, of European commodities and manufactures, the advantages and disadvantages attending their present regulations, the discoveries that are yet to be made, and the improvements which may still take place in the management of affairs in those countries: The singular inventions of the natives for passing great rivers, transporting their goods by the help of vessels of their own construction, their adroitness in some respects, and their stupidity in others.—From the due consideration of this part of the work, the reader will perceive, that in many things we have been imposed upon, in former accounts; and that other things, in a long course of years, are very much changed from what they were. But instead of old errors, we shall find many new truths, and some established from example and experience, that are of too great consequence not to be frequently remembered, and perfectly understood: Such as, that countries are not the better, and, which is still stranger, are not the richer, for producing immense quantities of gold and silver; since this prevents their being cultivated,

tivated, exposes the natives to pass their lives in
the severest drudgery, and, after all, makes the
digging of metal from the mine little more than
drawing water in a sieve; since, in such countries,
riches disappear almost as soon as they are revealed.
Industry alone, in the old world, and in the new, has
the power of acquiring and preserving wealth, and
this too without the trouble of mining.   Besides,
though not insisted upon, it will be evidently seen,
that severity in government, and superstition in religion, subvert both liberty and morals, and are consequently in all respects destructive of the happiness
of mankind.

THE account given by our authors, of the missions
which the Jesuits have established in Paraguay, is
as interesting as it is entertaining; and may be very
justly considered as one of the most curious and
best written parts of the whole performance; since,
at the same time that it breathes all the deference
and respect possible for the fathers, it informs us of
a great variety of facts of so much the more consequence, as, at the time it was written, nobody could
foresee that the courts of Madrid and Lisbon would
make so thorough a change as they have done in
their sentiments in regard to this order; and therefore the informations these gentlemen give us, are
the more to be relied on.   They shew us in what
manner, and under what specious pretences, the
Jesuits acquired a kind of independent possession
of so large a tract of country, and, except their
annual tribute, an almost absolute dominion over an
immense number of people.   They acquaint us,
that there is a civil government in every village, after
the model of the Spanish towns; but the magistrates are chosen by the people, subject only to the
approbation of the father Jesuit, who resides in,
                                                  and,

and, in reality, governs the village. We learn from them, that the Jesuits draw from the people all the commodities and manufactures that are fit for foreign commerce, which are vended by a commissary of their appointing, and the returns in European com-modities made to and distributed by them at their pleasure; they tell us, that the church in every village is spacious, and elegantly adorned; that, though they are styled villages, they are in effect large towns, and the houses in them neat, commodious, and, in comparison of the Spaniards', very well furnished. We learn from them, that, under a pretence of the excursions of the Portuguese, who used to seize these Indians and make them work in the mines, and of the savage Indians who surrounded them in a manner on all sides, the fathers have taught them the use of arms, make them spend their holidays in military exercises, have a large body of well-disciplined troops, magazines well furnished with military stores, together with mills and other necessary machines for making their own gunpowder. They likewise let us know, that, to prevent the manners of their disciples from being corrupted, the Jesuits exclude them entirely from all communication with strangers, whether Europeans or Indians, and suffer none to enter into their missions, who may report either the strength or the weakness of their condition, or penetrate into the mysteries of their policy.

ANOTHER point worthy of notice is, the account of the little island of Fernando de Norona. This, so inconsiderable in itself, so unfit for habitation, from its being sometimes three or four years without rain, was abandoned by the Portuguese; yet, being within sixty or seventy leagues of the coast of Brazil, was occupied by the French East-India company; which
induced

induced its former masters to repossess it, and to for-
tify it likewise, notwithstanding the aforementioned
inconvenience.   The building here no less than seven
forts, to cover and command three harbours, in the
largest of which there is a garrison of a thousand regular troops, relieved constantly once in six months
from Fernambuca, plainly proves, that either the force
of the Portuguese is much greater in those parts than
we commonly apprehend in Europe: or, which is
more probably the truth, that they are to the last
degree jealous and suspicious of that enterprizing
nation, who, in virtue of the great law of convenience, are for appropriating to their own use whatever they find themselves in a condition to seize and
to secure.   This gives us the true cause of that surprize
and uneasiness which the Portuguese, during the last
war, expressed, when a French squadron, with a body
of land-troops on board, intended against our settlements in the East Indies, touched there, on account
of an epidemic disease among their troops; which,
it seems, the Portuguese mistook for the epidemic
thirst of gold; and were so apprehensive of their
making a visit to their mines, that though they could
not well refuse them relief in their distress, yet they
took almost the same precautions as if declared enemies had landed in their country.

Another point of great utility, that will result
from the perusal of this work, is the obtaining a clear
and candid account of the flourishing state of the
French colonies in St. Domingo; which, considering
that the gentleman from whom we have it is a Spaniard, and consequently unexceptionable in his testimony, will suffice to give us a just idea of the prodigious advantages derived to France from their colonies in that island.   He observes, with great fair-
ness and freedom, that the French are well entitled to
the

the riches they acquire, from their industry and economy, since, though they occupy the worst part of the island, they are, out of all comparison, in a better condition than the Spanish inhabitants, who possess the better and more fertile part. He takes notice likewise, that though all correspondence between the two nations is forbidden under the severest penalties, it is notwithstanding open almost in the same degree as if there was no such prohibition; the reason is, because the French could scarce subsist, if they were not supplied with cattle from the Spaniards; and, on the other hand, the Spaniards must go naked, if they did not, by this means, obtain European commodities from the French; so idle a thing it is to think of making a law against necessity! By the balance of this trade the French acquire annually about two millions of pieces of eight, which returns in hard silver, with sugar, indigo, and the other commodities of the growth of their part of the island, which is admirably cultivated, to the ports of France; and is a very considerable addition to the value of their otherwise rich cargoes.

But the Portuguese and French are not the only strangers into whose circumstances, and management of affairs in America, our authors have enquired; the reader will find they took no less pains to make themselves well acquainted with the proceedings of the English. We have not only a full and distinct account of the taking of Louisbourg, and of the conquest of the island of Cape Breton in the war before the last; but we have also a very copious memorial, drawn from the papers of the marquis de la Maison Forte, of the colony of New England, which he had an opportunity of framing while he remained a prisoner at Boston. It would have been the more satisfactory if we had had the whole of his memoirs; for

VOL. I.   b   there

there is great reason to judge, from this specimen, that he must have made much deeper researches than he communicated to his Spanish friend, or, at least, than he judged expedient to communicate to the world; otherwise it is very hard to conceive upon what he grounded his notion, that, in the space of a century, the people of New, would be as numerous as those in Old England, and in a condition to give law to all the nations in North America. We have, besides, some very sensible remarks upon the cod-fishery, and the advantages arising from it, as well as our disputes with the French in that part of the world. These speculations, though not always critically right, deserve our notice, and even our strictest attention. For, however we may be at liberty to conduct our own affairs, we cannot limit the humour, or controul the politics, of our neighbours; and therefore it is of great service, to be well acquainted with their notions. The great importance of this subject might have been, and certainly was, discovered long ago, by our politicians of the first order; but it is now become apparent to all ranks of people, and, if the expression may be allowed, from being the object sometimes of discussions in the cabinet, is at present become the topic of vulgar politicians.

BESIDES those that have been already touched, there are a great variety of curious, instructive, and pleasing incidents, in this performance, that cannot fail of giving satisfaction to the ingenious and intelligent reader: Such as the discussing the causes, why it never rains naturally at Lima, or the country of Valles in Peru; the enquiries into the frequency of volcanoes in South America; the materials, if we may so speak, of those subterraneous fires, the accidents by which they are kindled, and the conse-
sequences

quences of their explosions; the author's sentiments as to earthquakes, their extent and direction, the circumstances preceding and attending them, and their different effects in different places.

THE value of this *Fourth Edition* is very much enhanced, by a number of curious, instructive, and explanatory *Notes.* These cannot fail of giving great satisfaction to the reader, as they serve to rectify some mistakes, and to set a variety of passages in a clearer light, from the writer's thorough acquaintance with the subjects treated in these voyages. He has also given a very accurate account of those parts of Brazil least known to us; and which may be therefore separated as a useful, as well as proper, supplement; and render the work, taken altogether, as complete as even a critical reader can desire.

CON-

CONTENTS

OF

VOLUME THE FIRST.

BOOK I.

b 3   BOOK

BOOK VI.

EXPLANATION

OF THE

P L A T E S.

b *A*

b *A lunar rainbow, observed by Don George Juan.*
c *The mountain of Cotopaxi, at the time of the eruption in* 1743.

FIG. 2.

View of a torrent, and the manner of passing it.
a *A bridge of bejucos.*
b *A tarabita, for the passage of horses.*
c *A tarabita, for men.*

FIG. 3.

Works of the ancient Indians, found in their tombs.
a *A tomb of the ancient Indians.*
b *Plan of a tomb, opened in the form of a cross.*
c *An idol of gold, or statue of some distinguished Indian.*
d *A hatchet, or axe, fixed in a javelin, and used by the Indians in war.*
e, f, g, *Hatchets of different forms.*
h, i, *Ynca rirpos, or mirroirs, formed out of Ynca stone.*
k *A large pot, used by the Indians for holding their chica.*
l *Sunga tirana, or knippers, used by the Indians for pulling out superfluous hairs from the face, &c.*
m *Pendant of gold or silver for the ear.*
n *A convex mirroir.*
o *A hatchet of stone.*
p, q, *Guainacaba cruches, or earthen jars, for holding their liquor.*
r *A tupu, or large pin, for fastening the anaco on their shoulders.*
f, t, *Tubus, a sort of needles, used by the Indians in fastening the plaits of the anaco.*

PLATE V.—Page 468.

FIG. 1.

a *A temple of the ancient Indians, near the village of Cayambe, in the province of Quito.*
b *Tombs of the ancient Indians.*
c *A fortification or retrenchment of the Indians on the top of a mountain.*
d *The village of Cayambe.*

FIG. 2.

View of the ruins of a palace of the Yncas, called Callo, near the town of Latacunga, in the province of Quito.

A *En-*

A *Entrance of the palace.*
B *Principal court of the palace.*
C *Apartments of the Yncas, divided into small chambers for the princes.*
D *Doors leading to the royal apartments.*
E *Parts, which were formerly subdivisions for the royal family.*
F *Others in the same taste, for the domestics.*
G *Offices for the use of the prince, with several small divisions for keeping curious and savage beasts.*
H *Apartments for the guards.*
K *A mount called Panacillo, which served as a watch-tower, when the monarch was in his palace.*
L *A river, which has its source in the desert of* Cotopaxi.

FIG. 3.

The balza.

a *The prow or head.*
b. *The stern.*
c *The awning or tilt.*
D *The poles or sheers, on which the sail is hoisted.*
e *A kind of bowsprit.*
g *A guara, drawn up.*
h *The fire hearth.*
i *The bowling of the sail.*
k *The back stays.*
l *The deck.*
1, 1, 1, *Other guaras for steering the balza.*

FIG. 4.

View of a palace and citadel of the Yncas, *near the Village of* Canar.
a *Entrance of the palace and fortress.*
b *The large court, or place of arms.*
c *The citadel in the form of a donjon.*
d *Barracks or apartments for the guards.*
e *The principal wall.*
f *Steps for mounting the walls.*
g *The apartments, having only one door to each.*
h *Lodgement for the soldiers.*
i, i, *A river running before the palace*
k, k, *Another river, which, joining the former, surrounds the whole structure.*
1, 1, 1, *Mountains surrounding the fortress.*

3

PLATE

PLATE VI.—Vol. II. Page 30.

A plan of Lima.
The port of Callao.
Different beasts, &c. in the kingdom of Peru.
Dress of the Peruvians.

PLATE VII.—Vol. II. Page 240.

FIG. 1.

Plan of the town of Cape François, in the Island of St.
Domingo.

1 *The church.*
2 *The college of Jesuits.*
3 *The place of arms.*
4 *Place of arms without the town.*
5 *The grand battery.*
6 *Caverns, or barracks.*
7 *The little battery.*
8 *Mouth of the river.*

FIG. 2.

Men of Chili.

FIG. 3.

Manner of killing the beasts in Chili.

1 *A mine, or quarry, of shells.*
2 *Manner of killing the beasts in Chili.*
3 *A pijaro nino.*
4 *Sea wolves.*
5 *The inclosure, in which they confine the beasts intended for the slaughter.*
6 *A Guaso on horseback, going to throw his noose at the beast.*

A VOYAGE

A

VOYAGE

TO

SOUTH AMERICA.

BOOK I.

Reasons for this Voyage; Navigation from the Bay of Cadiz to Carthagena in America, and a Description of the latter.

CHAP. I.

Motives of this Voyage to South America, with Remarks on the Navigation between Cadiz and Carthagena.

THE heart of man is naturally inclined to attempt things, the advantages of which appear to increase in proportion to the difficulties which attend them. It spares no pains, it fears no danger in attaining them; and instead of being diverted from its purpose, is animated with fresh vigour by opposition. The glory inseparable from arduous enterprises, is a powerful incentive, which raises the mind above itself; the hope of advantages determines the will, diminishes dangers, alleviates hardships, and le-

vels obfiacles, which otherwife would appear unfur-
mountable. Defire and refolution are not, however,
always fufficient to enfure fuccefs; and the beft-con-
certed meafures are not always profperous. Divine
Providence, whofe over-ruling and incomprehenfible
determinations direct the courfe of human actions,
feems to have prefcribed certain limits, beyond which
all our attempts are vain. The caufes his infinite wifdom has thought proper to conceal from us, and the
refult of fuch a conduct is rather an object of our re-
verence than fpeculation. The knowledge of the
bounds of human underftanding, a difcreet amufe-
ment and exercife of our talents for the demonftra-
tion of truths which are only to be attained by a continual and extenfive ftudy, which rewards the mind
with tranquillity and pleafure, are advantages worthy
of our higheft efteem, and objects which cannot be
too much recommended. In all times the defire of
enlightening others, by fome new difcovery, has roufed
the induftry of man, and engaged him in laborious
refearches, and by that means proved the principal
fource of the improvement of the fciences.

THINGS which have long baffled fagacity and application, have fometimes been difcovered by chance.
The firmeft refolution has often been difcouraged by
the infuperable precipices, which, in appearance, en-
circle his inveftigation. The reafon is, becaufe the obftacles are painted, by the imagination, in the moft
lively colours; but the methods of furmounting them
efcape our attention; till, fmoothed by labour and
application, a more eafy paffage is difcovered.

AMONG the difcoveries mentioned in hiftory, whether owing to accident or reflection, that of the Indies is not the leaft advantageous. Thefe parts were
for many ages unknown to the Europeans; or, at
leaft, the remembrance of them was buried in oblivion. They were loft through a long fucceffion of
time, and disfigured by the confufion and darknefs in
which

which they were found immerfed. At length the happy
æra arrived, when induftry, affifted by refolution,
was to remove all the difficulties exaggerated by ignorance. This is the epocha which diftinguifhed the
reign, in many other refpects fo glorious, of Ferdinand
of Arragon, and Ifabella of Caftile. Reafon and experience at once exploded all the ideas of rafhnefs and
ridicule which had hitherto prevailed. It feems as if
Providence permitted the refufal of other nations, to
augment the glory of our own; and to reward the
zeal of our fovereigns, who countenanced this important enterprife; the prudence of their fubjects in the
conduct of it, and the religious end propofed by both.
I mentioned accident or reflection, being not yet convinced, whether the confidence with which Chriftopher
Columbus maintained, that weftward there were lands
undifcovered, was the refult of his knowledge in cof-
mography and experience in navigation, or whether it
was founded on the information of a pilot, who had
actually difcovered them, having been driven on the
coafts by ftrefs of weather; and who, in return for
the kind reception he had met with at Columbus's
houfe, delivered to him, in his laft moments, the papers and charts relating to them.

The prodigious magnitude of this continent; the
multitude and extent of its provinces; the variety of
its climates, products, and curious particulars; and,
laftly, the diftance and difficulty of one part communicating with another, and efpecially with Europe, have
been the caufe, that America, though difcovered and
inhabited in its principal parts by Europeans, is but
imperfectly known by them; and at the fame time
kept them totally ignorant of many things, which would
greatly contribute to give a more perfect idea of fo confiderable a part of our globe. But though inveftiga-
tions of this kind are worthy the attention of a great
prince, and the ftudies of the moft piercing genius
among his fubjects; yet this was not the principal

B 2   intention

intention of our voyage. His majefty's wife refolution of lending us to this continent, was principally
owing to a more elevated and important defign.

The literary world are no ftrangers to the celebrated queftion that has lately produced fo many
treatifes on the figure and magnitude of the earth;
which had hitherto been thought perfectly fpherical.
The prolixity of later obfervations had given rife to
two oppofite opinions among philofophers. Both
fuppofed it to be elliptical; but one affirmed its tranfverfe diameter was that of the poles, and the other
that it was that of the equator. The folution of this
problem, in which not only geography and cofmo-
graphy are interefted, but alfo navigation, aftrono-
my, and other arts and fciences of public utility, was
what gave rife to our expedition. Who would have
imagined that thefe countries, lately difcovered,
would have proved the means of our attaining a perfect knowledge of the old world; and that, if the
former owed its difcovery to the latter, it would make
it ample amends by determining its real figure, which
had hitherto been unknown or controverted? who, I
fay, would have fufpected that the fciences fhould, in
that country, meet with treafures, not lefs valuable
than the gold of its mines, which has fo greatly enriched other countries? How many difficulties were
to be furmounted in the execution! what a feries of
obftacles were to be overcome in 'fuch long operations, flowing from the inclemency of the climates;
the difadvantageous fituation of the places where they
were to be made, and in fine, from the very nature of
the enterprife! All thefe circumftances infinitely heighten-
en the glory of the monarch, under whofe aufpices
the enterprife has been fo happily accomplifhed.
This difcovery was referved for the prefent age, and
for the two Spanifh monarchs, the late Philip V. and
Ferdinand VI. The former caufed the enterprife to
be carried into execution, the latter honoured it with
his

his countenance, and ordered the narrative of it to be publifhed; not only for the information and inftruction of his own fubjects, but alfo for thofe of other nations, to whom thefe accounts will prove equally advantageous. And, that this narrative may be the more inftructive, we fhall introduce the particular circumftances which originally gave occafion to our voyage, and were in a manner the bafis and rule of the other enterprifes, which will be mentioned in the fequel, each in its proper order.

THE attention of the Royal Academy of Sciences at Paris, for the improvement of human knowledge, and its continual ardour to difcover and apply the beft methods for that noble end, could not fit down contented under the uncertainty concerning the real figure and magnitude of the earth; the inveftigation of which had, for feveral years paft, employed the moft eminent geniufes of Europe. This learned affembly reprefented to their fovereign the neceffity of determining a point, the exact decifion of which was of fuch great moment, efpecially to geography and navigation; and at the fame time laid before him a method of doing it. This was, to meafure fome degree of the meridian near the equator; and (as was done with great propriety after our departure) by meafuring other degrees under the polar circle, in order to form a judgment of the different parts of its circumference, by their equality or inequality, and from thence to determine its magnitude and figure. No country feemed fo proper for this as the province of Quito in South America. The other countries under the equinoctial line, both in Afia and Africa, were either inhabited by favages, or not of an extent fufficient for thefe operations; fo that, after the moft mature reflection, that of Quito was judged to be the only place adapted to the plan in queftion.

HIS Moft Chriftian majefty Lewis XV. applied, by his minifters, to king Philip, that fome members

of his royal academy might pafs over to Quito, in
order to make there the neceffary obfervations; at
the fame time fhewing the intention and univerfal advantage of them, and how very remote they were
from any thing which tends to awaken a political jea-
loufy His majefty, perfuaded of the candour of
this application, and defirous of concurring in fo noble
a defign, as far as was confiftent with the dignity
of his crown and the fafety of his fubjects, referred
the matter to the council of the Indies: and, on their
favourable report, the licenfe was granted, with all
the neceffary recommendations and affurances of the
royal protection to the perfons who were to repair to
America to make thefe obfervations. The patents,
which were made out for them on the 14th and 20th
of Auguft 1734, contained the moft precife orders
to the viceroys, governors, &c. in the countries
through which they were to pafs, to aid and affift them,
to fhew them all friendfhip and civility, and to fee that
no perfons exacted of them for their carriages or labour more than the current price; to which his majefty was pleafed to add the higheft proofs of his royal
munificence, and of his zeal for the advancement of
the fciences, and efteem for their profeffors.

THIS general regard of his majefty was followed
by fome meafures, particularly defigned to promote
the honour of the Spanifh nation, and to give his
own fubjects a tafte for the fame fciences. He appointed two officers of his navy, well fkilled in mathematics, to join in the obfervations which were to
be made, in order to give them a greater dignity and
a more extenfive advantage; and that the Spaniards
might owe only to themfelves the fruits and improvements expected from them. His majefty alfo
conceived, that the French academicians having thefe
officers in their company, would be more regarded
by the natives; and, in the places through which they
were to pafs, all umbrage would be thus removed
from

from perfons who might not be fufficiently acquainted with the nature of their defign. Accordingly the commanders and directors of the academy of the royal Guardas Marinas received orders to recommend two perfons, whofe difpofitions not only promifed a perfect harmony and correfpondence with the French academicians, but who were capable of making, equally with them, the experiments and operations that might be neceffary in the courfe of the enterprife.

Don George Juan, commander of Aliaga, of the order of Malta, fub-brigadier in the Guardas Marinas, equally diftinguifhed by his application to the mathematics and his faithful fervices to the crown, was, with myfelf, propofed to his majefty, as qualified to contribute to the fuccefs of fuch an enterprife. We had commiffions given us as lieutenants of men of war, and, with all neceffary inftructions, were ordered to embark on board two fhips fitting out at Cadiz, for carrying to Carthagena, and thence to Porto Bello, the marquis de Villa-garcia, appointed viceroy of Peru. About the fame time the French academicians were to fail in a fhip of their nation, and, by way of St. Domingo, to join us at Carthagena, in order to proceed from thence in company.

The two men of war on board of which we had been ordered, were the Conquiftador of 64 guns, and the Incendio of 50; the former commanded by Don Francifco de Liano, of the order of Malta, commodore, and the latter by Don Auguftin de Iturriaga, by whom it was agreed that Don George Juan fhould go in the Conquiftador, and myfelf in the Incendio. We failed from Cadiz bay, May 26, 1735; but, the wind fhifting, were obliged to put back and come to an anchor about half a league without Las Puercas.

On the 28th, the wind coming about to the N. E. we again fet fail, and continued our courfe in the manner related in the two following Journals.

Journal of Don George Juan, on board the Conquiſtador.

THE ſecond of June 1735, ſaw the Canary iſlands; and the winds, which are uſually very variable in this paſſage, were either N. W. by N. or N. E. Don George Juan, by his reckoning, found the difference of longitude between Cadiz and the Pico of Teneriffe 10° 30′.

ACCORDING to father Feuillée's obſervations, made at Loratava, ſix minutes and a half eaſt of the Pico, the difference of the longitude betwixt the latter and the obſervatoay at Paris is 18° 51′. Subtracting therefore 8° 27′, which, according to the Connoiſſance des Tems, is the difference of longitude between that obſervatory and Cadiz; the difference of longitude between that city and the Pico is 10° 24′, and conſequently differs 6′ from Don George's reckoning.

ON the 7th we loſt ſight of the Canaries, and continued our courſe towards Martinico, ſteering ſouth between 42 and 45 degrees weſterly, increaſing the angle every day, till near the iſland, we ſteered due weſt under its parallel, and on the 26th of June diſcovered Martinico and Dominica.

THE difference of longitude between Cadiz and Martinico appeared, from our reckoning, to be 59° 55′, that is, 3° 55′ more than the chart of Antonio de Matos makes it; which is however generally followed in this voyage. According to the obſervations of father Laval, made at Martinico, the difference of longitude is 55° 8′ 45″; according to thoſe of father Feuillée, 55° 19′. This error in a great meaſure proceeds from a want of accuracy in the log-line; for had the pilot of the Conquiſtador, who found the ſame defect in his calculations, made the diſtance between the knots of the log-line 30 Engliſh feet, inſtead of 47 and a half,

half the difference of longitude, by account, would have been only 57°. This error in marking the log-line is common both to the pilots of Spain and other nations; and this, like many other faults in navigation, remains uncorrected for want of attention.

The distance between the knots on the log-line should contain $\frac{1}{120}$ of a mile, supposing the glass to run exactly half a minute: and though all agree in this respect, yet not in the true length of the mile, which ought to be determined by the most exact mensurations, as those of M. Cassini in France, ours in the province of Quito, or those of M. Maupertuis in Lapland. If the length of the degree be computed according to M. Cassini's measures, 57060 toises, a minute or geographical mile will contain 951 toises, or 5706 royal feet, of which $\frac{1}{120}$ is nearly equal to 47 feet 6¼ inches; and as the Paris foot is to that of London as 16 to 15 *; this, when reduced to English measure, makes near 50 feet 8¼ inches. And this is the true distance between each knot on the log-line.

This mensuration, which should have been hitherto the rule observed, is not exact, when compared to that which has been found from investigating the figure of the earth, which is discovered to be very different from what it has been imagined; so that it is not surprising that there should be found considerable differences in nautical calculations.

* According to the late regulation of the Royal Society of London, and the measures sent by it to the Academy of Sciences at Paris, and with which I was favoured by Martin Folkes, Esq. the worthy president of that society, the Paris foot is to that of London as 864 to 811, which shews how erroneous these are published by father Tosca .

ᵃ The Paris foot is divided into 12 inches, and each inch into 12 lines; wherefore, if we suppose each line to be divided in 310 parts,
    The Paris foot will be 1440 parts,
    The London,   1350.
These proportions were settled by the Royal Academy of Sciences at Paris, in their treatise of the figure and magnitude of the earth, Part xi. Chap. 5, which shews the erroneousness of the above. A.

The

The Author's Journal, on board the Incendio.

HAVING fet fail on the fame day, namely, the 28th of May 1735, and fteered S. between 52 and 56 deg. wefterly, we perceived, on June 2, about fix in the evening, the ifland of Savages, one of the Canaries; and on the 3d we faw Teneriffe. I found the difference of longitude between Cadiz and Naga-point to be 11° 6′, which agrees with the Englifh and Dutch charts, but differs a little from the true longitude determined by father Feuillée at Loratava, in the fame ifland of Teneriffe.

On the 4th, we had fight of the iflands of Palma, Gomera, and Fer; but again loft fight of them on the 5th. On the 29th, about noon, we made Martinico, and continuing our courfe, paffed between that ifland and Dominica. The difference of longitude between Martinico and Cadiz bay, according to my reckoning, was 57° 5′, one degree more than San Telmo's chart makes it. But it is proper to obferve, that, in order to eftimate my courfe, and avoid the danger of finding a great difference at making land, I followed two different calculations, one according to the meafures commonly given by pilots to the diftance between the knots on the log-line, of 47 Englifh feet and a half, and the other by reducing them to 47 royal feet; for though, in ftriétnefs, it ought to have been 47¼ of the latter, the difference being but fmall, I thought it beft to omit the half foot, that my reckoning might be before the fhip. According to the firft method, the difference of longitude between Cadiz and this ifland was between 60 and 61 degrees, which nearly agrees with the Journal of Don George.

From Martinico we continued our courfe towards Curafao, which we had fight of July 3d. The difference

ence of meridians between that and Martinico, Don George Juan found to be 6° 49′, whereas I made it 7° 56′. The cause of this disagreement was, that, finding a sensible difference in the latitudes, I regulated myself by the currents, imagining, according to the opinion of all our navigators, that they set to the N. W.; which Don George did not, and by that means his reckoning answered to the real distance betwixt these two islands, and mine was erroneous. But that the water was in motion, is not to be questioned; for in all the latitudes from June 30, to July 3, those found by observation exceeded those by account, 10′, 13′, and even 15 minutes; a sufficient proof that the currents run directly N. and not N. W.

From the 2d at six in the morning, till the day we made Curasao and Uruba, we had shallow water, of a greenish colour, which continued till about half past seven in the evening, when we entered the gulf.

Our course from Martinico to Curasao, during the two first days, was south 81° westerly; and the two last south 64° westerly. From thence to Carthagena we kept at a proper distance from the coast, so as to distinguish its most noted capes, and inhabited places.

On the 5th we discovered the mountains of St. Martha, so well known for their height, and being all covered with snow; and at six in the morning we crossed at the current of thick water, which issues with prodigious rapidity from the river de la Magdalena, and extends several leagues into the sea. About six in the evening found ourselves to the northward of Cape de Canoa, where we lay to, and continued till seven in the morning, when we set all our sails, which at eight in the evening brought us under fort Boca Chica, where we came to an anchor in 34 fathom water, the bottom muddy. On the 8th we endeavoured to get into Carthagena bay, but could not before the 9th securely moor our ship.

DURING

DURING our paſſage betwixt the Canary iſlands, we had faint and variable winds, with ſome ſhort calms; but, after we had loſt fight of them, the gales increaſed upon us, but moderate, and continued in this manner till we arrived within 170 or 180 leagues of Martinico, when we had ſqualls accompanied with violent rains. After paſſing the Canaries, at about twenty leagues from theſe iſlands, we had the wind at north-weſt, and at the diſtance of near 80 leagues it ſhifted to E. and E. N. E. We had nearly the ſame in the middle of the Atlantic ocean, and afterwards the wind came about to the E. with different degrees of velocity; but the variation was not ſuch as to occaſion any inconveniency.

THESE are the winds generally met with in this voyage. Sometimes it veers away to the W. and W. N. W. though it is very ſeldom known to continue on theſe points. Sometimes long calms intervene, which lengthen the voyage beyond the uſual time. All this depends on the ſeaſons; and according to the time of the voyage, the weather and winds are more or leſs favourable. The winds abovementioned are the moſt general; and the beſt time for making uſe of them, as they then are ſettled, is when the ſun approaches near the equator in his return from the tropic of Capricorn: for his approach to the autumnal equinox is the time when the calms moſt prevail.

FROM the iſlands of Martinico and Dominica to that of Curaſao and the coaſt of Carthagena, the winds continued the ſame as in the ocean, though more variable, and the weather leſs fair. I have ſaid, that about 170 leagues before we reached Martinico the winds were interrupted by ſqualls; and theſe are more common beyond thoſe iſlands, and are immediately ſucceeded by ſhort calms; after which the wind freſhens again for half an hour, an hour, two hours, and ſometimes longer. From what quarter theſe
                                        tornadoes

tornadoes or fqualls proceed, I cannot pofitively affirm; but this is certain, that when they are over,
the wind begins to blow from the fame point as
before, and nearly with the fame force. And here it
may be of ufe to obferve, that, on any appearance of
thefe fqualls in the atmofphere, the utmoft expedition
muft be ufed in getting the fhip in readinefs, their
impetuofity being fo fudden as to admit of no time
for preparatives; and therefore the leaft negligence
may be attended with the moft fatal confequences.

In the voyage from Cadiz to the Canaries, in fome
parts, though the winds are otherwife moderate, the
fea is agitated by thofe from the N. and N. W. fometimes in large and long waves; fometimes in fmall but
more frequent ones, which happens when the wind
blows ftrongly along the coaft of France and Spain;
for in the ocean the winds are fo mild, that the motion
of the fhip is hardly perceived, which renders the
paffage extremely quiet and agreeable. Within the
windward iflands, and even before we reach them, in
the parts where thefe terrible fqualls prevail, the fea is
agitated in proportion to their violence and duration;
but no fooner is the wind abated, than the water becomes again clear and fmooth.

The atmofphere of the ocean anfwers to the calm-
nefs of the winds and fea, fo that it is very feldom an
obfervation cannot be taken, either from the fun's
being obfcured, or the hazinefs of the horizon. This
is to be underftood of the fair feafon; for otherwife
here are dark days, when the air is filled with vapours,
and the horizon very hazy. At all times it is feen
filled with white and towering clouds, embellifhing
the fky with a variety of figures and ramifications,
which amufe the eye, tired with being fo long confined to two fuch fimilar objects as the fea and fky.
Within the windward iflands the variety is ftill
greater, the quantity of vapours profufely exhaled, filling it in fuch a manner, that fometimes nothing but
clouds

clouds are to be seen, though part of these are gradually dispersed by the heat of the sun, so that some parts are quite clear, others obscure; but a general darkness during the whole day is never known.

It is well known and allowed, that, through the whole extent of the ocean, not the least current is perceivable, till we arrive within the islands, where in some parts they are so strong and irregular, that, without the greatest vigilance and precaution, a ship will be in great danger among this archipelago. This subject, together with the winds peculiar to this coast, shall hereafter be considered more at large.

In the track to Martinico and Dominica there is a space where the water, by its white colour, visibly distinguishes itself from the rest of the ocean. Don George, by his estimate, found this space to terminate 100 leagues from Martinico; whereas, according to my reckoning, it reached only to within 108 leagues; it may therefore, at a medium, be placed at 104. This small difference, doubtless, proceeds from the difficulty of discovering where this whitish colour of the water terminates, towards Martinico. It begins at about 140 leagues from that island, which must be understood of the place where the different colours of the water are evident; for, if we reckon from where it begins to be just discernible, the distance is not less than 180 leagues. This tract of water is a certain mark for directing one's course; because, after leaving it, we have the satisfaction of knowing the remaining distance: it is not delineated on any map, except the new one lately published in France; though it would doubtless be of great use in them all.

Nothing farther remains, than to give an account of the variation of the needle in different parts in which we found the ship by her latitude and longitude; a point of the utmost consequence in navigation, not only with regard to the general advantage to mariners in knowing the number of degrees intercepted
                                                            between

between the magnetic and true north of the world; but alſo as, by repeated obſervations of this kind, the longitude may be found, and we may know within a degree, or a degree and a half, the real place of the ſhip; and this is the neareſt approximation to which this has been carried by thoſe who revived it at the beginning of this century. Among theſe the chief was that celebrated Engliſhman, Dr. Edmund Halley: in emulation of whom, many others of the ſame nation, as alſo ſeveral Frenchmen, applied themſelves to the improvement of it. We already enjoy the fruits of their labours in the variation charts lately publiſhed, though they are principally uſeful only in long voyages; where the difference of two or of even three degrees is not accounted a conſiderable error, when there is a certainty that it cannot exceed that number. This ſyſtem, though new with regard to the uſe it is now applied to, is far from being ſo among the Spaniards and Portugueſe, very plain veſtiges of it remaining in their old treatiſes of navigation. Maniel de Figueyredo, coſmographer to the king of Portugal, in his Hydrographia, or Examin de Pilotos, printed at Liſbon in 1608, chap. ix. and x. propoſes a method for finding, from the variation of the needle, the diſtance run in ſailing eaſt and weſt. And Don Lazaro de Flores, in his Arte de Navegar, printed in 1672, chap. i. part ii. quotes this author, as an authority to confirm the ſame remark made by himſelf; adding (chap. ix.) that the Portugueſe, in all their regulations concerning navigation, recommend it as a certain method. It muſt however be acknowledged, that thoſe ancient writers have not handled this point with the penetration and accuracy of the Engliſh and French, aſſiſted by a greater number of more recent obſervations. And that the obſervations made in this voyage may be of the moſt general uſe, I ſhall inſert them in the two following tables; previouſly informing the reader, that the longitudes correſponding with each are true, the

error

error of the courfe with regard to the difference of meridians being corrected from the obfervations of the fathers Laval and Feuillée.

Variations obferved by Don George Juan, the Longitude being reckoned weft from Cadiz.

Deg. of Lat.		Deg. of Long.		Variation obferved.		Variation by the chart.		Differ- ence.	
27	30	11	00	8	00 W.	9	00 W.	1	00
25	30	14	30	6	20	7	20	1	00
24	00	17	00	4	30	6	00	1	30
23	20	18	30	3	30	5	00	1	00
22	30	20	00	2	30	4	30	2	00
21	50	22	00	1	30	4	00	2	30
21	35	26	00	0	30	3	00	2	30
16	20	43	00	4	00 E.	2	30 E.	2	00
15	40	45	00	5	00	3	20	1	40
Off Martinico				6	00	5	00	1	00

Variations obferved by the Author, the Longitude being reckoned from the former Meridian.

Deg. of Lat.		Deg. of Long.		Variation obferved.		Variation by the chart.		Differ- ence.	
36	20	00	25	9	30 W.	13	00 W.	3	30
31	23	08	22	7	00	10	30	3	30
30	11	10	21	6	00	9	30	3	30
26	57	14	54	4	00	7	00	3	00
25	52	15	59	3	40	6	30	2	50
16	28	43	46	0	30 E.	2	00 E.	1	30
15	20	47	32	2	30	4	00	1	30
Off Cape de la Vela				6	00	7	30	1	30

To the above obfervations on the variation of the needle, compared with thofe on the variation chart, firft publifhed by the great Dr. Halley in 1700, and corrected

rected in 1744, from other obfervations and journals by Mc r Montaine and Dodfon of London, I fhall add fome reflections, in order to expofe the negligence in conftructing the magnetic needles.   1. It appears that the variations obferved by Don George Juan do not agree with mine, which is not to be attributed to a defect in the obfervations.   This is fufficiently evident from comparing them.   The differences between thofe obferved by Don George and thofe on the chart, are nearly every where uniform; the moft confiderable being a degree and 30 minutes; one making the variation 2° 30′, and the other a degree only.   This probably arofe from the motion of the fhip, which hinders the needle from being entirely at reft; or from the difk of the fun, by reafon of intervening vapours, not being accurately determined, or fome other unavoidable accident; the error, when the difference is lefs than a degree, being fcarce perceivable in thefe obfervations. Thus, on a medium, the rational conclufion is, that the needle ufed in thefe obfervations varied a degree and 40 minutes lefs than thofe when the map was conftructed.

The fame uniformity appears in the differences between my obfervations and the chart; but it muft be obferved, that having ufed two different needles, the particulars of each nearly correfpond, fo that between the five firft, the greateft difference is of 40 minutes, which intervene between the fmalleft difference of 2° 50′, and the greateft of 3° 30′.   Hence, taking the medium between both, the difference between my obfervations and the chart will be 3° 16′, the latter being fo much lefs than the former.   The three laft do not want this operation, the difference of 1° 30′ being equal in all, and the variations refulting from thefe obfervations are alfo lefs than thofe delineated on the map; the variation having paffed to a different fpecies; namely, from N. W. to N. E.   This demonftrates, that the firft needle I made ufe of, whether it had been ill touched or the fteel not accurately placed, varied 1° 30′ wefterly lefs than that ufed by Don George

Juan; and as this officer continued his obfervations, to the end of the voyage with the fame needle, the difference, which at firft was negative, on the variation changing its denomination became pofitive; and from my changing inftruments, the difference on my fide continued always negative. The reafon of this is, that the difference of the five firft obfervations proceeded lefs from a real difference in the variation, than from the poles of the needle, which was fo far from anfwering exactly with the meridian-line on the compafs-card, that it inclined towards the N. W.; the contrary happened in the fecond compafs made ufe of, its inclination being towards the N. E.; confequently, whatever the angle of that inclination was, it occafioned a proportionate diminution in the variation of a contrary fpecies.

THESE obfervations, thus compared, fhew the errors to which navigators are liable, for want of attention in making choice of proper needles, which they fhould be careful to procure, not only well made and exact, but alfo ftrictly tried with regard to their inclination to the true meridian, before they venture to depend upon them in any voyage. In this point Spain is guilty of a notorious neglect, notwithftanding it is evidently the fource of a thoufand dangerous errors; for a pilot, in correcting the courfe he has fteered, in making ufe of a compafs whofe variation is different from the true, will confequently find a difference between the latitude by account and the latitude obferved; and to make the neceffary equation according to the rules commonly received in failing on points near the meridian, he muft either increafe or diminifh the diftance, till it agrees with the latitude, whereas in this cafe the principal error proceeded from the rhomb. The fame thing happens in parts where it is apprehended there may be currents; which often occur in failing when the latitude by account, and that by obfervation, difagree; though in reality the water has no motion, the difference proceeding entirely from making ufe of another variation

tion in the courfe, than that of the needle by which the
fhip is fteered; as was the cafe with me in failing from
Martinico to Curafao, and likewife of all the artifts on
board the fhip. Another error incident to navigators,
though not fo much their own, is to fteer the fhip by one
needle, and obferve the variation by another; for
though they have been compared, and their differences
carefully obferved, their motions being unequal, though
at the beginning of the voyage the difference was only
a certain number of degrees, the continual friction of
the former on the pivot, renders the point of the needle,
on which it is fufpended, more dull than the other,
which is only hung when they make obfervations, being
at all other times kept with the greateft care; and hence
proceeds the change obfervable in their differences.
In order to remedy this evil, all needles intended to be
ufed at fea fhould be equally proper for obferving the
variation; and the obfervation made with thofe before
placed in the bittacle; and, to improve the charts of
variation, fhould be touched in the fame manner, and
adjufted to the meridian of a place, where the exact
variation is known. Thus obfervations made in the
fame places by different fhips, would not be found fo
confiderably to vary; unlefs the interval of time between two obfervations be fuch as to render fenfible
that difference in the variation, which has been obferved
for many years paft, and is allowed of by all nations.

THESE are the caufes of the manifeft difference between needles; there may be others, but this is not
the proper place for enumerating them.

CHAP. II.

Defcription of Carthagena.

ON the 9th of July 1735, we landed, and Don
George Juan and myfelf immediately waited on
the governor of the place. We were informed that the

French

French academicians were not yet arrived, nor was there any advice of them. Upon this information, and being by our inftruĉtions obliged to wait for them, we agreed to make the beſt uſe of our time ; but were unhappily deſtitute of inſtruments, thoſe ordered by his majeſty from Paris and London not being finiſhed when we left Cadiz, but were forwarded to us at Quito ſoon after our arrival. We however fortunately heard that there were ſome in the city, formerly belonging to Brigadier Don Juan de Herrera, engineer of Carthagena ; by theſe we were enabled to make obſervations on the latitude, longitude, and variation of the needle. We alſo drew plans of the place and the bay from thoſe of this engineer, with the neceſſary additions and alterations.

In theſe operations we employed ourſelves till the middle of November 1735, impatient at the delay of the French academicians. At length, on the 15th, a French armed veſſel came to an anchor, during the night, under Boca Chica ; and to our great ſatisfaĉtion we learned, that the long-expeĉted gentlemen were on board. On the 16th we viſited them, and were received with all imaginable politeneſs by Mr. de Ricour, captain of a man of war, and king's lieutenant of Guarico, in the iſland of St. Domingo ; and Meſſ. Godin, Bouguer, and de la Condamine, academicians, who were accompanied by Meſſ. Juſſieu, botaniſt ; Seniergues, ſurgeon ; Verguin, Couplet, and Deſſordonais, aſſociates ; Morenvile, draughtſman ; and Hugot, clockmaker.

Our intention being to go to the equator with all poſſible expedition, nothing remained but to fix on the moſt convenient and expeditious route to Quito. Having agreed to go by the way of Porto Bello, Panama, and Guayaquil, we prepared to ſail ; in the mean time, by help of the inſtruments brought by the academicians, we repeated our obſervations on the latitude, weight of the air, and the variation of the needle ; the reſult of which will appear in the following deſcription.

THE

THE city of Carthagena ftands in 10 deg. 25 min. 48 ½ fec. north latitude; and in the longitude of 282 deg. 28 min. 36 fec. from the meridian of Paris; and 301 deg. 19 min. 36 fec. from the meridian of Pico Teneriffe; as appeared from our obfervations. The variation of the needle we alfo, from feveral obferva- tions, found to be 8 deg. eafterly.

THE bay, and the country, before called Calamari, were difcovered in 1502 by Roderigo de Baftidas; and in 1504 Juan de la Cofa and Chriftopher Guerra began the war againft the Indian inhabitants, from whom they met with greater refiftance than they expeacted; thofe Indians being a martial people, and valour fo na- tural to them, that even the women voluntarily fhared in the fatigues and dangers of the war. Their ufual arms were arrows, which they poifoned with the juice of certain herbs; whence the flighteft wounds were mortal. Thefe were fucceeded by Alonfo de Ojeda, who fome years after landed in the country, attended by the fame Juan de la Cofa, his chief pilot, and Amerco Vefpucio, a celebrated geographer of thofe times; but made no greater progrefs than the others, though he had feveral encounters with the Indians. Nor was Gregorio Hernandez de Oviedo more fortu- nate. But, at length, the conqueft of the Indians was accomplifhed by Don Pedro de Heredia, who, after gaining feveral victories over them, peopled the city in the year 1533, under the title of a government.

THE advantageous fituation of Carthagena, the ex- tent and fecurity of its bay, and the great fhare it at- tained of the commerce of that fouthern continent, foon caufed it to be erected into an epifcopal fee. The fame circumftances contributed to its prefervation and increafe, as the moft efteemed fettlement and ftaple of the Spaniards; but at the fame time they drew on it the hoftilities of foreigners, who, thirfting after its riches, or induced by the importance of the place, have feveral times invaded, taken, and plundered it.

C 3   THE

THE firft invafion was made foon after its eftablifh-
ment in 1544, by certain French adventurers, conducted by a Corfican pilot, who, having fpent fome time
there, gave them an account of its fituation, and the
avenues leading to it, with every other particular neceffary to the fuccefsful conduct of their enterprife;
which they accordingly effected. The fecond invader
was Francis Drake, termed the deftroyer of the new
conquefts, who, after giving it up to pillage, fet it on
fire, and laid half the place in afhes ; and its fatal deftruction was only prevented by a ranfom of a hundred
and twenty thoufand filver ducats paid him by the
neighbouring colonies.

IT was invaded a third time in 1597, by the French,
commanded by M. de Pointis, who came before the
place with a large armament, confifting partly of Fli-
buftiers, little better than pirates: but, as fubjects to
the king of France, were protected by that monarch.
After obliging the fort of Boca Chica to furrender,
whereby the entrance of the bay was laid open, he
landed his men, and befieged Fort Lazaro, which was
followed by the furrender of the city. But the capitulation was no fecurity againft the rage of avarice,
which had configned it to pillage.

THIS eafy conqueft has by fome been attributed to
a private correfpondence between the governor and
Pointis; and what increafes the fufpicion is, that he
embarked on board the French fquadron at its departure, together with all his treafures and effects, none
of which had fhared in the general calamity.

THE city is fituated on a fandy ifland, which forming
a narrow paffage on the S. W opens a communication
with that part called Tierra Bomba, as far as Boca
Chica. The neck of land which now joins them, was
formerly the entrance of the bay; but it having been
clofed up by orders from Madrid, Boca Chica became
the only entrance; and this alfo has been filled up fince
the attempt of the Englifh in 1741, who, having made
themfelves

themfelves mafters of the forts which defended it, entered the bay with an intent of taking the city; but they mifcarried in their attempt, and retired with confiderable lofs. This event caufed orders to be difpatched for opening the old entrance, by which all fhips now enter the bay. On the north fide the land is fo narrow, that, before the wall was begun, the diftance from fea to fea was only 35 toifes; but afterwards enlarging, forms another ifland on this fide, and the whole city is, excepting thefe two places which are very narrow, entirely furrounded by the fea. Eaftward it communicates, by means of a wooden bridge, with a large fuburb called Xexemani, built on another ifland, which has alfo a communication with the continent by means of another wooden bridge. The fortifications both of the city and fuburb are conftructed in the modern manner, and lined with free-ftone. The garrifon in times of peace confifts of ten companies of regulars, each containing, officers included, 77 men; befides feveral companies of militia.

In the fide of Xexemani, at a fmall diftance from that fuburb, on a hill, is a fort called St. Lazaro, commanding both the city and fuburb. The height of the hill is between 20 and 21 toifes, having been geometrically meafured. It is joined to feveral higher hills, which run in an eaftern direction. Thefe terminate in another hill of confiderable height, being 84 toifes, called Monte de la Popa, and on the top of it is a convent of bare-footed Auguftines, called Nueftra Senora de la Popa. Here is an enchanting profpect, extending over the country and coaft to an immenfe diftance.

The city and fuburbs are well laid out, the ftreets being ftraight, broad, uniform, and well paved. The houfes are built of ftone, except a few of brick; but confift chiefly of only one ftory above the ground-floor; the apartments well contrived. All the houfes have balconies and lattices of wood, as more durable in this climate than iron, the latter being foon corroded and

C 4

deftroyed

deftroyed by the moifture and acrimonious quality of the nitrous air; from whence, and the fmoky colour of the walls, the outfide of the buildings makes but an indifferent appearance.

The churches and convents of this city are the cathedral, that of the Trinity in the fuburbs, built by bifhop Don Gregory de Molleda, who alfo in 1734 founded a chapel of eafe dedicated to St. Toribio. The orders which have convents at Carthagena are thofe of St. Francis, in the fuburbs, St Dominic, St. Auguftin, La Merced, alfo the Jacobines, and Recollets; a college of Jefuits and an hofpital of San Juan de Dios. The nunneries are thofe of St. Clara and St. Terefa. All the churches and convents are of a proper architecture, and fufficiently capacious; but there appears fomething of poverty in the ornaments, fome of them wanting what even decency requires. The communities, particularly that of St. Francis, are pretty numerous, and confift of Europeans, white Creoles, and native Indians.

Carthagena, together with its fuburbs, is equal to a city of the third rank in Europe. It is well peopled, though moft of its inhabitants are defcended from the Indian tribes. It is not the moft opulent in this country, for, befides the pillages it has fuffered, no mines are worked here; fo that moft of the money feen in it is fent from Santa Fe and Quito, to pay the falaries of the governor and other civil and military officers, and the wages of the garrifon; and even this makes no long ftay here. It is not however unfrequent to find perfons who have acquired handfome fortunes by commerce, whofe houfes are fplendidly furnifhed, and who live in every refpect agreeable to their wealth. The governor refides in the city, which till 1739 was independent of the military government. In civil affairs, an appeal lies to the audience of Santa Fe; and a viceroy of Santa Fe being that year created, under the title of viceroy of New Granada, the government of Carthagena became

fubject

fubject to him alfo in military affairs. The firft who filled this viceroyalty was lieutenant-general Don Sebaftian de Eflava ; who defended Carthagena againft the powerful invafion of the Englifh in 1741.

CARTHAGENA has alfo a bifhop, whofe fpiritual jurifdiction is of the fame extent as the military and civil government. The ecclefiaftical chapter is compofed of the bifhop and prebends. There is alfo a court of inquifition, whofe power reaches to the three provinces of Ifla Efpanola (where it was firft fettled), Terra Firma, and Santa Fe.

BESIDES thefe tribunals, the police and adminiftration of juftice in the city is under a fecular magiftracy, confifting of regidores, from whom every year are chofen two alcaldes, who are generally perfons of the higheft efteem and diftinction. There is alfo an office of revenue, under an accomptant and treafurer : here all taxes and monies belonging to the king are received ; and the proper iffues directed. A perfon of the law, with the title of auditor de la gente de guerra, determines proceffes.

THE jurifdiction of the government of Carthagena reaches eaftward to the great river de la Magdalena, and along it fouthward, till, winding away, it borders on the province of Antioquia ; from thence it ftretches weftward to the river of Darien ; and from thence northward to the ocean, all along the coafts between the mouths of thefe two rivers. The extent of this government from E. to W. is generally computed at 53 leagues.; and from S. to N. 85. In this fpace are feveral fruitful vallies, called by the natives favannahs ; as thofe of Zamba, Zenu, Tolu, Mompox, Baranca, and others ; and in them many fettlements large and fmall, of Europeans, Spanifh Creoles, and Indians. There is a tradition, that all thefe countries, together with that of Carthagena, whilft they continued in their native idolatry, abounded in gold ; and fome veftiges of the old mines of that metal are ftill to be feen, in the neighbourhoods of Simiti San Lucas, and Guamaco ;

but

but they are now neglected, being, as imagined, exhausted. But what equally contributed to the richnefs of this country was the trade it carried with Choco and Darien; from whence they brought, in exchange for this metal, the feveral manufactures and works of art they ftood in need of. Gold was the moft common ornament of the Indians, both men and women.

CHAP. III.

Defcription of Carthagena Bay.

CARTHAGENA bay is one of the beft, not only on the coaft, but alfo in all the known parts of this country. It extends 2½ leagues from north to fouth; has a fufficient depth of water and good anchorage; and fo fmooth, that the fhips are no more agitated than on a river. The many fhallows indeed, at the entrance, on fome of which there is fo little water that even fmall veffels ftrike, render a careful fteerage neceffary. But this danger may be avoided, as it generally is, by taking on board a pilot; and for further fecurity, his majefty maintains one of fufficient experience, part of whofe employment is to fix marks on the dangerous places.

The entrance to the bay, as I have already obferved, was through the narrow ftrait called Boca Chica, a name very properly adapted to its narrownefs, fignifying in Spanifh Little Mouth, admitting only one fhip at a time, and even fhe muft be obliged to keep clofe to the fhore. This entrance was defended on the E. by a fort called St. Lewis de Boca Chica, at the extremity of Tierra Bomba, and by Fort St. Jofeph on the oppofite fide in the ifle of Baru. The former, after fuftaining, in the laft fiege by the Englifh, a vigorous attack both by fea and land, and a cannonading of eleven days, its defences ruined, its parapets beat down, and all its artillery difmounted, was relinquifhed. The enemy being thus mafters of it, cleared the entrance,

and

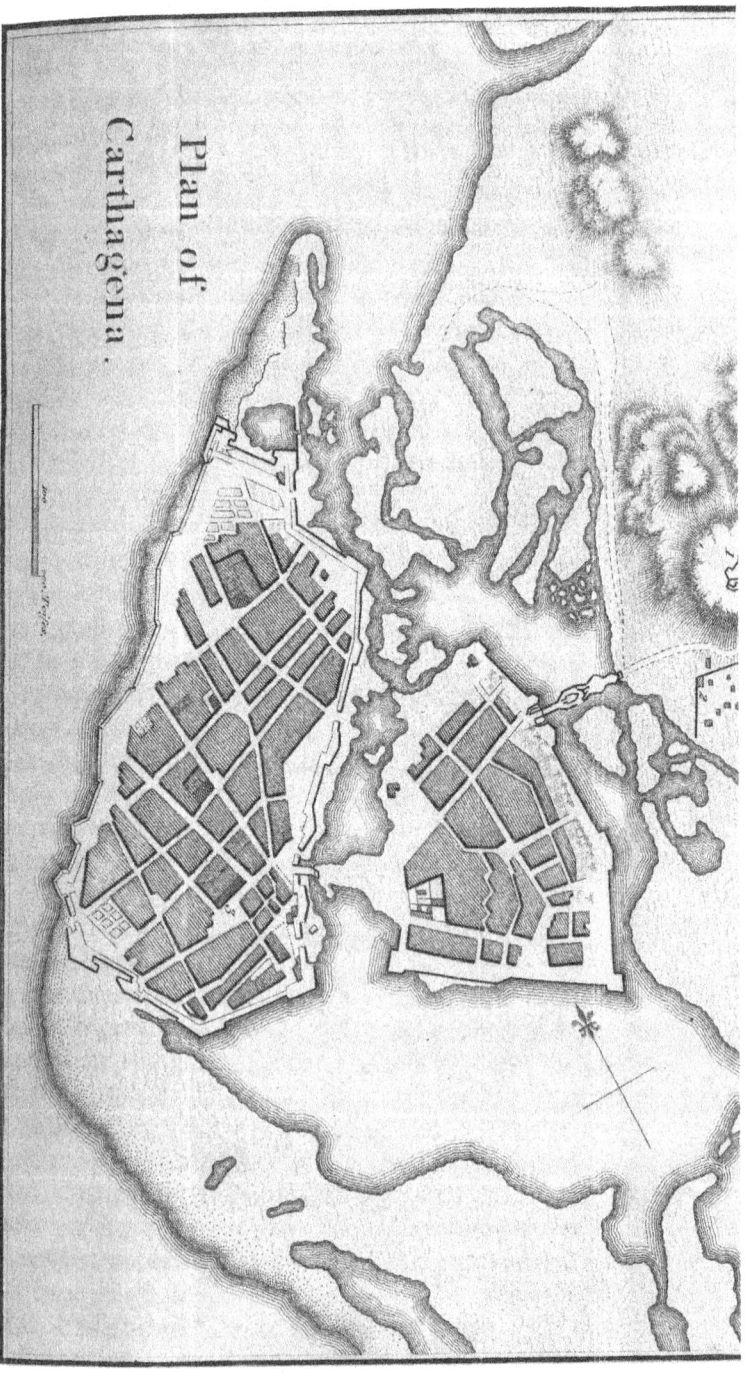

Plan of
Carthagena.

1. Fort of St Philip de Barras. 2. Hospital of St Lazaro. 3. The Cathedral.

Mode. cosp. Brauel.

Bay of Carthagena.

The Town

Forte de St Cruz

Scale of Toises.

Boca Grande

Boca Chica

and with their whole squadron and armaments moved
to the bottom of the bay.   But, by the diligence and
industry of our people, they found all the artillery of
fort Santa Cruz nailed up.   This fort was also, from
its largeness, called Castillo Grande, and commanded
all the ships which anchor in the bay.   This, together
with that of Boca Chica, St. Joseph, and two others,
which defended the bay, called Manzanillo and Paste-
lillo, the enemy, enraged at their disappointment, demolished when they quitted the bay.   The promising
beginning of this invasion, as I have already observed,
gave occasion to the shutting up and rendering impracticable the entrance of Boca Chica, and of opening and
fortifying the former strait; so that an enemy would
now find it much more difficult to force a passage.

Tʜᴇ tides in this bay are very irregular, and the same
may nearly be said of the whole coast.   It is often seen
to flow a whole day, and afterwards ebbs away in four
or five hours; yet the greatest alteration observed in its
depth is two feet, or two feet and a half. Sometimes it
is even less sensible, and only to be perceived by the
current or flow of the water.   This circumstance increases the danger of striking, though a serenity continually reigns there.   The bottom also being composed
of a gravelly ooze, whenever a ship is aground, it often
happens that she must be lightened before she can be
made to float.

Tᴏᴡᴀʀᴅs Boca Chica, and two leagues and a half
distant from it seawards, there is a shoal of gravel and
coarse sand, on many parts of which there is not above
a foot and a half of water. In 1735, the Conquistador
man of war, bound from Carthagena to Porto Bello,
struck on this shoal, and owed her safety entirely to a
very extraordinary calm. Some pretended to say that the
shoal was before known by the name of Salmedina; but
the artists on board affirmed the contrary, and that the
shoal on which she struck had never been heard of before.   From the observations of the pilots and others,
Nueſtra

Nueſtra Senora de la Popa bore E. N. E. two degrees north, diſtance two leagues; the caſtle of St. Lewis de Boca Chica, E. S. E. diſtance three leagues and a half, and the north part of Iſla Voſaria, ſouth one quarter weſterly. It muſt, however, be remembered that theſe obſervations were made on the apparent rhombs of the needle.

The bay abounds with great variety of fiſh both wholeſome and agreeable to the palate; the moſt common are the ſhad, the taſte of which is not indeed the moſt delicate. The turtles are large and well taſted. But it is greatly infeſted with ſharks, which are extremely dangerous to ſeamen, as they immediately ſeize every perſon they diſcover in the water, and ſometimes even venture to attack them in their boats. It is a common diverſion for the crews of thoſe ſhips who ſtay any time in the bay, to fiſh for theſe rapacious monſters, with large hooks faſtened to a chain; though, when they have caught one, there is no eating it, the fleſh being as it were a kind of liquid fat. Some of them have been ſeen with four rows of teeth; the younger have generally but two. The voracity of this fiſh is ſo prodigious, that it ſwallows all the filth either thrown out of ſhips, or caſt up by the ſea. I myſelf ſaw in the ſtomach of one, the entire body of a dog, the ſofter parts only having been digeſted. The natives affirm that they have alſo ſeen alligators; but this being a freſh-water animal, if any were ever ſeen in the ſea, it muſt be ſomething very extraordinary.

In the bay the galleons from Spain wait the arrival of the Peru fleet at Panama; and on the firſt advice of this, ſail away for Porto Bello; at the end of the fair held at that town, they return into this bay, and, after taking on board every neceſſary for their voyage, put to ſea again as ſoon as poſſible. During their abſence the bay is little frequented; the country veſſels, which are only a few bilanders and feluccas, ſtay no longer than is neceſſary to careen and fit out for proſecuting their voyage. CHAP.

CHAP. IV.

Of the Inhabitants of Carthagena.

THE inhabitants may be divided into different casts or tribes, who derive their origin from a coalition of Whites, Negroes, and Indians. Of each of these we shall treat particularly.

THE Whites may be divided into two classes, the Europeans, and Creoles, or Whites born in the country. The former are commonly called Chapetones, but are not numerous; most of them either return into Spain after acquiring a competent fortune, or remove up into inland provinces in order to increase it. Those who are settled at Carthagena, carry on the whole trade of that place, and live in opulence; whilst the other inhabitants are indigent, and reduced to have recourse to mean and hard labour for subsistence. The families of the White Creoles compose the landed interest ; some of them have large estates, and are highly respected, because their ancestors came into the country invested with honourable posts, bringing their families with them when they settled here. Some of these families, in order to keep up their original dignity, have either married their children to their equals in the country, or sent them as officers on board the galleons; but others have greatly declined. Besides these, there are other Whites, in mean circumstances, who either owe their origin to Indian families, or at least to an intermarriage with them, so that there is some mixture in their blood ; but when this is not discoverable by their colour, the conceit of being Whites alleviates the pressure of every other calamity.

AMONG the other tribes which are derived from an intermarriage of the Whites with the Negroes, the first are the Mulattos. Next to these the Tercerones, produced from a White and a Mulatto, with some approximation to the former, but not so near as to obliterate

their

their origin. After thefe follow the Quarterones, proceeding from a White and a Terceron. The laft are
the Quinterones, who owe their origin to a White and
Quarteron. This is the laft gradation, there being no
vifible difference between them and the Whites, either
in colour or features ; nay, they are often fairer than
the Spaniards. The children of a White and Quinteron are alfo called Spaniards, and confider themfelves
as free from all taint of the Negro race. Every perfon
is fo jealous of the order of their tribe or caft, that if,
through inadvertence, you call them by a degree lower
than what they actually are, they are highly offended,
never fuffering themfelves to be deprived of fo valuable
a gift of fortune.

BEFORE they attain the clafs of the Quinterones,
there are feveral intervening circumftances which
throw them back ; for between the Mulatto and the
Negro there is an intermediate race, which they
call Sambos, owing their origin to a mixture between
one of thefe with an Indian, or among themfelves.
They are alfo diftinguifhed according to the cafts their
fathers were of. Betwixt the Tercerones and the
Mulattos, the Quarterones and the Tercerones, &c.
are thofe called Tente en el Ayre, fufpended in the
air, becaufe they neither advance nor recede. Children, thofe parents are a Quarteron or Quinteron,
and a Mulatto or Terceron, are Salto atras, retrogrades,
becaufe, inftead of advancing towards being Whites,
they have gone backwards towards the Negro race.
The children between a Negro and Quinteron are
called Sambos de Negro, de Mulatto, de Terceron, &c.

THESE are the moft known and common tribes or
Caftas ; there are indeed feveral others proceeding
from their intermarriages ; but, being fo various, even
they themfelves cannot eafily diftinguifh them ; and
thefe are the only people one fees in the city, the
eftancias *, and the villages ; for if any Whites, efpecially

* Eftancia properly fignifies a manfion, or place where one ftops
to reft; but at Carthagena it implies a country-houfe, which, by reafon

cially women, are met with, it is only accidental; thefe generally refiding in their houfes; at leaft, if they are of any rank or character.

These cafts, from the Mulattos, all affect the Spanifh drefs, but wear very flight ftuffs on account of the heat of the climate. Thefe are the mechanics of the city; the Whites, whether Creoles or Chapitones, difdaining fuch a mean occupation, follow nothing below merchandife. But it being impoffible for all to fucceed, great numbers not being able to procure fufficient credit, they become poor and miferable from their averfion to thofe trades they follow in Europe; and, inftead of the riches which they flattered themfelves with poffeffing in the Indies, they experience the moft complicated wretchednefs.

The clafs of Negroes is not the leaft numerous, and is divided into two parts; the free and the flaves. Thefe are again fubdivided into Creoles and Bozares, part of which are employed in the cultivation of the haziandes *, or eftancias. Thofe in the city are obliged to perform the moft laborious fervices, and pay out of their wages a certain quota to their mafters, fubfifting themfelves on the fmall remainder. The violence of the heat not permitting them to wear any clothes, their only covering is a fmall piece of cotton ftuff about their waift; the female flaves go in the fame manner. Some of thefe live at the eftancias, being married to the flaves who work there; while thofe in the city fell in the markets all kind of eatables, and dry fruits, fweetmeats, cakes made of the maize, and caffava, and feveral other things about the ftreets. Thofe who have children fucking at their breaft, which is the cafe of the generality, carry them on their fhoulders, in order to have their arms at liberty; and when the infants are hungry,

___

fon of the great number of flaves belonging to it, often equals a confiderable village.

* Hazianda in this place fignifies a country-houfe, with the lands belonging to it.

they

they give them the breaſt either under the arm or over the ſhoulder, without taking them from their backs. This will perhaps appear incredible ; but their breaſts, being left to grow without any preſſure on them, often hang down to their very waiſt, and are not therefore difficult to turn over their ſhoulders for the convenience of the infant.

The dreſs of the Whites, both men and women, differs very little from that worn in Spain. The perſons in grand employments wear the ſame habits as in Europe ; but with this difference, that all their clothes are very light, the waiſtcoats and breeches being of fine Bretagne linen, and the coat of ſome other thin ſtuff. Wigs are not much worn here ; and during our ſtay, the governor and two or three of the chief officers only appeared in them. Neckcloths are alſo uncommon, the neck of the ſhirt being adorned with large gold buttons, and theſe generally ſuffered to hang looſe. On their heads they wear a cap of very fine and white linen. Others go entirely bareheaded, having their hair cut from the nape of the neck *. Fans are very commonly worn by men, and made of a very thin kind of palm in the form of a creſcent, having a ſtick of the ſame wood in the middle. Thoſe who are not of the White claſs, or of any eminent family, wear a cloak and a hat flapped ; though ſome Mulattos and Negroes dreſs like the Spaniards and great men of the country.

The Spaniſh women wear a kind of petticoat, which they call pollera, made of a thin ſilk, without any lining; and on their body, a very thin white waiſtcoat; but even this is only worn in what they call winter, it being inſupportable in ſummer. They however always lace in ſuch a manner as to conceal their breaſts. When they go abroad, they wear a mantelet; and on the days of

---

* Here, and in moſt parts of South America, they have their hair cut ſo ſhort, that a ſtranger would think every man had a wig, but did not wear it on account of the heat.—A.

precept, they go to mafs at three in the morning, in
order to difcharge that duty, and return before the violent heat of the day, which begins with the dawn *.

WOMEN wear over their pollera a taffety petticoat,
of any colour they pleafe, except black; this is pinked
all over, to fhew the other they wear under it. On
the head is a cap of fine white linen, covered with
lace, in the fhape of a mitre, and, being well ftarched,
terminates forward in a point. This they call panito,
and never appear abroad without it, and a mantelet
on their fhoulders. The ladies, and other native
Whites, ufe this as their undrefs, and it greatly becomes them; for having been ufed to it from their infancy, they wear it with a better air. Inftead of
fhoes, they only wear, both within and without doors,
a kind of flippers, large enough only to contain the
tip of their feet. In the houfe their whole exercife
confifts in fitting in their hammocks †, and fwinging
themfelves for air. This is fo general a cuftom, that
there is not a houfe without two or three, according
to the number of the family. In thefe they pafs the
greater part of the day; and often men, as well as
women, fleep in them, without minding the inconveniency of not ftretching the body at full length.

BOTH fexes are poffeffed of a great deal of wit and
penetration, and alfo of a genius proper to excel in
all kinds of mechanic arts. This is particularly confpicuous in thofe who apply themfelves to literature,
and who, at a tender age, fhew a judgment and perfpicacity, which, in other climates, is attained only by
a long feries of years and the greateft application.
This happy difpofition and perfpicacity continues till
they are between twenty and thirty years of age, after

* The heat is inconfiderable, compared with that of the afternoon, till half an hour after funrife. A.

† Thefe hammocks are made of twifted cotton, and commonly
knit in the manner of a net, and make no fmall part of the traffick
of the Indians, by whom they are chiefly made. A.

which

which they generally decline as fast as they rose; and frequently, before they arrive at that age, when they should begin to reap the advantage of their studies, a natural indolence checks their farther progress, and they forsake the sciences, leaving the surprising effects of their capacity imperfect.

The principal cause of the short duration of such promising beginnings, and of the indolent turn so often seen in these bright geniuses, is doubtless the want of proper objects for exercising their faculties, and the small hopes of being preferred to any post answerable to the pains they have taken. For as there is in this country neither army nor navy, and the civil employments very few, it is not at all surprising that the despair of making their fortunes, by this method, should damp their ardour for excelling in the sciences, and plunge them into idleness, the sure forerunner of vice; where they lose the use of their reason, and stifle those good principles which fired them when young and under proper subjection. The same is evident in the mechanic arts, in which they demonstrate a surprising skill in a very little time; but soon leave these also imperfect, without attempting to improve on the methods of their masters. Nothing indeed is more surprising than the early advances of the mind in this country, children of two or three years of age conversing with a regularity and seriousness that is rarely seen in Europe at six or seven; and at an age when they can scarce see the light, are acquainted with all the depths of wickedness.

The genius of the Americans being more forward than that of the Europeans, many have been willing to believe that it also sooner decays; and that at sixty years, or before, they have outlived that solid judgment and penetration, so general among us at that time of life; and it has been said that their genius decays, while that of the Europeans is hastening to its maturity and perfection. But this is a vulgar prejudice,

dice, confuted by numberlefs inftances, and particularly by the celebrated father Fr. Benito Feyjoo,
Têatro Critico, vol. iv. effay 6. All who have travelled
with any attention through thefe countries, have obferved in the natives of every age a permanent capacity, and uniform brightnefs of intellect; if they were
not of that wretched number, who diforder both their
minds and bodies by their vices. And indeed one
often fees here perfons of eminent prudence and extenfive talents, both in the fpeculative and practical
fciences, and who retain them, in all their vigour, to
a very advanced age.

CHARITY is a virtue in which all the inhabitants of
Carthagena, without exception, may be faid particularly to excel; and did they not liberally exert it towards European ftrangers, who generally come hither
to feek their fortune, fuch would often perifh with
ficknefs and poverty. This appears to me a fubject
of fuch importance, though well known to all who
have vifited this part of the world, that I fhall add a
word or two on it, in order to undeceive thofe, who,
not contented with perhaps a competent eftate in their
own country, imagine that is only fetting their foot in
the Indies, and their fortune is made.

THOSE who on board the galleons are called Puli-
zones, as being men without employment, ftock, or
recommendation; who, leaving their country as fugi-
tives, and, without licenfe from the officers, come to
feek their fortune in a country where they are utterly
unknown; after traverfing the ftreets till they have
nothing left to procure them lodging or food, they
are reduced to have recourfe to the laft extremity,
the Francifcan hofpital; where they receive, in a
quantity fufficient barely to keep them alive, a kind
of pap made of cafava; of which as the natives themfelves will not eat, the difagreeablenefs to wretched
mortals never ufed to fuch food, may eafily be conceived.

ceived *. This is their food; their lodging is the entrance of the fquares and the porticos of churches, till their good fortune throws them in the way of hiring themfelves to fome trader going up the country, who wants a fervant. The city merchants, ftanding in no need of them, difcountenance thefe adventurers. Affected by the difference of the climate, aggravated by bad food, dejected and tortured by the entire difappointment of their romantic hopes, they fall into a thoufand evils, which cannot well be reprefented; and among others, that diftemper called Chapetonada, or the diftemper of the Chapetones, without any other fuccour to fly to, than Divine Providence; for none find admittance into the hofpital of St. Juan de Dios, but thofe who are able to pay, and, confequently, poverty becomes an abfolute exclufion. Now it is that the charity of thefe people becomes confpicuous. The Negro and Mulatto free women, moved at their deplorable condition, carry them to their houfes, and nurfe them with the greateft care and affection. If any one die, they bury him by the charity they procure, and even caufe maffes to be faid for him. The general iffue of this endearing benevolence is, that the Chapetone, on his recovery, during the fervour of his gratitude, marries either his Negro or Mulatto benefactrefs, or one of her daughters; and thus he becomes fettled, but much more wretchedly than he could have been in his own country, with only his own labour to fubfift on.

THE difintereftednefs of thefe people is fuch, that their compaffion towards the Chapetones muft not be imputed to the hopes of producing a marriage, it being very common for them to refufe fuch offers,

* This is called Mandioc by the natives, and is the chief fubftitute the poorer people have for bread; and fo far from being rejected even by the richer, that many prefer it to bread made from the beft European flour, much more to bifcuit, which after fuch a voyage generally begins to be full of weevils. A.

either

either with regard to themſelves or their daughters, that their miſery may not be perpetual, but endeavour to find them maſters whom they may attend up the country, to Santa Fe, Popyan, Quito, and Peru, whither their inclination or proſpects lead them.

They who remain in the city, whether bound by one of the above marriages, or, which is but too common, are in another condition very dangerous to their future happineſs, turn Pulperos *, Canoeros, or ſuch-like mean occupations : in all which, they are ſo haraſſed with labour, and their wages ſo ſmall, that their conditon in their own country muſt have been miſerable indeed, if they have not reaſon to regret quitting it. The height of their enjoyment, after toiling all day and part of the night, is to regale with bananas, a cake of maize or caſava, which ſerves for bread, and a ſlice of caſajo, or hung-beef; without taſting wheat bread during the whole year.

Others, not a few, equally unfortunate, retire to ſome ſmall eſtancia, where, in a Bujio, or ſtraw hut, they live little different from beaſts, cultivating, in a very ſmall ſpot, ſuch vegetables as are at hand, and ſubſiſting on the ſale of them.

What has been obſerved with regard to the Negro and Mulatto women, and which may alſo be extended to the other caſts, is, as to the charitable part, applicable to all the women and whites; who, in every tribe, are of a very mild and amiable diſpoſition ; and from their natural ſoftneſs and ſympathy excel the men in the practice of that Chriſtian virtue.

Among the reigning cuſtoms here, ſome are very different from thoſe of Spain, or the moſt known parts of Europe. The principal of theſe are the uſe of brandy, chocolate, honey, ſweetmeats, and ſmoking tobacco ; all which ſhall be taken notice of.

* Pulperos are men who work in a kind of tent, called in Spaniſh Pulperios, and the Canoeros are watermen who carry goods in Pirogues or canoes.

The

THE ufe of brandy is fo common, that the moft regular and fober perfons never omit drinking a glafs of it every morning about eleven o'clock; alleging that this fpirit ftrengthens the ftomach, weakened by copious and conftant perfpiration, and fharpens the appetite. Hazer las onze, to take a whet at eleven, that is, to drink a glafs of brandy, is the common invitation. This cuftom, not efteemed pernicious by thefe people, when ufed with moderation, has degenerated into vice; many being fo fond of it, that, during the whole day, they do nothing but hazer las onze. Perfons of diftinction ufe Spanifh brandy; but the lower clafs and Negroes very contentedly take up with that of the country, extracted from the juice of the fugar cane, and thence called Agoa ardente de canna, or cane brandy, of which fort the confumption is much the greateft.

CHOCOLATE, here known only by the name of cacao, is fo common, that there is not a Negro flave but conftantly allows himfelf a regale of it after breakfaft; and the Negro women fell it ready made about the ftreets, at the rate of a quarter of a real (about five farthings fterling) for a difh. This is however fo far from being all cacao, that the principal ingredient is maize; but that ufed by the better fort is neat, and worked as in Spain. This they conftantly repeat an hour after dinner, but never ufe it fafting, or without eating fomething with it.

THEY alfo make great ufe of fweetmeats and honey; never fo much as drinking a glafs of water without previoufly eating fome fweetmeats. Honey is often preferred, as the fweeter, to conferves or other fweetmea s either wet or dry. Their fweetmeats are eaten with wheat bread, which they ufe only with thefe and chocolate; the honey they fpread on cafava cakes.

THE paffion for fmoking is no lefs univerfal, prevailing among perfons of all ranks in both fexes. The

ladies

ladies and other white women fmoke in their houfes, a decency not obferved either by the women of the other cafts, nor by the men in general, who regard neither time nor place. The manner of ufing it is, by flender rolls compofed of the leaves of that plant; and the women have a particular manner of inhaling the fmoke. They put the lighted part of the roll into their mouths, and there continue it a long time without its being quenched, or the fire incommoding them. A compliment paid to thofe for whom they profefs an intimacy and efteem, is, to light their tobacco for them, and to hand them round to thofe who vifit them. To refufe the offer would be a mark of rudenefs not eafily digefted; and accordingly they are very cautious of paying this compliment to any but thofe whom they previoufly know to be ufed to tobacco. This cuftom the ladies learn in their child-hood from their nurfes, who are Negro flaves; it is fo common among perfons of rank, that thofe who come from Europe eafily join in it, if they intend to make any confiderable ftay in the country.

One of the moft favourite amufements of the natives here, is a ball, or Fandango. Thefe are the diftinguifhed rejoicings on feftivals and remarkable days. But while the galleons, guarda coftas, or other Spanifh fhips are here, they are moft common, and at the fame time conducted with the leaft order; the crews of the fhips forcing themfelves into their ball-rooms. Thefe diverfions, in houfes of diftinction, are conducted in a very regular manner; they open with Spanifh dances, and are fucceeded by thofe of the country, which are not without fpirit and grace-fulnefs. Thefe are accompanied with finging, and the parties rarely break up before daylight.

The Fandangos, or balls, of the populace, confift principally in drinking brandy and wine, intermixed with indecent and fcandalous motions and geftures; and thefe continual rounds of drinking foon give rife

D 4   to

to quarrels, which often bring on misfortunes. When strangers of rank visit the city, they are generally at the expense of these balls; as the entrance is free, and no want of liquor, they need give themselves no concern about the want of company.

THEIR burials and mournings are something singular; as in this particular they endeavour to display their grandeur and dignity, too often at the expense of their tranquillity. If the deceased be a person of condition, his body is placed on a pompous Catafalco, erected on the principal apartment of the house, amidst a blaze of tapers. In this manner the corpse lies twenty-four hours or longer, for friends to visit it at all hours; as also the lower class of women, among whom it is a custom to lament over the deceased.

THESE women, who are generally dressed in black, come in the evening, or during the night, into the apartment where the corpse lies; and having approached it, throw themselves on their knees, then rise and extend their arms as to embrace it; after which, they begin their lamentations, in a doleful tone, mixed with horrid cries, which always conclude with the name of the deceased; afterwards they begin, in the same disagreeable vociferations, his history, rehearsing all his good and bad qualities, not even omitting his amours of any kind, and in so circumstantial a narrative, that a general confession could hardly be more full; at length, quite spent, they withdraw to a corner stored with brandy and wine, on which they never fail plentifully to regale themselves. As these depart from the body, others succeed, till they have all taken their turn. The same, afterwards, is repeated by the servants, slaves, and acquaintance of the family, which continues, without intermission, during the remainder of the night; whence may easily be imagined the confusion and noise occasioned by this dismal vociferous ceremony.

THE

THE funeral alſo is accompanied with the like noiſy lamentations; and even after the corpſe is depoſited in the grave, the mourning is continued in the houſe for nine days, during which time the Pacientes or mourners, whether men or women, never ſtir from the apartment, where they receive the Peſanes, or compliments of condolence. During nine nights, from ſun-ſet to ſun riſing, they are attended by their relations and intimate acquaintances; and it may be truly ſaid of them, that they are all ſincerely ſorrowful; the mourners for the loſs of the deceaſed, and the viſitors from the uneaſineſs and fatigue of ſo uncomfortable an attendance.

CHAP. V.

Of the Climate of Carthagena, and the Diſeaſes incident to Natives and Foreigners.

THE climate of Carthagena is exceſſively hot, for, by obſervations we made on the 19th of November 1735, by a thermometer conſtructed according to Mr. Reaumur, the ſpirit was elevated to 1025½; and in our ſeveral experiments, made at different hours, varied only from 1024 to 1026. By experiments made the ſame year at Paris on a thermometer of the ſame gentleman, the ſpirit roſe on the 16th of July at 3 in the afternoon, and on the 10th of Auguſt at half an hour after 3, to 1025½, and this was the greateſt degree of heat felt at Paris during that year; conſequently the degree of heat in the hotteſt day at Paris, is continual at Carthagena.

BUT the nature of this climate chiefly diſplays itſelf from the month of May to the end of November, the ſeaſon they call winter; becauſe, during that time, there is almoſt a continual ſucceſſion of thunder, rain, and tempeſts; the clouds precipitating the rain with
ſuch

such impetuosity, that the streets have the appearance of rivers, and the country of an ocean. The inhabitants make use of this opportunity, otherwise so dreadful, for filling their cisterns; this being the only sweet water they can procure. Besides the water saved for private uses, there are large reservoirs on the bastions, that the town may not be reduced to the shocking consequence of wanting water. There are indeed wells in most houses; but the water being thick and brackish, is not fit to drink, but serves for other uses.

From the middle of December to the end of April, the rains cease, and the weather becomes agreeable, the heat being something abated by the N E. winds which then set in. This season they call summer; besides which, there is another called the little summer of St. John, as, about the festival of that saint, the rains are intermitted, and refreshing gales begin to blow, and continue about a month.

The invariable continuance of such great heats, without any sensible difference between night and day, occasions such profuse perspiration, that the wan and livid complexion of the inhabitants would make a stranger suspect they were just recovered from some terrible distemper. Their actions are conformable to their colour; in all their motions there is something lax and sluggish; it even affects their speech, which is soft and slow, and their words generally broken. But notwithstanding all these appearances of sickness and debility, they enjoy a good state of health. Strangers from Europe retain their strength and ruddy colour possibly for three or four months; but afterwards both suffer such decays from the excessive perspiration, that these new comers are no longer to be distinguished by their countenances from the old inhabitants. Young persons are generally most affected by the climate, which spares the more aged, who preserve their vivid countenance, and so confirmed a state of health as even to reach their 80th year and upwards;

upwards : this is common among all the claffes of inhabitants.

THE fingularity of the climate, in all probability,
occafions the fingularity of fome of the diftempers
which here affect the human race ; and thefe may be
confidered in two different lights ; one, as only attack-
ing the Europeans newly landed, and the other, as
common both to Creoles and Chapitones.

THOSE of the firft kind are, in the country, commonly called Chapetonadas, alluding to the name
given there to the Europeans. Thefe diftempers are
fo very deleterious, that they carry off a multitude of
people, and thin the crews of European fhips ; but
they feldom laft above three or four days, in which
time the patient is either dead or out of danger. The
nature of this diftemper is but little known, being
caufed in fome perfons by cold, and in others by in-
digeftions ; it foon brings on the vomito prieto, or
black vomit, which is the fatal fymptom ; very few
being ever known to recover. Some, when the vomit
attacks them, are feized with fuch a delirium, that,
were they not tied down, they would tear themfelves
to pieces, and thus expire in the midft of their furious
paroxyfms. It is remarkable, that only the new-
comers from Europe are fubject to this diftemper, and
that the natives, and thofe who have lived fome time
here, are never affected by it ; but enjoy an uninterrupted ftate of health. amidft the dreadful havock it makes
among others. It is alfo obferved to rage more among
the common feamen, than thofe who have been able
to live on more wholefome food ; whence, falt meat
has been confidered as pernicious in bringing on this
diftemper, and that the humours it produces, together
with the labour and hardfhips of the feamen, incline
their blood to putrefaction, and from this putrefaction
the vomito prieto is fuppofed to have its origin. Not
that the failors are its only victims, for even paffengers, who poffibly have not tafted any falt meat
during

during the voyage, often feel its effects. Another remarkable circumstance is, that persons who have been once in this climate are never after, upon their return again, subject to this distemper; but enjoy the same state of health with the natives, even though they do not lead the most temperate lives.

The investigation of the cause of this strange distemper has exercised the attention of all the surgeons in the galleons, as well as the physicians of the country; and the result of their researches is, that they impute it to the food, labour, and hardships of the seamen. Doubtless these are collateral causes; but the principal question is, why persons exempt from those inconveniences, frequently die of the distemper? Unhappily, after all the experiments that have been made, no good method of treatment has been discovered; no specific for curing it, nor preservative against it. The symptoms are so vague, as sometimes not to be distinguished from those of slight indispositions; and though the vomit be the determinate symptom, the fever preceding it is observed to be very oppressive, and extremely affecting to the head.

This distemper does not shew itself immediately after the arrival of the European ships in the bay, nor has it been long known here; for what was formerly called Chapetonadas, were only indigestions, which, though always dangerous in these climates, were, with little difficulty, cured by remedies prepared by the women of the country, and which are still used with success, especially if taken in the beginning. The ships afterwards going to Porto Bello, were there first attacked by this terrible disease, which has always been attributed to the inclemency of the climate, and the fatigue of the seamen in unloading the ships, and drawing the goods during the fair.

The vomito prieto was unknown at Carthagena and all along the coast, till the years 1729 and 1730.

In

In 1729, Don Domingo Juſtiniani, commodore of the guarda coſtas, loſt ſo conſiderable a part of his ſhips' companies at Santa Martha, that the ſurvivors were ſtruck with aſtoniſhment and horror at the havock made among their comrades. In 1730, when the galleons under Don Manuel Lopez Pintado came to Carthagena, the ſeamen were ſeized with the ſame dreadful mortality; and ſo ſudden were the attacks of the diſeaſe, that perſons walking about one day, were the next carried to their graves.

The inhabitants of Carthagena, together with thoſe in the whole extent of its government, are very ſubject to the mal de San Lazaro, or leproſy, which ſeems ſtill to gain ground. Some phyſicians attribute the prevalence of it to pork, which is here a very common food; but it may be objected, that in other countries, where this fleſh is as frequently eaten, no ſuch effects are ſeen, whence it evidently appears that ſome latent quality of the climate muſt alſo contribute to it. In order to ſtop the contagion of this diſtemper, there is without the city, an hoſpital called San Lazaro, not far from the hill on which is a caſtle of the ſame name. In this hoſpital all perſons of both ſexes labouring under this diſtemper are confined, without any diſtinction of age or rank; and if any refuſe to go, they are forcibly carried thither. But here the diſtemper increaſes among themſelves, they being permitted to intermarry, by which means it is rendered perpetual. Beſides, their allowance being here too ſcanty to ſubſiſt on, they are permitted to beg in the city; and from their intercourſe with thoſe in health, the number of lepers never decreaſes, and is at preſent ſo conſiderable, that their hoſpital reſembles a little town. Every perſon at his entering this ſtructure, where he is to continue during life, builds a cottage called in the country bujio, proportional to his ability, where he lives in the ſame manner as before in his houſe, the prohibition of not going beyond the limits preſcribed him,

him, unlefs to afk alms in the city, only excepted. The ground on which the hofpital ftands is furrounded by a wall, and has only one gate, and that always carefully guarded.

AMIDST all the inconveniences attending this diftemper, they live a long time under it, and fome even attain to an advanced age. It alfo greatly increafes the natural defire of coition, and intercourfe of the fexes: fo that, to avoid the diforders which would refult from indulging this paffion, now almoft impoffible to be controlled, they are permitted to marry.

IF the leprofy be common and contagious in this climate, the itch and herpes are equally fo, efpecially among Europeans, who are not feafoned to the climate; and, if neglected in the beginning, it is dangerous to attempt a cure when cuftom has rendered them natural. The remedy againft them, in the firft ftage, is a kind of earth called maquimaqui, found in the neighbourhood of Carthagena, and, on the account of this virtue, exported to other parts.

ANOTHER very fingular diftemper, though not fo common, is the cobrilla, or little fnake, being, as the moft fkilful think, a tumour caufed by certain malignant humours, fettled longitudinally between the membrane of the fkin, and daily increafing in length, till the fwelling quite furrounds the part affected, which is ufually the arm, thigh, and leg; though fometimes it has been known to fpread itfelf all over thefe parts. The external indications of it are, a round inflamed tumour, of the thicknefs of a quarter of an inch, attended with a flight pain, but not vehement, and a numbnefs of the part, which often terminates in a mortification. The nátives are very fkilful in removing it by the following procefs. They firft examine where (according to their phrafe) the head is, to which they apply a fmall fuppurative plafter, and gently foment the whole tumour with oil. The next day the fkin under the plafter is found divided, and

through

through the orifice appears a kind of white fibre, about the fize of a coarfe fewing thread; and this, according to them, is the cobrilla's head, which they carefully faften to a thread of filk, and wind the other end of it about a card, rolled up like a cylinder. After this they repeat the fomentation with oil, and the following day continue to wind about the cylindric card the part of this fmall fibre which appears in fight. Thus they proceed till the whole is extracted, and the patient entirely cured. During this operation their chief care is not to break the cobrilla; becaufe, they fay, it would then caufe a humour to fpread through the body, and produce a great quantity of fuch little fnakes, as they will have them to be, when the cure would become extremely difficult. It is a current notion among them, that when it has, for want of care in the beginning, completed the circle, and, according to them, joined its head with its tail, the difeafe generally proves fatal. But this is very feldom the cafe; the pain warning the patient immediately to apply a remedy, which fhould be accompanied with emollients for difperfing the humour.

These people firmly believe it to be a real cobrilla or fmall fnake, and accordingly have called it by that name. At its firft appearance, a fmall flow motion may indeed be perceived; but this is foon over, and poffibly proceeds from the compreffion or extenfion of the nervous fibres which compofe it, without its having any animal life. I do not, however, pretend to determine abfolutely on this point.

Besides thefe, another diftemper common in this country is the fpafm, or convulfion, which always proves mortal, and feldom comes alone. And of this I fhall fpeak when I defcribe other parts of America, where it is equally dangerous, and more common.

CHAP.

CHAP. VI.

Description of the Country, and of the Trees and Vegetables in the Neighbourhood of Carthagena.

THE country about Carthagena is so luxuriant, that it is impossible to view without admiration the rich and perpetual verdure of the woods, and plants it naturally produces. But these are advantages of which the natives make little use; their innate sloth and indolence not allowing them to cultivate the gifts of nature, which seem to have been dealt out with a lavish hand. The interwoven branches of the trees form a shelter impenetrable both to heat and light.

THE trees here are large and lofty, their variety admirable, and entirely different from those of Europe. The principal of these for dimensions are, the caobo or acajou, the cedar, the maria, and the balsam tree. Of the first are made the canoes and champanes used for fishing, and the coast and river trade, within the jurisdiction of this government. These trees produce no eatable fruit; but their wood is compact, fragrant, and beautiful. The cedar is of two kinds, white and reddish; but the last most esteemed. The maria and the balsam trees, besides the usefulness of their timber, distil those admirable balsams called maria oil, and balsam of Tolu, so called from a village in the neighbourhood of which it is found in the greatest quantity, and of a peculiar excellency.

BESIDES these trees, here are also the tamarind, the medlar, the sapote, the papayo, the guayabo, the cannafistulo or caffia, the palm, the mançanillo, and several others, most of them producing a wholesome and palatable fruit, with a durable and variegated wood. The mançanillo is particularly remarkable; its name is derived from the Spanish word mançan,

an

an apple, which the fruit of this tree exactly refembles in fhape, colour, and flavour; but contains, under this beautiful appearance, fuch a fubtle poifon, that its effects are perceived before it is tafted. The tree is large, and its branches form near the top a kind of crown; its wood hard, and of a yellowifh tinct. On being cut, it iffues out a white juice, but not unlike that of the fig-tree, lefs white and of a thinner con-fiftence; but equally poifonous with the fruit itfelf; for if any happens to drop on any part of the flefh, it immediately caufes an ulcer and inflammation, and, unlefs fpeedy application be ufed, foon fpreads through all the other parts of the body *; fo that it is necef-fary, after felling it, to leave it till thoroughly dried, in order to its being worked without danger; and then appears the beauty of this wood, which is exquifitely variegated and veined like marble on its yellow ground. Upon tafting its fruit, the body immediately fwells, till the violence of the poifon, wanting fuf-ficient room, burfts it; as has been too fully con-firmed by feveral melancholy inftances of European failors who have been fent on fhore to cut wood. The fame unhappy confequence alfo attended great num-bers of Spaniards at the conqueft of thefe countries, till, according to Herrera, common oil was found to be the powerful antidote to this fubtle poifon.

But fuch is the malignity of the mançanillo, that if a perfon happens to fleep under it, he is foon awaked, and finds his body fwelled almoft as much as if he had actually eaten the fruit †; and continues in great

* The juice dropping on the flefh generally caufes an inflamma-tion; but I do not remember ever to have feen an ulcer produced, or any very bad effects, the hot burning pain excepted. A.

† The author is here mifinformed. Indeed perfons, who have flept under the tree, have afterwards complained of an head-ach. Thofe who happen to take fhelter under it in a fhower, generally feel the fame effect from the dropping of the leaves, as though the juice had dropt on them. A.

danger

danger and tortures; till relieved by repeated anointings and the use of cooling draughts. The very beasts themselves, by their natural instinct, are so far from eating its fruit, that they never approach the tree.

The palm-trees, rising with their tufted heads above the branches of the others, form a grand perspective on the mountains. These, notwithstanding the difference is scarce perceivable, are really of different kinds, as is evident from the diversity of their fruit. They distinguish four principal species: the first produce coco; the second dates, of a very pleasant taste; the third, called palma-real, whose fruit, though of the same figure, but something less than the date, is not at all palatable, but has a very disagreeable taste; and the fourth, which they call corozo, has a fruit larger than dates, of an exquisite taste, and proper for making cooling and wholesome draughts. The palmitos, or branches of the palma-real, are agreeably tasted, and so large as frequently to weigh from two to three arrobas *. The other species also produce them, but neither in such plenty, nor so succulent. Palm-wine is also extracted from all the four; but that from the palma-real and corozo is much the best. The manner of making it, is either by cutting down the palm-tree, or boring a hole in the trunk, in which is placed a tap, with a vessel under it for receiving the liquor, which, after five or six days fermentation, becomes fit for drinking. The colour of it is whitish; the taste racy: it bears a greater head than beer, and is of a very inebriating quality. The natives, however, reckon it cooling, and it is the favourite liquor of the Indians and Negroes. The guaiacum and ebony trees are equally common; and their hardness almost equal to that of iron. These species of wood are sometimes carried into Spain, where they are greatly esteemed, but here they are disregarded from their great plenty.

* The arroba is 25 pounds.

Among

Among the variety of vegetables, which grow under the fhade of the trees, and along the funny borders of the woods, the moft common is the fenfitive; on touching one of the leaves of which, all thofe on the fame branch immediately clofe againft each other. After a fhort interval, they begin gradually to open and feparate from each other, till they are entirely expanded. The fenfitive is a fmall plant about a foot and a half or two feet in height, with a flender ftem, and the branches proportionally weak and tender. The leaves are long, and ftand fo clofe together, that all on one branch may be confidered as a fingle leaf, four or five inches in length, and ten lines in breadth; which, being fubdivided into he other ftill fmaller, forms in each of them the true leaf, which is about four or five lines in length, and not quite one in breadth. On touching one of thefe fmall leaves, all of them immediately quit their horizontal pofition, and fly into a perpendicular direction, clofing their inward fuperficies, fo that thofe, which before this fenfitive motion made two leaves, now feem as but one. The vulgar name of this plant at Carthagena being improper to be mentioned here, we fhall omit it; in other parts it is more decently called la vergonoza, the bafhful, and la donçella, the maiden. The common people imagine that this effect is caufed by pronouncing its name at the inftant of the touch; and are amazed that a plant fhould have the wifdom of fhewing its obedience to what was ordered, or that it was too much affected by the injury offered it to conceal its refentment.

We afterwards meet with this plant at Guayaquil, where the climate feems to be better adapted to it than that of Carthagena; for it is not only more common, but grows to three or four feet in height, the leaves and every part in proportion.

In the woods about Carthagena are found a great quantity of bejucos of a different magnitude, figure, and colour, and fome of the ftems flat. One fpecies is

particularly

particularly known on account of its fruit called habilla de Carthagena, the bean of Carthagena.   It is
about an inch broad, and nine lines in length, flat, and
in the shape of a heart.   The shell, though thin, is
hard, and on the outside scabrous. It contains a kernel
resembling an almond, but less white, and extremely
bitter.   This is one of the most effectual antidotes
known in that country against the bites of vipers and
serpents ; for a little of it being eaten immediately after the bite, it presently stops the effects of the poison ;
and accordingly all who frequent the woods, either for
felling trees or hunting, never fail to eat a little of this
habilla fasting, and repair to their work without any
apprehension.   I was informed by an European, who
was a famous hunter, and by several other persons
worthy of credit, that, with this precaution, if any one
happened to be bit by a serpent, it was attended with
no ill consequence.   The natives tell you, that, this
habilla being hot in the highest degree, much of it
cannot be eaten, that the common dose of it is less than
the fourth part of a kernel, and that no hot liquor, as
wine, brandy, &c. must be drunk immediately after
taking it.   In this case they doubtless derive their
knowledge from experience.   This valuable habilla is
also known in other parts of America near Carthagena,
and goes every where by its name, as being the peculiar product of its jurisdiction.

CHAP. VII.

*Of the Beasts, Birds, Reptiles, and Insects, in the
Territories of Carthagena.*

FROM the trees and plants in this jurisdiction, we
shall proceed to the different kinds of animals ;
some of which are tame for the use and pleasure of its
inhabitants ; others wild, and of such different qualities

ties and kinds, as wonderfully diſplay the diverſity which the Author of nature has ſhewn in the multitude of his works.   The quadrupeds and reptiles frequent the dry and deſert places, and are diſtinguiſhed by an endleſs variety of ſpots, whilſt the vivid plumage of the feathered race glows with exquiſite beauty; and the brilliant ſcales of another kind conceal the moſt active poiſons.

The only tame eatable animals are the cow and the hog, of which there are great plenty.   The beef, though not abſolutely bad, cannot be ſaid to be palatable.   The conſtant heat of the climate preventing the beaſts from fattening, deprives their fleſh of that ſucculency it would otherwiſe have acquired: the pork is delicate, and allowed not only to be the beſt in all America, but even to exceed any in Europe.   This, which is the uſual food of Europeans and Creoles at Carthagena, beſides its palatableneſs, is alſo looked upon to be ſo wholeſome, that even ſick perſons are allowed it preferably to poultry, which is here very good and in great abundance.

I must not omit a ſingular ſtratagem practiſed here for taking wild geeſe, the extreme cheapneſs of which naturally inclined us to aſk how they caught them in ſuch quantities: in anſwer to our queſtion, we received the following account.   Near Carthagena, to the eaſtward of Monte de la Popa, is a large lake called la Cienega de Teſcas, abounding with fiſh, but reckoned unwholeſome.   The water of this lake, communicating with the ſea, is ſalt, but without increaſe or decreaſe, the difference of the tides here being inſignificant.   Every evening vaſt flights of geeſe retire hither from all the neighbouring countries, as their natural place of reſt during the night.   The perſons who catch theſe birds, throw into the lake about 15 or 20 large calabaſhes, which they call totumos; and the geeſe, being accuſtomed to ſee theſe calabaſhes floating on the water, never avoid them.   In three or four days the

perſons

perfons return early in the morning to the lake, with another calabafh, having holes in it for feeing and breathing. This calabafh he places on his head, and walks in the water, with only the calabafh above the furface. In this manner, with all poffible ftillnefs, he moves towards the geefe, pulling them under water with one hand, and then feizing them with the other. When he has thus taken as many as he is able to carry, he returns towards the fhore, and delivers them to his companion, who waits for him at a certain diftance in the water. This done, he renews his fport, either till he has taken as many as he defires, or the birds begin to difperfe over the country.

OTHER perfons make it their bufinefs to procure different kinds of game, as deer, rabbits, and wild boars, called here fajones; but thefe are eaten only by the country Negroes and Indians, except the rabbits, which meet with a good market in the city.

THE wild beafts are alfo of various kinds; as tigers, which make a great havock, not only among the cattle, but among the human fpecies. Their fkin is very beautiful, and fome are as large as little horfes*. Here are alfo leopards, foxes, armadillos, a kind of fcaly lizard; ardillas, or fquirrels, and many others; befides innumerable kinds of monkies living in the woods, fome remarkable for their fize, others for their colour The artifice generally obferved by the fox, in defending itfelf againft dogs or other animals, by whom it is purfued, by voiding its urine on its own tail and fprinkling it on them, effectually here anfwers the intention; the fmell of it being fo ftrong and fetid, that it throws the dogs into diforder, and thus the fox efcapes. The ftench of this urine is fo great, that it may be fmelt a quarter of a league from the place; and very often for half an hour after. The fox here is not much bigger than a large cat; but delicately fhaped; has a very fine coat, and of a cinnamon

* They are not larger than maftiff dogs. A.

colour;

colour; but no large brush on its tail. The hair however is spungy, and forms a bunch proper for the above-mentioned method of defence.

NATURE, which has furnished the fox with such an effectual defence, has not forgot the armadillo, the name of which partly describes it. The size of it is about that of a common rabbit, though of a very different shape; the snout, legs, and tail, resembling those of a pig. His whole body is covered with a strong shell, which, answering exactly every where to the irregularities of its structure, protects it from the insults of other animals, without affecting its activity. Besides this, he has another, as a helmet, connected by a joint to the former; this guards his head, and thus he is every way safe.

THESE shells are variegated with several natural relievos, as it were, in chiaro oscuro, so that they are at once his defence, and a beautiful ornament. The Negroes and Indians, who eat its flesh, give a high character of it.

AMONG the monkies of this country, the most common are the micos, which are also the smallest. They are generally about the size of a cat, of a brownish colour; and too well known to need any further description. The larger kind, which are less known, I shall describe in another place.

THE birds seen in this hot climate are so numerous, that it is impossible to give a distinct representation of them; particularly of the beauty and brilliancy of their various plumage. The cries and croakings of some, mixed with the warblings of others, disturb the pleasure which would flow from the melody of the latter, and render it impossible to distinguish the different cries of the former; and yet in this instance we may observe the wisdom of Nature in distributing her favours; the plumage of those birds being the most beautiful, whose croakings are the most offensive; while, on the other hand, those whose appearance has

E 4   nothing

nothing remarkable excel in the fweetnefs of their notes. This is particularly evident in the guacamayo, the beauty and luftre of whofe colours are abfolutely inimitable by painting; and yet there is not a more fhrill and difagreeable found than the noife it makes: this is in a great meafure common to all other birds, whofe bills are hard and crooked, and their tongue thicker than ufual, as the parrots, the cotorras, and the periquitos. All thefe birds fly in troops, fo that the air often founds with their cries.

But of all the fingularities among the feathered race, nothing is more remarkable than the bill of the tulcan, or preacher. This bird is about the fize of a common pigeon, but its legs much larger; its tail is fhort, and its plumage of a dark colour, but fpotted with blue, purple, yellow, and other colours; which have a beautiful effect on the dark ground. Its head is beyond all proportion to its body, but otherwife he would not be able to fupport his bill, which, from the root to the point, is at leaft fix or eight inches, and the upper mandible has, at its root, a bafe of at leaft an inch and a half, of a triangular figure, whofe apex is at the point of the bill. The two lateral fuperficies form a kind of elevation on the upper part; and the third receives the lower mandible, which clofes with the upper through the whole length; fo that the two parts are every where perfectly equal, and from their root narrows infenfibly, till neai the top, where it fuddenly becomes incurvated, and terminates in a ftrong and fharp point. The tongue is formed like a feather, and of a deep red colour, like the whole infide of its mouth. The bill is variegated with all thofe bright colours which adorn the plumage of other birds. At the bafe, and alfo at the convexity, it is generally of a light yellow, forming a kind of riband half an inch in breadth. The reft is of a fine deep purple, except two ftreaks near the root, of a rich fcarlet, an inch diftant from each other. The inward

ward flefhy parts, which touch when the bill is clofed, are furnifhed with teeth, which form the furface of its two ferrated mandibles. The name of preacher has been given to this bird from its cuftom of perching on the top of a tree above his companions, while they are afleep, and making a noife refembling ill-articulated founds, moving his head to the right and left, in order to keep off the birds of prey from feizing on the others. They are eafily rendered fo very tame, as to run about in houfes, and come when called. Their ufual food is fruit ; but the tame eat other things, and in general whatever is given them.

To defcribe all the other extraordinary birds would engage me in a prolixity of little entertainment or ufe; but I hope a word or two on the gallinazos will be excufed. This bird is about the fize of a pea-hen, but the neck and head fomething larger. From the crop to the bafe of the bill, inftead of feathers, it has a wrinkled glandulous and rough fkin, covered with fmall warts and tubercles. Its feathers are black, which is alfo the colour of this fkin, but ufually with fomething of a brownifh tinct. Its bill is well pro-portioned, ftrong, and a little crooked. They are fo numerous and tame in the city, that it is not uncom-mon to fee the ridges of the houfes covered with them. They are alfo very ferviceable, for they clean the city from all kinds of filth and ordure, greedily devouring any dead animal, and, when thefe are wanting, feek other filth. They have fo quick a fcent, that they will fmell at the diftance of three or four leagues * a dead carcafe, and never leave it till they have entirely re-duced it to a fkeleton †. The infinite number of thefe

* The author fhould have faid *miles*. A.
† It is furprifing to fee what numbers of thefe birds gather round the carcafe of a dead whale, which is no uncommon thing on thefe coafts. The carcafe fhall be covered with them; and yet their number fhall be nothing in comparifon to that hovering about, waiting for their turn, for which they often fight They
are

thefe birds found in fuch hot climates, is an excellent
provifion of nature, as otherwife the putrefaction
caufed by the conftant and exceffive heat would render
the air infupportable to human life. At firft they fly
heavily, but afterwards dart up out of fight. On the
ground they hop along with a kind of torpor, though
their legs are ftrong and well proportioned. They
have three toes forward turning inwards, and one in
the infide, turned a little backwards; fo that, the feet
interfering, they cannot walk with any agility, but are
obliged to hop or fkip. Each toe has a long and
thick claw.

WHEN the gallinazos find no food in the city, their
hunger drives them into the country, among the beafts
in the paftures; and on feeing any one with a fore on
the back, they immediately alight on it, and attack the
part affected. It is in vain for the poor beaft to endeavour to free itfelf from thefe devourers, either by
rolling on the ground, or hideous cries; for they
never quit their hold, but with their bills fo widen the
wound that the creature foon expires.

THERE is another kind of gallinazos, fomewhat
larger than thefe, only to be met with in the country.
In fome of thefe the head and part of the neck are
white, in fome red, and in others a mixture of both
thefe colours. A little above the beginning of the
crop, they have a ruff of white feathers. Thefe are
equally fierce and carnivorous with the former; and
called the kings of the gallinazos; probably becaufe
the number of them is but few: and it is obferved, that
when one of thefe has faftened on a dead beaft, none
of the others approach till he has eaten the eyes, with
which he generally begins, and is gone to another
part, when they all flock to the prey.

BATS are very common all over the country; but
Carthagena is infefted with fuch multitudes of them,

are feldom above a fortnight in making a fkeleton of a large
whale. A.

that

that after funfet, when they begin to fly, they may, without any hyperbole, be faid to cover the ftreets like clouds *. They are the moft dexterous bleeders both of men and cattle; for the inhabitants being obliged, by the exceffive heats, to leave open the doors and windows of the chambers where they fleep, the bats get in, and if they happen to find the foot of any one bare, they infinuate their tooth into a vein, with all the art of the moft expert furgeon, fucking the blood till they are fatiated, and withdraw their tooth; after which the blood flows out at the orifice. I have been affured, by perfons of the ftricteft veracity, that fuch an accident has happened to them; and that, had they not providentially awaked foon, their fleep would have been their paffage into eternity; they having loft fo large a quantity of blood, as hardly to be able to bind up the orifice. The puncture not being felt is (befides the great precaution with which it is made) attributed to the gentle and refrefhing agitation of the air by the bat's wings, hindering the perfon from feeling this flight puncture by throwing him into a deeper fleep. Nearly the fame thing happens to horfes, mules, and affes; but beafts of a thick and hard fkin are not expofed to this inconveniency.

We fhall next proceed to the infects and reptiles, in which nature has no lefs difplayed its infinite power. The great number of them is not only an inconvenience to the inhabitants, but health and even life itfelf often fuffers from the malignity of their poifon. The principal are the fnakes, the cientopies †, the fcorpions, and the fpiders; of all which there are different kinds, and their poifons of different activity.

* They are almoft as large as rats; and the infide of the roofs of the outhoufes are generally lined with them. A.
† Or hundred feet. They are very common throughout the warmer regions of America. Common falt is a fpecific againft their bite, as alfo againft the fting of the fcorpion. A.

Of the fnakes, the moft common, and at the fame time the moft poifonous, are the corales, or coral-fnakes, the cafcabeles, or rattle-fnakes, and the culebras de bejuco *. The firft are generally between four and five feet in length, and an inch in diameter. They make a very beautiful appearance, their fkin being all over variegated with a vivid crimfon, yellow, and green. The head is flat and long, like that of the European viper. Each mandible is furnifhed with a row of pointed teeth, through which, during the bite, they infinuate the poifon; the perfon bit, immediately fwells to fuch a degree, that the blood gufhes out through all the organs of fenfe, and even the coats of the veins at the extremities of the fingers burft, fo that he foon expires. The cafcabel or rattle-fnake feldom exceeds two feet, or two feet and a half, in length; though there are fome of another fpecies, which are three and a half. Its colour is brown, variegated with deeper fhades of the fame tinct; at the end of its tail is the cafcabel or rattle, in the form of a garvanzo or French-bean pod, when dried on the plant, and like that has five or fix divifions, in each of which are feveral fmall round bones; thefe, at every motion of the fnake, rattle, and thence give rife to its name. Thus nature, which has painted the coral fnake with fuch fhining colours, that it may be perceived at a diftance, has formed the latter in fuch a manner, that, as its colours render it difficult to diftinguifh it from the ground, the rattle might give notice of its approach.

The culebras de bejuco, which are very numerous, have their name from their colour and fhape refembling the branches of the bejuco, and, as they hang down from that plant, appear as real parts of the bejuco, till a too near approach unhappily difcovers the miftake; and though their poifon be not fo active as

* They are called Cobras by the natives, which is their common name for all kinds of ferpents. A.

that

that of the others, without a fpeedy application of fome
fpecific, it proves mortal.   Thefe remedies are perfeCtly known to the Negroes, Mulattos, and Indians
frequenting the woods, and called curanderos.   But
the fafeft antidote is the habilla, already mentioned.

It is not, however, often that thefe dangerous ferpents bite any one, unlefs, from inadvertence or defign,
he has been the aggreffor.   Befides, they ate fo far
from having any extraordinary agility, that they are
remarkably torpid, and, as it were, half dead; fo that,
were it not for their motion in retiring to hide
themfelves among the leaves, it would be difficult to
determine whether they were dead or alive.

There are few parts of Europe which do not produce the cientopies or fcolopendra; but at Carthagena
they not only fwarm, but are of a monftrous fize, and
the more dangerous, as breeding more commonly in
houfes than in the fields.   They are generally a yard
in length, fome a yard and a quarter, the breadth about
five inches, more or lefs, according to the length.
Their figure is nearly circular, the back and fides covered with hard fcales, of a mufk colour, tinged with
red; but thefe fcales are fo articulated, as not in the
leaft to impede their motion, and at the fame time fo
ftrong as to defend them againft any blow; fo that
the head is the only place where you can ftrike them
to any purpofe.   They are alfo very nimble, and their
bite, without timely application, proves mortal: nor
is the patient free from confiderable torture, till the
medicine has deftroyed the malignity of the poifon.

The alacranes, or fcorpions, are not lefs common,
and of different kinds, as black, red, mufk colour, and
fome yellow.   The firft generally breed in dry rotten
wood, and others in the corners of houfes, in clofets
and cupboards.   They are of different fizes, the largeft
about three inches long, exclufive of the tail.   The
fting alfo of fome is lefs dangerous than that of
others; that of the black is reckoned the moft malignant,

lignant, though timely care prevents its being fatal. The stings of the other kinds produce fevers, numbnesses in the hands and feet, forehead, ears, nose and lips, tumours in the tongue, and dimness of sight; these disorders last generally 24 or 48 hours, when, by degrees, the patient recovers. The natives imagine, that a scorpion falling into the water purifies it, and therefore drink it without any examination. They are so accustomed to these insects, that they do not fear them, but readily lay hold of them, taking care not to touch them only in the last vertebræ of the tail, to avoid being stung; sometimes they cut their tails off and play with them. We more than once entertained ourselves with an experiment of putting a scorpion into a glass vessel, and injecting a little smoke of tobacco, and immediately by stopping it found that its aversion to this smell is such, that it falls into the most furious agitations, till, giving itself several repeated stings on the head, it finds relief by destroying itself. Hence we see that its poison has the same effect on itself as on others.

Here is also another insect called caracol soldado, or the soldier-snail. From the middle of the body to the posterior extremity it is shaped like the common snail, of a whitish colour and a spiral form: but the other half of the body resembles a crab, both in size and the disposition of its claws. The colour of this, which is the principal part of its body, is of a light brown. The usual length, exclusive of the tail, is about two inches, and the breadth one and a half. It is destitute both of shell and scale, and the body every where flexible. Its resource against injuries is to seek a snail-shell of a proper size, in which it takes up its habitation. Sometimes it drags this snail-shell with it, and at other times quits it, while it goes out in quest of food; but, on the least appearance of danger, it hastens back to the shell, and thrusts itself into it, beginning with its hind part, so that the fore part fills
the

the entrance, while the two claws are employed in its
defence, the gripe of which is attended with the same
symptoms as the sting of a scorpion. In both cases the
patient is carefully kept from drinking any water,
which has been known to bring on convulsions; and
these always prove fatal.

The inhabitants relate, that when this creature grows
too large for making its way into the shell which was
its retreat, it retires to the sea coast, in order to find
there a larger, where killing the wilk, whose shell best
suits him, he takes possession of it, which is indeed
the same method it took to obtain its first habitation.
This last circumstance, and the desire of seeing the
form of such a creature, induced Don George Juan
and myself to desire the inhabitants to procure us
some; and upon examination, we found all the above-
mentioned particulars were really true; except the
bite, which we did not choose to experience.

There are several other sorts of insects remaining,
which, though smaller, yet afford equal reason for admiration to a curious examiner; particularly the infinite variety of mariposas, or butterflies, which,
though differing visibly in figure, colours, and decorations, we are at a loss to determine which is the most
beautiful.

If these are so entertaining to the sight, there are
others no less troublesome; so that it would be more
eligible to dispense with the pleasure of seeing the
former, than to be continually tortured by the latter;
as the moschetos, of which large clouds may be seen,
especially among the savannahs and manglares, or
plantations of mangrove-trees, so that the one, as
affording the herbage on which they feed, and the
other, as the places where they produce their young,
are rendered impassable.

There are four principal species of this insect: the
first called zancudos, which are the largest; the second
the moschetos, differing little or nothing from those
of

of Spain * ; the third gegenes, which are very small and of a different shape, resembling the weevil, about the size of a grain of muftard-feed, and of an ash-colour. The fourth are the mantas blancas, or white cloaks, and so very minute that the inflammation of their bite is felt before the infect that caused it is seen. Their colour is known by the infinite numbers of them which fill the air, and from thence they had their name. From the two former, few houses are free. Their sting is followed by a large tumour, the pain of which continues about two hours. The two last, which chiefly frequent fields and gardens, raise no tumour, but cause an insupportable itching. Thus, if the extreme heat renders the day troublesome, these imperceptible infects disturb the repose of the night. And though the mosquiteros, a kind of gauze curtains, in some measure defend us from the three former, they are no safeguard against the latter, which make their way between the threads ; unless the stuff be of a closer texture, in which case the heat becomes insupportable.

The infect of Carthagena called nigua, and in Peru pique, is shaped like a flea, but almost too small for sight. It is a great happiness that its legs have not the elasticity with those of fleas ; for, could this infect leap, every animal body would be filled with them ; and, consequently, both the brute and human species be soon extirpated by the multitudes of these infects. They live amongst the dust, and therefore are most common in filthy places. They insinuate themselves into the legs, the soles of the feet, or toes, and pierce the skin with such subtilty, that there is no being aware of them, till they have made their way into the flesh †. If they are perceived at the beginning, they are extracted with little pain ; but if the head only has

* Or the gnat of England. A.
† They seldom insinuate themselves into the legs. A.

pierced

pierced through the skin, an incision must be made before it can be taken out. If they are not soon perceived, they make their way through the skin, and take up their lodging between that and the membrane of the flesh; and sucking the blood, form a nidus or nest, covered with a white and fine tegument, resembling a flat pearl; and the insect is, as it were, inchased in one of the faces, with its head and feet outwards, for the convenience of feeding, while the hinder part of the body is within the tunic, where it deposits its eggs; and as the number of these increases, the nidus enlarges, even to the diameter of a line and a half, or two lines, to which magnitude it generally attains in four or five days. There is an absolute necessity for extracting it; for otherwise it would burst of itself, and by that means scatter an infinite number of germs, resembling nits, in size, shape, and colour, which becoming niguas, would, as it were, undermine the whole foot. They cause an extreme pain, especially during the operation of extracting them; for sometimes they penetrate even to the bone; and the pain, even after the foot is cleared of them, lasts till the flesh has filled up the cavities they had made, and the skin is again closed.

The manner of performing this operation is both tedious and troublesome; the flesh contiguous to the membrane, where the eggs of the insect are lodged, is separated with the point of a needle, and those eggs so tenaciously adhere to the flesh and this membrane, that, to complete the operation without bursting the tegument, and putting the patient to the most acute pain, requires the greatest dexterity. After separating on every side the small and almost imperceptible fibres, by which it was so closely connected with the membranes and muscles of the part, the perilla, as they term it, is extracted, the dimensions of which are proportional to the time it has existed. If unfortunately it should burst, the greatest care must be used clear away all the roots, particularly not to leave the principal nigua ; as, before the wound could be healed, there would be a new brood, further within the flesh; and consequently the cure much more difficult and painful.

THE cavity left by the removal of the nidus, must be immediately filled either with tobacco ashes, chewed tobacco, or snuff; and, in hot countries, as Carthagena, great care must be taken not to wet the foot for the first two days, as convulsions would ensue; a distemper seldom got over: this consequence has possibly been observed in some, and from thence considered as general *.

THE first entrance of this insect is attended with no sensible pain; but, the next day, it brings on a fiery itching, extremely painful, but more so in some parts than in others. This is the case in extracting it, when the insect gets between the nails and the flesh, or at the extremity of the toes. In the sole of the foot, and other parts where the skin is callous, they cause little or no pain.

THIS insect shews an implacable hatred to some animals, particularly the hog; which it preys on with such voracity, that when their feet come to be scalded, after being killed, they are found full of cavities made by this corroding insect.

MINUTE as this creature is, there are two kinds of it; one venomous, and the other not. The latter perfectly resembles the flea in colour, and gives a whiteness to the membrane where it deposits its eggs. This causes no pain, but what is common in such cases. The former is yellowish, its nidus of an ash-colour, and its effects more extraordinary; as, when lodged at the extremity of the toes, it violently in-

* There is no necessity for this precaution, as is well known to the honest tar. The tobacco ashes, &c. entirely destroy the nits or ovaria, if any be left. A.

flames

flames the glands of the groin, and the pain continues, without abatement, till the nigua is extracted, that being the only remedy; after which, the swelling subsides, and the pain ceases, those glands corresponding with the foot, where the cause of the pain resided. The true cause of this apparently strange effect I shall not undertake to investigate; the general opinion is, that some small muscles extending from those glands to the feet, being affected by the poison of the bite, communicate it to the glands, whence proceed the pain and inflammation. All I can affirm is, that I have often experienced it, and at first with no small concern; till having frequently observed, that these effects ceased on extracting the nigua, I thence concluded it to be the true cause of the disorder. The same thing happened to all the French academicians, who accompanied us in this expedition, and particularly to M. de Juffieu, botanist to the king of France, whom frequent experience of these kinds of accidents taught to divide these insects into two kinds.

As the preceding animals and insects chiefly exercise their malignant qualities on the human species, so there are others which damage and destroy the furniture of houses, particularly all kinds of hangings, whether of cloth, linen, silk, gold or silver stuffs, or laces; and indeed every thing, except those of solid metal, where their voracity seems to be wearied out by the resistance. This insect, called comegen, is nothing more than a kind of moth or maggot; but so expeditious in its depredations, that in a very short time it entirely reduces to dust one or more bales of merchandise where it happens to fasten; and, without altering the form, perforates it through and through, with a subtility which is not perceived till it comes to be handled, and then, instead of thick cloth or linen, one finds only small shreds and dust. At all times the strictest attention is requisite to prevent such accidents, but chiefly at the arrival of the galleons; for

F 2   then

then it may do immenfe damage among the vaft quantity of goods landed for warehoufes, and for fale in the fhops. The beft, and indeed the only method, is, to lay the bales on benches, about half a yard from the ground, and to cover the feet of them with alquitran, or naphtha, the only prefervative againft this fpecies of vermin; for, with regard to wood, it eats into that as eafily as into the goods, but will not come near it when covered with naphtha as above.

NEITHER would this precaution be fufficient for the fafety of the goods, without a method of keeping them from touching the walls; and then they are fufficiently fecured. This infect is fo fmall, as to be fcarcely vifible to the naked eye; but of fuch activity, as to deftroy all the goods in a warehoufe, where it has got footing, in one night's time. Accordingly it is ufual that in running the rifks of commerce, in goods configned to Carthagena, the circumftances are fpecified, and in thefe are underftood to be included the loffes that may happen in that city by the comegen. This infect infefts neither Porto Bello, nor even places nearer Carthagena, though they have fo many other things in common with that city; nor is it fo much as known among them.

WHAT has been faid, will, I hope, be fufficient to give an adequate idea of this country, without fwelling the work with trivial obfervations, or fuch as have been already publifhed by others. We fhall now proceed to treat diftinctly of other equally wonderful works of Omnipotence, in this country.

CHAP.

CHAP. VIII.

Of the esculent Vegetables produced in the Terri-
tories of Carthagena, and the Food of the In-
habitants of that City.

THOUGH Carthagena has not the convenience
of being furnished by its soil with the different
kinds of European vegetables, it does not want for
others, far from being contemptible, and of which the
inhabitants eat with pleasure. Even the Europeans,
who at their first coming cannot easily take up with
them, are not long before they like them so well as to
forget those of their own country.

THE constant moisture and heat of this climate
will not admit of barley, wheat, and other grain of
that kind; but produces excellent maize and rice in
such abundance, that a bushel of maize, sown, usually
produces an hundred, at harvest. From this grain
they make the bollo, or bread, used in all this country; they also use it in feeding hogs and fattening
poultry. The maize bollo has no resemblance to the
bread made of wheat, either in shape or taste. It is
made in form of a cake; is of a white colour, and
an insipid taste. The method of making it is, to soak
the maize, and afterwards bruise it between two
stones; it is then put into large bins filled with water,
where, by rubbing and shifting it from one vessel
into another, they clear it from its husk; after this it
is ground into a paste, of which the bollos are made.
These bollos, being wrapped up in plantane or
vijahua leaves, are boiled in water, and used as bread;
but, after twenty-four hours, become tough and of a
disagreeable taste. In families of distinction the
bollo is kneaded with milk, which greatly improves
it; but, being not thoroughly penetrated by the li-
quids, it never rises, nor changes its natural colour;

so

fo that, inftead of a pleafing tafte, it has only that of
the flour of maize.

BESIDES the bollo * here is alfo the cafava bread,
very common among the Negroes, made from the
roots of yuca, names, and moniatos. After carefully
taking off the upper fkin of the root, they grate it,
and fteep it in water, in order to free it from a ftrong
acrid juice, which is a real poifon, particularly that of
the moniato. The water being feveral times fhifted,
that nothing of this acrimony may remain, the
dough is made into round cakes, about two feet
diameter, and about three or four lines in thicknefs.
Thefe cakes are baked in ovens, on plates of copper,
or a kind of brick made for that purpofe. It is a nourifhing and ftrengthening food, but very infipid. It
will keep fo well, that at the end of two months it
has the fame tafte as the firft day, except being more
dry.

WHEAT bread is not entirely uncommon at Carthagena; but, as the flour comes from Spain, the
price of it may well be conceived to be above the
reach of the generality. Accordingly it is ufed only
by the Europeans fettled at Carthagena, and fome
few Creoles; and by thefe only with their chocolate
and conferves. At all other meals, fo ftrong is the
force of a cuftom imbibed in their infancy, they prefer
bollos to wheat bread, and eat honey with cafava.

THEY alfo make, of the flour of maize, feveral
kinds of paftry, and a variety of foods equally palatable and wholefome; bollo itfelf being never known
to difagree with thofe who ufe it.

BESIDES thefe roots, the foil produces plenty of
camiótes, refembling, in tafte, Malaga potatoes; but
fomething different in fhape, the camiotes being ge-

* Or cake made of mandioc yams, and fweet potatoes (or camiotes), which they grate and mix together. The bollo is far from
infipid, when a proper quantity of the camiote is put in. A.

nerally

nerally roundifh and uneven. They are both pickled and ufed as roots with the meat; but, confidering the goodnefs and plenty of this root, they do not improve it as they might.

PLANTATIONS of fugar-canes abound to such a degree, as extremely to lower the price of honey; and a great part of the juice of thefe canes is converted into fpirit for the difpofing of it. They grow fo quick as to be cut twice in a year. The variety of their verdure is a beautiful ornament to the country.

HERE are alfo great numbers of cotton-trees, fome planted and cultivated, and thefe are the beft; others fpontaneoufly produced by the great fertility of the country. The cotton of both is fpun, and made into feveral forts of ftuffs, which are worn by the Negroes of the Haciendas, and the country Indians.

CACAO trees alfo grow in great plenty on the banks of the river Magdalena, and in other fituations which that tree delights in; but thofe in the jurifdiction of Carthagena excel thofe of the Caracas, Maracaybo, Guayaquil, and other parts, both in fize and the goodnefs of the fruit. The Carthagena cacao or chocolate is little known in Spain, being only fent as prefents; for, as it is more efteemed than that of other countries, the greater part of it is confumed in this jurifdiction, or fent to other parts of America. It is alfo imported from the Caracas, and fent up the country, that of the Magdalena not being fufficient to anfwer the great demand there is for it in thefe parts. Nor is it amifs to mix the former with the latter, as correcting the extreme oilinefs of the chocolate, when made only with the cacao of the Magdalena. The latter, by way of diftinction from the former, is fold at Carthagena by millares, whereas the former is difpofed of by the bufhel, each weighing 110 pounds; but that of Maracaybo weighs only 96 pounds. This is the moft valuable treafure which nature could have beftowed on this country; though it has carried its

F 4   bounty

bounty still farther, in adding a vast number of delicious fruits which evidently display the exuberance of the soil. Nothing strikes a spectator with greater admiration, than to see such a variety of pompous trees, in a manner emulating each other, through the whole year, in producing the most beautiful and delicious fruits. Some resemble those of Spain; others are peculiar to the country. Among the former, some are indeed cultivated, the latter flourish spontaneously.

Those of the same kind with the Spanish fruits are melons, water-melons, called by the natives Blanciac, grapes, oranges, medlars, and dates. The grapes are not equal to those of Spain; but the medlars as far exceed them: with regard to the rest, there is no great difference.

Among the fruits peculiar to the country, the preference, doubtless, belongs to the pine-apple; and accordingly its beauty, smell, and taste, have acquired it the appellation of queen of fruits. The others are the papayas, guanabanas, guayabas, sapotes, mameis, platanos, cocos, and many others, which it would be tedious to enumerate, especially as these are the principal; and therefore it will be sufficient to confine our descriptions to them.

The ananas or pine-apple, so called from its resembling the fruit or the cones of the European pine-tree, is produced by a plant nearly resembling the aloe, except that the leaves of the pine apple are longer, but not so thick, and most of them stand near the ground in a horizontal position; but as they approach nearer the fruit, they diminish in length, and become less expanded. This plant seldom grows to above three feet in height, and terminates in a flower resembling a lily, but of so elegant a crimson, as even to dazzle the eye. The pine-apple makes its first appearance in the centre of the flower, about the size of a nut; and as this increases, the lustre of the flower fades, and the leaves

expand

expand themfelves to make room for it, and fecure it both as a bafe and ornament. On the top of the apple itfelf, is a crown or tuft of leaves, like thofe of the plant, and of a very lively green. This crown grows in proportion with the fruit, till both have attained their utmoft magnitude, and hitherto they differ very little in colour. But as foon as the crown ceafes to grow, the fruit begins to ripen, and its green changes to a bright ftraw colour: during this gradual alteration of colour, the fruit exhales fuch a fragrancy as difcovers it, though concealed from fight. While it continues to grow, it fhoots forth on all fides little thorns, which, as it approaches towards maturity, dry and foften, fo that the fruit is gathered without the leaft inconvenience. The fingularities which concentre in this product of nature, cannot fail of ftriking a contemplative mind with admiration. The crown, which was to it a kind of apex, while growing in the woods, becomes itfelf, when fown, a new plant; and the ftem, after the fruit is cut, dies away, as if fatisfied with having anfwered the intention of nature in fuch a product; but the roots fhoot forth frefh ftalks, for the farther increafe of fo valuable a fpecies.

THE pine-apple, though feparated from the plant, retains its fragrancy for a confiderable time, when it begins to decay. The odour of it not only fills the apartment where the fruit is kept, but even extends to the contiguous rooms. The general length of this delicious fruit is from five to feven inches, and the diameter near its bafis three or four, diminifhing regularly, as it approaches to its apex. For eating, it is peeled and cut into round flices, and is fo full of juice, that it entirely diffolves in the mouth. Its flavour is fweet, blended with a delightful acidity. The rind, infufed in water, after a proper fermentation, produces a very cooling liquor, and ftill retains all the properties of the fruit.

THE

THE other fruits of this country are equally valuable in their feveral kinds; and fome of them alfo diftinguifhed for their fragrancy, as the guayaba, which is, befides, both pectoral and aftringent.

THE moft common of all are, the platanos, the name of which, if not its figure and tafte, is known in all parts of Europe *. Thefe are of three kinds. The firft is the banana, which is fo large as to want but little of a foot in length. Thefe are greatly ufed, being not only eaten as bread, but alfo an ingredient in many made difhes. Both the ftone and kernel are very hard; but the latter has no noxious quality. The fecond kind are the dominicos, which are neither fo long nor fo large as the bananas, but of a better tafte; they are ufed as the former.

THE third kind are the guineos, lefs than either of the former, but far more palatable, though not reckoned fo wholefome by the natives, on account of their fuppofed heat. They feldom exceed four inches in length; and their rind, when ripe, is yellower, fmoother, and brighter, than that of the two other kinds. The cuftom of the country is to drink water after eating them; but the European failors, who will not be confined in their diet, but drink brandy with every thing they eat, make no difference between this fruit and any other; and to this intemperance may, in fome meafure, be attributed the many difeafes with which they are attacked in this country, and not a few fudden deaths; which are, indeed, apt to raife, in the furvivors, concern for their companions for the pre-

* The plantane and banana are, I believe, little known in Europe by name. The firft two forts the Author defcribes, are better known by the names of the long and fhort plantane, and the laft by the name of banana, than by thofe he has given them. They have neither ftone nor kernel, but a very fmall feed, as fmall as that of thyme, which lies in the fruit in rows like that of a cucumber, to which the banana bears the greateft refemblance of any thing in England; only it is fmooth and not fo large. A.

fent;

fent; but they foon return to the fame exceffes, not remembering, or rather choofing to forget, the melancholy confequences.

By what we could difcover, it is not the quality of the brandy which proves fo pernicious, but the quantity; fome of our company making the experiment of drinking fparingly of this liquor after eating the guineos, and.repeating it feveral times without the leaft inconvenience. One method of dreffing them, among feveral others, is to roaft them in their rind, and afterwards flice them, adding a little brandy and fugar to give them a firmnefs. In this manner we had them every day at our table, and.the Creoles themfelves approved of them.

·The papayas are from fix to eight inches in length, and refemble a lemon, except that towards the ftalk they are fomewhat lefs than at·the other extremity. Their rind is green, the pulp white, very juicy, but ftringy, and the tafte a gentle acid, not pungent. This is the fruit of a tree, and not, like the pine-apple and platano, the product of a plant. The guayaba and the following are alfo the fruit of trees.

The guanabana approaches very near the melon, but its rind is much fmoother, and of a greenifh colour. Its pulp is of a yellowifh caft, like that of fome melons, and not very different in tafte. But the greateft diftinction between thefe two fruits is a naufeous fmell in the guanaba. The feed is round, of a fhining dark colour, and about two lines in diameter. It confifts of a very fine tranfparent pellicle, and a kernel folid and juicy. The fmell of this little feed is much ftronger and more naufeous. The natives fay, that, by eating this feed, nothing is to be apprehended from the fruit, which is otherwife accounted heavy and hard of digeftion; but, though the feed has no ill tafte, the ftomach is offended at its fmell:

The fapotes are round, about two inches in circumference, the rind thin and eafily feparated from the
fruit;

fruit; the colour brown, ſtreaked with red. The fleſh is of a bright red, with little juice, viſcid, fibrous, and compact. It cannot be claſſed among delicious fruits, though its taſte is not diſagreeable. It contains a few ſeeds, which are hard and oblong.

The mameis are of the ſame colour with the ſapotes, except that the brown is ſomething lighter. Their rind alſo requires the aſſiſtance of a knife, to ſeparate it. The fruit is very much like the brunion plum, but more ſolid, leſs juicy, and, in colour, more lively. The ſtone is proportioned to the largeneſs of the fruit, which is betwixt three and four inches in diameter, almoſt circular, but with ſome irregularities. The ſtone is an inch and a half in length, and its breadth in the middle, where it is round, one inch. Its external ſurface is ſmooth, and of a brown colour, except on one ſide, where it is vertically croſſed by a ſtreak reſembling the ſlice of a melon in colour and ſhape. This ſtreak has neither the hardneſs nor ſmoothneſs of the reſt of the ſurface of the ſtone, which ſeems in this place covered and ſomething ſcabrous.

The coco is a very common fruit, and but little eſteemed; all the uſe made of it being to drink the juice whilſt fluid, before it begins to curdle. It is, when firſt gathered, full of a whitiſh liquor, as fluid as water, very pleaſant and refreſhing. The ſhell which covers the coco nut, is green on the outſide, and white within; full of ſtrong fibres, traverſing it on all ſides in a longitudinal direction, but eaſily ſeparated with a knife. The coco is alſo whitiſh at that time, and not hard; but, as the conſiſtency of its pulp increaſes, the green colour of its ſhell degenerates into yellow. As ſoon as the kernel has attained its maturity, this dries and changes to brown; then becomes fibrous, and ſo compact, as not to be eaſily opened and ſeparated from the coco, to which ſome of thoſe fibres adhere. From the pulp of theſe

cocos

cocos is drawn a milk like that of almonds, and at Carthagena is ufed in dreffing rice.

Though lemons, of the kind generally known in Europe, and of which fuch quantities are gathered in fome parts of Spain, are very fcarce; yet there are fuch numbers of another kind, called futiles or limes, that the country is, in a manner, covered with the trees that produce them, without care or culture. But the tree and its fruit are both much lefs than thofe of Spain, the height of the former feldom exceeding eight or ten feet; and from the bottom, or a litle above, divides into feveral branches, whofe regular expanfion forms a very beautiful tuft: The leaf, which is of the fame fhape with that of the European lemon, is lefs, but fmoother; the fruit does not exceed a common egg in magnitude; the rind very thin; and it is more juicy, in proportion, than the lemon of Europe, and infinitely more pungent and acid; on which account the European phyficians pronounce it detrimental to health; though, in this country, it is a general ingredient in their made difhes. There is one fingular ufe to which this fruit is applied in cookery. It is a cuftom with the inhabitants not to lay their meat down to the fire above an hour at fartheft, before dinner or fupper; this is managed by fteeping it for fome time in the juice of thefe limes, or fqueezing three or four, according to the quantity of meat, into the water, if they intend boiling; by which means the flefh becomes fo foftened as to admit of being thoroughly dreffed in this fhort fpace. The people here value themfelves highly on this preparative, and laugh at the Europeans for fpending a morning about what they difpatch fo expeditioufly.

This country abounds in tamarinds; a large branchy tree, the leaf of a deep green; the pods of a middle fize, and flat; the pulp of a dark brown, a pleafant tafte, very fibrous, and is called by the fame name as the tree itfelf. In the middle of the pulp is a hard

feed,

feed, or ftone, fix or eight lines in length, to two in breadth. Its tafte is an acid fweetnefs, but the acid predominates; and it is only ufed when diffolved in water as a cooling liquor, and then but moderately, and not for many days fucceffively; its acidity and extreme coldnefs weakening and debilitating the ftomach.

Another fruit, called mani, is produced by a fmall plant. It is of the fize and fhape of a pinecone; and eaten either roafted, or as a conferve. Its quality is directly oppofite to that of the former, being hot in the higheft degree; and, confequently, not very wholefome in this climate.

The products which are not natural here, befides wheat, barley, and other grain, are grapes, almonds, and olives: confequently the country is deftitute of wine, oil, and raifins, with which they are fupplied from Europe: this neceffarily renders them very dear; fometimes they are not to be had at any price. When this is the cafe with regard to wine, great numbers fuffer in their health; for, as all thofe who do not accuftom themfelves to drink brandy at their meals, which are far the greater number, except the Negroes, being ufed to this wine, their ftomach, for want of it, lofes the digeftive faculty, and thence are produced epidemical diftempers. This was an unhappy circumftance at our arrival, when wine was fo extremely fcarce, that mafs was faid only in one church.

The want of oil is much more tolerable; for, in dreffing either fifh or flefh, they ufe hog's lard, of which they have fo great a quantity, as to make it an ingredient in their foup, which is very good, and, confidering the country, not at all dear: inftead of lamps too, they ufe tallow candles: fo that they want oil only for their falads.

From fuch plenty of flefh, fowl, and fruits, an idea may be formed of the luxuriancy of the tables in this country; and, indeed, in the houfes of perfons

fons of wealth and diftinction, they are ferved with the greateft decency and fplendour. Moft of the difhes are dreffed in the manner of this country, and differ confiderably from thofe of Spain; but fome of them are fo delicate, that foreigners are no lefs pleafed with them, than the gentlemen of the country. One of their favourite difhes is the agi-aco, there being fcarcely a genteel table without it. It is a mixture of feveral ingredients, which cannot fail of making an excellent ragout. It confifts of pork fried, birds of feveral kinds, plantanes, maize pafte, and feveral other things highly feafoned with what they call pimento, or aji.

The inhabitants of any figure generally make two meals a-day, befides another light repaft. That in the morning, their breakfaft, is generally compofed of fome fried difh, paftry of maize flour, and things of that nature, followed by chocolate. Their dinner confifts of a much greater variety; but at night the regale is only of fweetmeats and chocolate. Some families, indeed, affect the European cuftom of having regular fuppers, though they are generally looked upon at Carthagena as detrimental to health. We found, however, no difference as to ourfelves; and poffibly the ill effects flow from excefs in the other meals.

CHAP. IX.

Of the Trade of Carthagena, and other Countries of America, on the Arrival of the Galleons and other Spanifh Ships.

THE bay of Carthagena is the firft place in America at which the galleons are allowed to touch; and thus it enjoys the firft fruits of commerce,

by

by the public fales made there. Thefe fales, though
not accompanied with the formalities obferved at
Porto Bello fair, are very confiderable. The traders
of the inland provinces of Santa Fe; Popayan, and
Quito, lay out not only their own ftocks, but alfo the
monies-intrufted to them by commiffions, for feveral
forts of goods, and thofe fpecies of provifions which
are moft wanted in their refpective countries. The
two provinces of Santa Fe and Popayan have no other
way of fupplying themfelves with the latter, than from
Carthagena. Their traders bring gold and filver in
fpecie, ingots, and duft, and alfo emeralds; as, befides the filver mines worked at Santa Fe, and which
daily increafe by frefh difcoveries, there are others
which yield the fineft emeralds. But the value of
thefe gems being now fallen in Europe, and particularly in Spain, the trade of them, formerly fo confiderable, is now greatly leffened, and confequently the
reward for finding them. All thefe mines produce
great quantities of gold, which is carried to Choco,
and there pays one fifth to the king, at an office
erected for that purpofe.

This commerce was for fome years prohibited, at
the folicitation of the merchants of Lima, who complained of the great damages they fuftained by the
tranfportation of European merchandifes from Quito
to Peru; which being thus furnifhed, while the traders
of Lima were employed at the fairs of Panama and
Porto Bello, at their return, they found, to their
great lofs, the price of goods very much lowered. But
it being afterwards confidered, that reftraining the
merchants of Quito and other places from purchafing
goods at Carthagena, on the arrival of the galleons,
was of great detriment to thofe provinces; it was
ordered, in regard to both parties, that, on notice
being given in thofe provinces, of the arrival of the
galleons at Carthagena, all commerce, with regard to
European commodities, fhould ceafe between Quito
and

and Lima, and that the limits of the two audiences
fhould be thofe of their commerce: that is, that Quito
fhould not trade beyond the territories of Loja and
Zamora; nor Lima, beyond thofe of Piura, one of the
jurifdiclions of its audience. By this equitable expedient, thofe provinces were, in time, fupplied with the
goods they wanted, without any detriment to the trade
of Peru. This regulation was firft executed in 1730,
on the arrival of the fquadron commanded by Don
Manuel Lopez Pintado, who had orders, from the
king, to place commerce on this footing, provided it
bid fair to anfwer the intentions of both parties, and
that no better expedient could be found. Accordingly
this was carried into execution; being not only well
adapted to the principal end, but alfo, during the ftay
of the galleons at Carthagena, procured bufinefs for
the Cargadores *, in the fale of their goods; and thus
made themfelves ample amends for their expenfes.

During the prohibition, the merchants of Carthagena were obliged to have recourfe to the Flotila of
Peru, in their courfe from Guayaquil to Panama; or
to wait the return of the galleons to Carthagena, and,
confequently, purchafe only the refufe of Porto Bello
fair; both which were, doubtlefs, confiderable griev-
ances to them. If they purfued the firft, they were
obliged to travel acrofs the whole jurifdiction of Santa
Fe to Guayaquil, which was a journey of above four
hundred leagues, with confiderable fums of money,
which having difpofed of in merchandifes, the charges
of their return were ftill greater. In fine, the loffes
inevitable in fuch a long journey, where rapid rivers,
mountains, and bridges, were to be croffed, and their
merchandifes expofed to a thoufand accidents, rendered this method utterly impracticable; fo that they
were obliged to content themfelves with the remains
of the fair; though it was very uncertain whether

* Perfons who bring European goods for fale.

　　　　thefe

thefe would be fufficient to anfwer the demand. Be-fides, the inland merchants ran the hazard of not meet-ing at Carthagena with goods fufficient, in quality and quantity, to anfwer their charges; and were fome-times actually obliged to return with the money, and the vexation of a fruitlefs, though expenfive, journey. Thefe inconveniences produced a repeal of the pro-hibition, and commerce was placed on the prefent equitable footing.

This little fair at Carthagena, for fo it may be called, occafions a great quantity of fhops to be opened, and filled with all kinds of merchandife; the profit partly refulting to Spaniards who come in the galleons, and are either recommended to, or are in partnerfhip with, the Cargadores; and partly to thofe already fettled in that city. The Cargadores furnifh the former with goods, though to no great value, in order to gain their cuftom; and the latter, as perfons whom they have already experienced to be good men; and both in proportion to the quicknefs of their fale. This is a time of univerfal profit; to fome by letting lodgings and fhops, to fome by the increafe of their refpective trades, and to others by the labour of their Negro flaves, whofe pay alfo is proportionally in-creafed, as they do more work in this bufy time. By this brifk circulation through all the feveral ranks, they frequently get a furplus of money beyond what is fufficient for providing themfelves with neceffaries. And it is not uncommon for flaves, out of their fav-ings, and after paying their mafters the daily tribute, to purchafe their freedoms.

This affluence extends to the neighbouring villages, eftancias, and the moft wretched chacaras, of this ju-rifdiction; for, by the increafe of ftrangers to a fourth, third, and fometimes one half, of the ufual number of people, the confumption, and confequently the price of provifions, advances, which is, of courfe, no fmall advantage to thofe who bring them to market.

This

This commercial tumult lasts while the galleons continue in the bay : for they are no sooner gone, than silence and tranquillity resume their former place. This the inhabitants of Carthagena call tiempo muerto, the dead time; for, with regard to the trade carried on with the other governments, it is not worth notice. The greater part of it consists in some bilanders from La Trinidad, the Havannah, and St. Domingo, bringing leaf-tobacco, snuff, and sugars; and returning with Magdalena cacao, earthen-ware, rice, and other goods wanted in those islands. And even of these small vessels, scarcely one is seen for two or three months. The same may be said of those which go from Carthagena to Nicaragua, Vera-Cruz, Honduras, and other parts; but the most frequent trips are made to Porto Bello, Chagra, or Santa Martha. The reason why this commerce is not carried on more briskly is, that most of these places are naturally provided with the same kind of provisions; and consequently are under no necessity of trafficking with each other.

Another branch of the commerce of Carthagena, during the tiempo muerto, is carried on with the towns and villages of its jurisdiction, from whence are brought all kinds of necessaries and even the luxuries of life, as maize, rice, cotton, live hogs, tobacco, plantanes, birds, casava, sugar, honey, and cacao, most of which is brought in canoes and champanas, a sort of boats proper for rivers. The former are a kind of coasters, and the latter come from the rivers Magdalena, Sinu, and others. Their returns consist of goods for apparel, with which the shops and warehouses furnish themselves from the galleons, or from prizes taken on the coast by the king's frigates, or privateers.

No eatable pays any duty to the king; and every person may, in his own house, kill any number of pigs he thinks he shall sell that day; no salted pork is eaten, because it is soon corrupted by the excessive

heat

heat of the place. All imports from Spain, as brandy, wine, oil, almonds, raifins, pay a duty, and are afterwards fold without any farther charge, except what is paid by retailers, as a tax for their fhop or ftall.

Besides thefe goods, which keep alive this flender inland commerce, here is an office for the affiento of Negroes, whither they are brought, and, as it were, kept as pledges, till fuch perfons as want them on their eftates come to purchafe them; Negroes being generally employed in hufbandry and other laborious country works. This indeed gives fome life to the trade of Carthagena, though it is no weighty article. The produce of the royal revenues in this city not being fufficient to pay and fupport the governor, garrifon, and a great number of other officers, the deficiency is remitted from the treafurers of Santa Fe and Quito, under the name of Situado, together with fuch monies as are requifite for keeping up the fortifications, furnifhing the artillery, and other expenfes, neceffary for the defence of the place and its forts.

BOOK II.

Voyage from Carthagena to Porto Bello.

CHAP. I.

General Winds and Currents between Carthagena and Porto Bello.

WHEN the French frigate had watered, and was ready for sailing, we embarked on board her, on the 24th of November 1735; the next day we put to sea, and on the 29th of the same month, at half an hour after five in the evening, came to an anchor at the mouth of Porto Bello harbour, in fourteen fathom water; Castle Todo Fierro, or the iron castle, bearing N. E. four degrees northerly; and the south point of the harbour east one quarter northerly. The difference of longitude between Carthagena and Punta de Nave, we found to be 4° 24'.

We had steered W. N. W. and W. one quarter northerly, till the ship was observed to be in the eleventh degree of latitude, when we stood to the west. But when our difference of longitude from Carthagena was 3° 10', we altered our course to S. W. and S. a quarter westerly, which, as already observed, on the 29th of November, at 5 in the evening, brought us in sight of Punta de Nave, which being south of us, we were obliged to make several tacks before we could get into the harbour.

In this passage we met with fresh gales. The two first days at north quarter easterly, and the other days till we made the land at N. E.; a high sea running the

G 3   whole

whole time. But we were no fooner in fight of Punta
de Nave, than it became calm, and a breeze from the
land fprung up, which hindered us from getting that
day into the harbour. It alfo continued contrary on
the 30th ; but by the help of our oars, and being
towed, we got at laft to the anchoring place, where
we went on fhore, with our baggage and inftruments
neceffary for beginning our obfervations. But this
being the moft proper place for mentioning the winds
which prevail in this paffage, along the coaft, and
that of Carthagena, we fhall beftow fome paragraphs
on them.

THERE are two forts of general winds on thefe
coafts; the one called brifas, which blow from the
N. E. and the other called vendabales, which come from
the W. and W. S. W. The former fet in about the
middle of November, but are not fettled till the beginning or middle of December, which is here the
fummer, and continue blowing frefh and invariable till
the middle of May; they then ceafe, and are fucceeded by the vendabales, but with this difference, that
thefe do not extend farther than 12 or 12½ degrees of
latitude ; beyond which the brifas conftantly reign,
though with different degrees of ftrength, and veer
fometimes to the eaft, and at other times to the north.

THE feafon of the vendabales is attended with violent ftorms of wind and rain ; but they are foon over,
and fucceeded by a calm equally tranfitory ; for the
wind gradually frefhens, efpecially near the land,
where thefe phenomena are more frequent. The fame
happens at the end of October and beginning of November, the general winds not being fettled.

IN the feafon of the brifas, the currents as far as 12°
or 12° 30′ of latitude, fet to the weftward, but with lefs
velocity than ufual at the changes of the moon, and
greater at the full. But beyond that latitude, they
ufually fet N. W. Though this muft not be underftood without exception ; as, for inftance, near iflands
or

or fhoals, their courfe becomes irregular : fometimes they flow through long channels ; and fometimes they are met by others ; all which proceeds from their feveral directions, and the bearings of the coafts ; fo that the greateft attention is neceffary here, the general accounts not being fufficient to be relied on ; for, though they have been given by pilots who have for twenty or thirty years ufed this navigation, in all kinds of veffels, and therefore have acquired a thorough knowledge, they themfelves confefs that there are places where the currents obferve no kind of regularity, like thofe we have mentioned.

WHEN the brifas draw near their period, which is about the beginning of April, the currents change their courfe, running to the eaftward for eight, ten, or twelve leagues from the coaft, and thus continue during the whole feafon of the vendabales; on which account, and the winds being at this feafon contrary for going from Carthagena to Porto Bello, it is neceffary to fail to 12 or 13 degrees of latitude, or even fometimes farther; when, being without the verge of thofe winds, the voyage is eafily performed.

WHILE the brifas blow ftrongeft, a very impetuous current fets into the gulf of Darien ; and out of it during the feafon of the vendabales. This fecond change proceeds from the many rivers which difcharge themfelves into it, and at that time being greatly fwelled by the heavy rains, peculiar to the feafon ; fo that they come down with fuch rapidity, as violently to propel the water out of the gulf. But in the feafon of the brifas thefe rivers are low, and fo weak, that the current of the fea overcomes their refiftance, fills the gulf, and returns along the windings of the coaft.

CHAP. II.

Description of the Town of St. Philip de Porto Bello.

THE town of St. Philip de Porto Bello, according to our obfervations, ftands in 9° 34′ 35″ north latitude; and by the obfervations of father Feuillée, in the longitude of 277° 50′ from the meridian of Paris, and 296° 41′ from the Pico of Teneriffe. This harbour was difcovered on the 2d of November 1502 by Chriftopher Columbus, who was fo charmed with its extent, depth, and fecurity, that he gave it the name of Porto Bello, or the fine harbour. In the profecution of his difcoveries, he arrived at that which he called Baftimentos, where, in 1510, was founded by Diego de Niqueza the city of Nombre de Dios, " the name " of God;" fo called from the commander having faid to his people on his landing, "Here we will make " a fettlement in the name of God," which was accordingly executed. But this place was, in its infancy, entirely deftroyed by the Indians of Darien. Some years after, the fettlement was repaired, and the inhabitants maintained their ground till 1584, when orders arrived from Philip II. for their removing to Porto Bello; as much better fituated for the commerce of that country.

PORTO BELLO was taken and plundered by John Morgan, an Englifh adventurer, who infefted thofe feas; but, in confideration of a ranfom, he fpared the forts and houfes.

THE town of Porto Bello ftands near the fea, on the declivity of a mountain which furrounds the whole harbour. Moft of the houfes are built of wood. In fome the firft ftory is of ftone, and the remainder of wood. They are about 130 in number; moft of them large and fpacious. The town is under the jurifdic-

2   tion

tion of a governor, with the title of lieutenant-
general; being such under the president of Panama,
and the term of his post is without any specified limitation. He is always a gentleman of the army, having
under him the commandants of the forts that defend
the harbour; whose employments are for life.

It consists of one principal street, extending along
the strand, with other smaller crossing it, and running
from the declivity of the mountain to the shore, together with some lanes, in the same direction with the
principal street, where the ground admits of it. Here
are two large squares; one opposite to the custom-
house, which is a structure of stone, contiguous to the
quay; the other opposite the great church, which is
of stone, large, and decently ornamented, considering
the smallness of the place. It is served by a vicar and
other priests, natives of the country.

Here are two other churches, one called Nuestra
Signora de la Merced, with a convent of the same
order; the other St. Juan de Dios, which, though it
bears the title of an hospital, and was founded as such,
is very far from being so in reality. The church
of la Merced is of stone, but mean, and ruinous,
like the convent, which is also decayed; so that,
wanting the proper conveniencies for the religious to
reside in, they live in the town dispersed in private
houses.

That of St. Juan de Dios is only a small building
like an oratory, and not in better condition than that
of la Merced. Its whole community consists of a
prior, chaplain, and another religious, and sometimes
even of less: so that its extent is very small, since,
properly speaking. it has no community; and the
apartment intended for the reception of patients consists only of one chamber, open to the roof, without
beds or other necessaries. Nor are any admitted but
such as are able to pay for their treatment and diet.
It is therefore of no advantage to the poor of the
place;

place; but ferves for lodging fick men belonging to the men of war which come hither, being provided with neceffaries from the fhips, and attended by their refpective furgeons, lodging-room being the only thing afforded them by this nominal hofpital.

At the eaft end of the town, which is the road to Panama, is a quarter called Guiney, being the place where all the Negroes of both fexes, whether flaves or free, have their habitations. This quarter is very much crowded when the galleons are here, moft of the inhabitants of the town entirely quitting their houfes for the advantage of letting them, while others content themfelves with a fmall part in order to make money of the reft. The Mulattoes and other poor families alfo remove, either to Guiney, or to cottages already erected near it, or built on this occafion. Great numbers of artificers from Panama likewife, who flock to Porto Bello to work at their refpective callings, lodge in this quarter for cheapnefs.

Towards the fea, in a large tract between the town and Gloria caftle, barracks are alfo erected, and principally filled with the fhips' crews; who keep ftalls of fweetmeats, and other kind of eatables brought from Spain. But at the conclufion of the fair, the fhips put to fea, and all thefe buildings are taken down, and the town returns to its former tranquillity and emptinefs.

By an experiment we made with the barometer in a place a toife above the level of the fea, the height of the mercury was 27 inches 11 lines and a half.

CHAP.

CHAP. III.

Defcription of Porto Bello Harbour.

THE name of this port indicates its being com- modious for all forts of fhips or veffels, great or fmall ; and though its entrance is very wide, it is well defended by Fort St. Philip de Todo Fierro. It ftands on the north point of the entrance, which is about 600 toifes broad, that is, a little lefs than the fourth part of a league ; and the fouth fide being full of rifes of rocks, extending to fome diftance from the fhore, a fhip is obliged to ftand to the north, though the deepeft part of the channel is in the middle of the entrance, and thus continues in a ftraight direction, having 9, 10, or 15 fathom water, and a bottom of clayey mud, mixed with chalk and fand.

On the fouth fide of the harbour, and oppofite to the anchoring-place, is a large caftle, called Sant Jago de´la Gloria, to the eaft of which, at the diftance of about 100 toifes, begins the town, having before it a point of land projecting into the harbour. On this point ftood a fmall fort called St. Jerome, within ten toifes of the houfes. All thefe were demolifhed by the Englifh admiral Vernon, who, with a numerous naval force *, in 1739, made himfelf mafter of this port ; having found it fo unprovided with every thing, that the greater part of the artillery, efpecially that of the caftle de Todo Fierro, or iron caftle, was dif- mounted for want of carriages, part of the few mili- tary ftores unferviceable, and the garrifon fhort of its complement even in time of peace. The gover- nor of the city, Don Bernardo Gutierrez de Bucane- gra, was alfo abfent at Panama, on fome accufation brought againft him. Thus the Englifh, meeting no

---

* The numerous naval force, mentioned by our author, confifted, we know, of fix fhips only.

refiftance,

refiſtance, eaſily fucceeded in their deſign upon this city, which furrendered by capitulation.

The anchoring-place for the large ſhips is N. W. of Gloria caſtle, which is nearly the centre of the harbour; but leſſer veſſels, which come farther up, muſt be careful to avoid a fand-bank, lying 150 toiſes from St. Jerome's fort, or point, bearing from it W. one quarter northerly; and on which there is only a fathom and a half, or, at moſt, two fathom water.

N. W. of the town is a little bay, called la Caldera, or the kettle, having four fathom and a half water; and is a very proper place for careening ſhips and veſſels, as, befides its depth, it is perfectly defended from all winds. In order to go into it, you muſt keep pretty cloſe to the weſtern ſhore till about a third part of the breadth of the entrance, where you will have five fathom water (whilſt on the eaſtern ſide of the fame entrance there is not above two or three feet), and then ſteer directly towards the bottom of the bay. When the ſhips are in, they may moor with four cables eaſt and weſt, in a ſmall baſon, formed by the Caldera; but care muſt be taken to keep them always on the weſtern ſide.

N. E. of the town is the mouth of a river called Caſcajal, which affords no freſh water within a quarter of a league or upwards from its mouth; and it is not uncommon to fee in it Caymanes, or alligators.

The tides are here irregular; and in thisparticular, as well as that of the winds, there is no difference between this harbour and that of Carthagena; except that here the ſhips muſt always be towed in, being either becalmed, or the wind directly againſt them.

From obfervations we made, both by the pole-ſtar and the fun's azimuth, we found the variation of the needle in this harbour to be 8° 4' eaſterly.

Among the mountains which furround the whole harbour of Porto Bello, beginning from St. Philip de Todo Fierro, or the iron caſtle (which is ſituated on

their

1. A Spanish Lady of Quito.
2. An Indian woman of Distinction.

3. An Indian Barber.
4. A Monk of Quito.

5. An Indian Peasant.
6. An Indian Woman of the common sort.

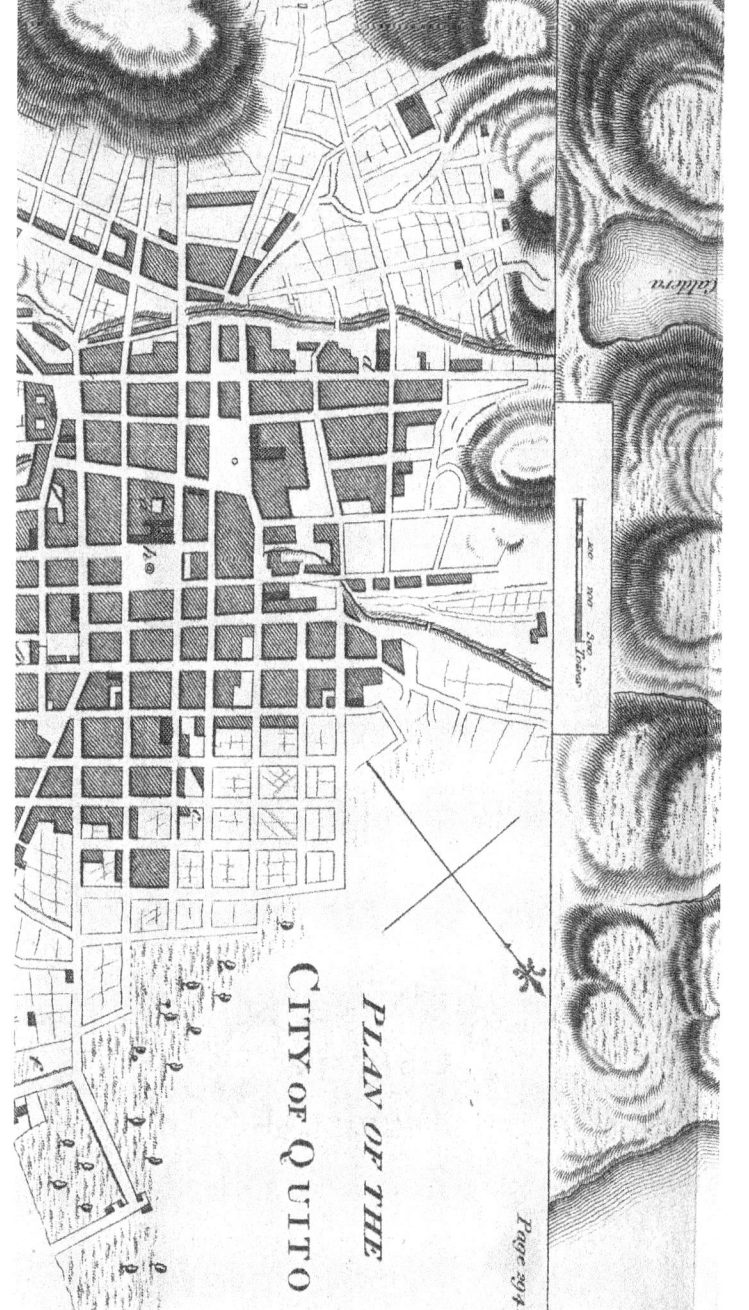

PLAN OF THE CITY OF QUITO

Page 204.

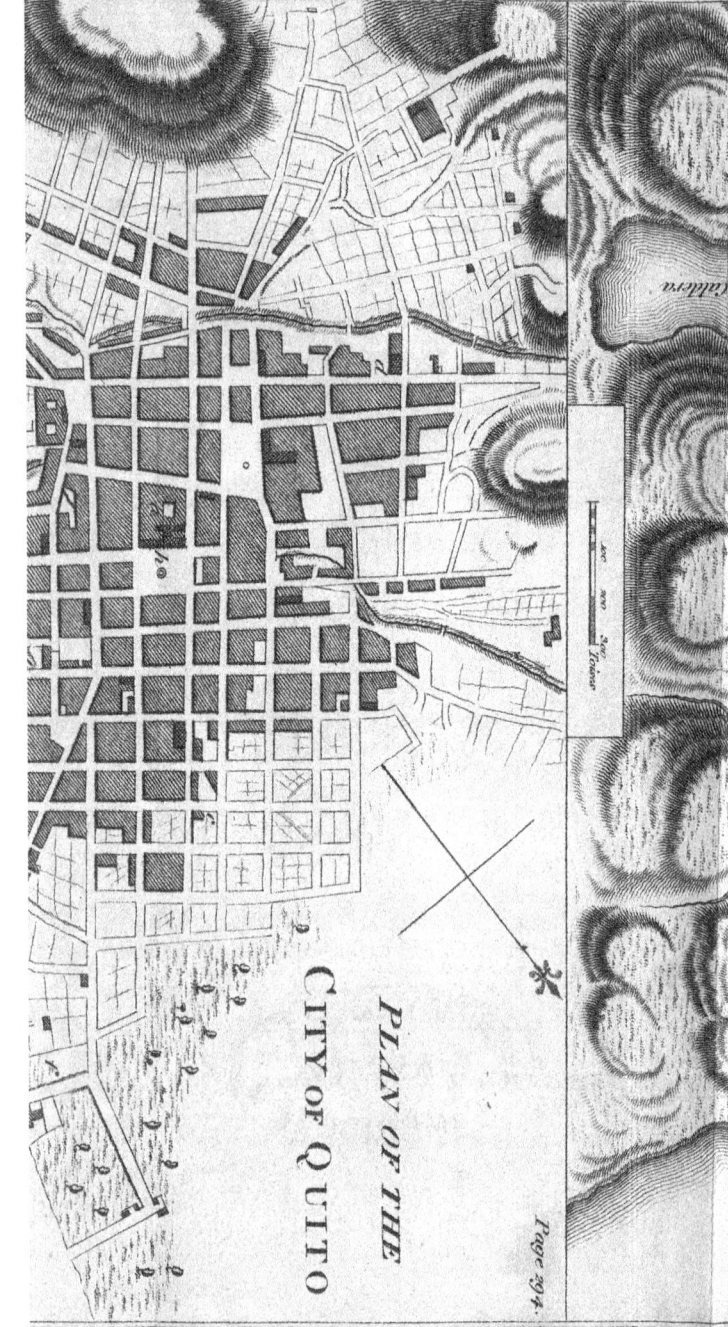

PLAN OF THE CITY OF QUITO

Page 294.

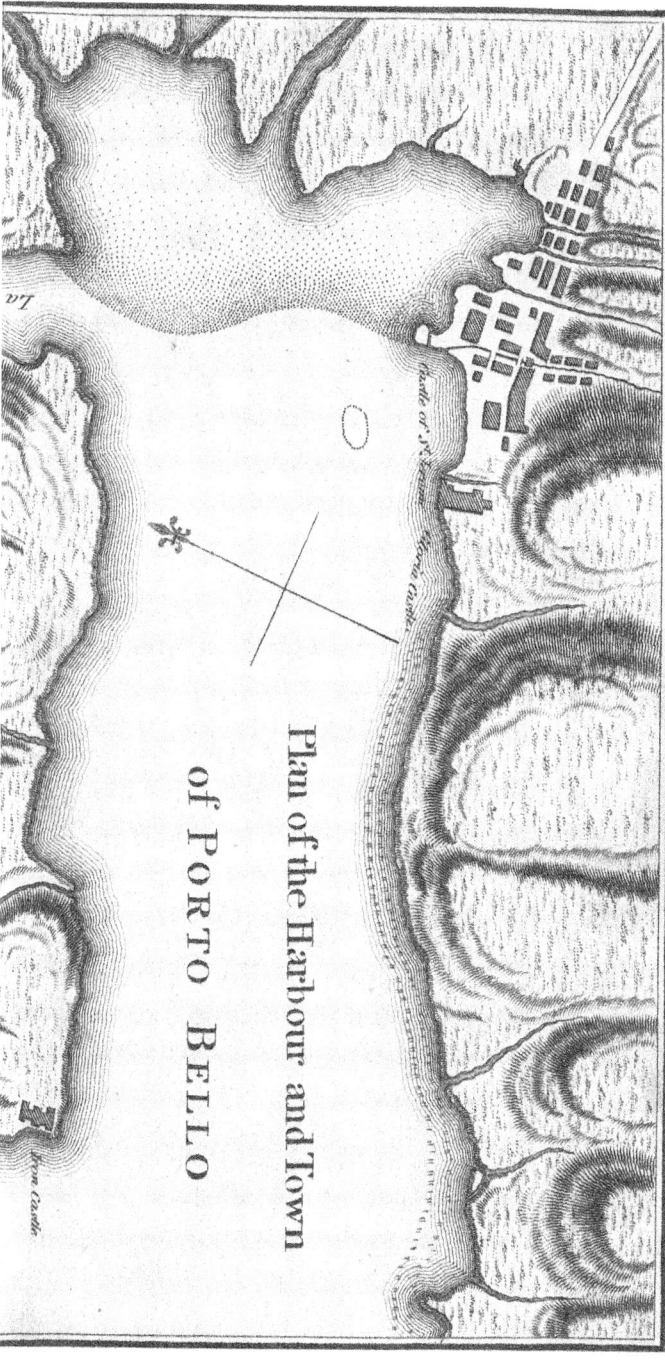

Plan of the Harbour and Town
of PORTO BELLO

their declivity), and, without any decreafe of height, extends to the oppofite point, one is particularly remarkable by its fuperior loftinefs, as if defigned to be the barometer of the country, by foretelling every change of weather. This mountain, diftinguifhed by the name of Capiro, ftands at the utmoft extremity of the harbour, in the road to Panama. Its top is always covered with clouds of a denfity and darknefs feldom feen in thofe of this atmofphere; and from thefe, which are called the capillo or cap, has poffibly been corruptly formed the name of Monte Capiro. When thefe clouds thicken, increafe their blacknefs, and fink below their ufual ftation, it is a fure fign of a tempeft; while, on the other hand, their clearnefs and afcent as certainly indicate the approach of fair weather. It muft however be remembered, that thefe changes are very frequent and very fudden. It is alfo feldom that the fummit is ever obferved clear from clouds; and when this does happen, it is only, as it were, for an inftant.

The jurifdiction of the governor of Porto Bello is limited to the town and the forts; the neighbouring country, over which it might be extended, being full of mountains covered with impenetrable forefts, except a few vallies, in which are thinly fcattered fome farms or Aaciendas; the nature of the country not admitting of farther improvements.

CHAP. IV.

Of the Climate of Porto Bello, and the Diftempers which prove fo fatal to the Crews of the Galleons.

THE inclemency of the climate of Porto Bello is fufficiently known all over Europe. Not only ftrangers who come thither are affected by it, but even the natives themfelves fuffer in various manners. It deftroys

deftroys the vigour of nature, and often untimely cuts the thread of life. It is a current opinion, that formerly, and even not above twenty years fince, parturition was here fo dangerous, that it was feldom any women did not die in childbed. As foon therefore as they had advanced three or four months in their pregnancy, they were fent to Panama, where they continued till the danger of delivery was paft. A few indeed had the firmnefs to wait their deftiny in their own houfes; but much the greater number thought it more advifable to undertake the journey, than to run fo great a hazard of their lives.

The exceffive love which a lady had for her hufband, blended with a dread that he would forget her during her abfence, his employment not permitting him to accompany her to Panama, determined her to fet the firft example of acting contrary to this general cuftom. The reafons for her fear were fufficient to juftify her refolution to run the rifk of a probable danger, in order to avoid an evil which fhe knew to be certain, and muft have embittered the whole remainder of her life. The event was happy; fhe was delivered, and recovered her former health; and the example of a lady of her rank did not fail of infpiring others with the like courage, though not founded on the fame reafons; till, by degrees, the dread which former melancholy cafes had impreffed on the mind, and gave occafion to this climate's being confidered as fatal to pregnant women, was entirely difperfed.

Another opinion equally ftrange is, that the animals from other climates, on their being brought to Porto Bello, ceafe to procreate. The inhabitants bring inftances of hens brought from Panama or Carthagena, which immediately on their arrival grew barren, and laid no more eggs; and even at this very time the horned cattle fent from Panama, after they have been here a fhort time, lofe their flefh fo as not to be eatable; though they do not want for plenty of
good

good pasture. It is certain that there are no horses or asses bred here, which tends to confirm the opinion that this climate checks the generation of creatures produced in a more benign or less noxious air. However, not to rely on the common opinion, we inquired of some intelligent persons, who differed but very little from the vulgar, and even confirmed what they asserted, by many known facts, and experiments performed by themselves.

The liquor in Mr. Reaumur's thermometer, on the 4th of December 1735, at six in the morning, stood at 1021, and at noon rose to 1023.

The heat here is excessive, augmented by the situation of the town, which is surrounded with high mountains, without any interval for the winds, whereby it might be refreshed. The trees on the mountains stand so thick, as to intercept the rays of the sun ; and, consequently, hinder them from drying the earth under their branches : hence copious exhalations, which form large clouds, and precipitate in violent torrents of rain ; these are no sooner over, than the sun breaks forth afresh, and shines with its former splendour ; though scarce has the activity of his rays dried the surface of the ground not covered by the trees, when the atmosphere is again crowded by another collection of thick vapours, and the sun again concealed. Thus it continues during the whole day : the night is subject to the like vicissitudes ; but without the least diminution of heat in either.

These torrents of rain, which, by their suddenness and impetuosity, seem to threaten a second deluge, are accompanied with such tempests of thunder and lightning, as must daunt even the most resolute : this dreadful noise is prolonged by repercussions from the caverns of the mountains, like the explosion of a cannon, the rumbling of which is heard for a minute after. To this may also be added the howlings and shrieks of the multitudes of monkies of all kinds, which live in
the

the foreſts of the mountains, and which are never louder than when a man of war fires the morning and evening gun, though they are ſo much uſed to it.

This continual inclemency, added to the fatigue of the ſeamen in unloading the ſhips, carrying the goods on ſhore in barges, and afterwards drawing them along on ſledges, cauſes a very profuſe tranſpiration, and conſequently renders them weak and faint; and they, in order to recruit their ſpirits, have recourſe to brandy, of which there is, on theſe occaſions, an incredible conſumption. The exceſſive labour, immoderate drinking, and the inclemency and unhealthfulneſs of the climate, muſt jointly deſtroy the beſt conſtitutions, and produce thoſe deleterious diſeaſes ſo common in this country. They may well be termed deleterious; for the ſymptoms of all are fatal, the patients being too much attenuated to make any effectual reſiſtance; and hence epidemics and mortal diſtempers are ſo very common.

It is not the ſeamen alone who are ſubject to theſe diſeaſes; others, ſtrangers to the ſeas, and not concerned in the fatigues, are attacked by them; and, conſequently, is a ſufficient demonſtration that the other two are only collateral, though they tend both to ſpread and inflame the diſtemper; it being evident, that when the fluids are diſpoſed to receive the ſeeds of the diſtemper, its progreſs is more rapid, and its attacks more violent. On ſome occaſions, phyſicians have been ſent for from Carthagena, as being ſuppoſed to be better acquainted with the propereſt methods of curing the diſtempers of this country, and conſequently more able to recover the ſeamen; but experience has ſhewn, that this intention has been ſo little anſwered, that the galleons or other European ſhips, which ſtay any time here, ſeldom depart, without burying half, or, at leaſt, a third of their men; and hence this city has, with too much reaſon, been termed the grave of the Spaniards; but it may,

with

with much greater propriety, be applied to thofe of other nations who vifit it. This remark was fufficiently confirmed by the havoc made among the Englifh, when their fleet, in 1726, appeared before the port, with a view of making themfelves mafters of the treafure brought thither from all parts to the fair held at the arrival of the galleons, which, at that time, by the death of the marquis Grillo, were commanded by Don Francifco Cornejo, one of thofe great officers whofe conduct and refolution have done honour to the navy of Spain. He ordered the fhips under his command to be moored in a line within the harbour; and erected, on the entrance, a battery, the care of which he committed to the officers of the fhips; or rather, indeed, fuperintended it himfelf, omitting no precaution, but vifiting every part in perfon. Thefe preparatives ftruck fuch a confternation into the Englifh fleet, though of confiderable force, that, inftead of making any attempt, they formed only a blockade, depending on being fupplied with provifions from Carthagena, and that famine would at length oblige the Spaniards to give up what they at firft intended to acquire by force; but when the admiral thought himfelf on the point of obtaining his ends, the inclemency of the feafon declared itfelf among his fhips' companies, fweeping away fuch numbers, that in a fhort time he was obliged to return to Jamaica, with the lofs of above half his people.

But, notwithftanding the known inclemency of the climate of Porto Bello, and its general fatality to Europeans, the fquadron of 1730 enjoyed there a good ftate of health, though the fatigues and irregularities among the feamen were the fame: nor was there any perceivable change in the air. This happy fingularity was attributed to the ftay of the fquadron at Carthagena, where they paffed the time of the epidemia, by which their conftitutions were better adapted to this climate; and hence it appears, that

the principal caufe of thefe diftempers flows from the
conftitutions of the Europeans not being ufed to it;
and thus they either die, or become habituated to it,
like the natives, Creoles, and other inhabitants.

CHAP. V.

Account of the Inhabitants and Country about
Porto Bello.

IN feveral particulars there is no effential difference
between Carthagena and Porto Bello; fo that I
fhall only mention thofe peculiar to the latter; and
add fome obfervations, tending to convey a more exact
knowledge of this country.

THE number of the inhabitants of Porto Bello, by
reafon of its fmallnefs, and the inclemency of its climate, is very inconfiderable, and the greater part of
thefe, Negroes and Mulattos, there being fcarce thirty
White families; thofe, who by commerce or their
eftates are in eafy circumftances, removing to Panama.
So that thofe only ftay at Porto Bello, whofe employments oblige them to it; as the governor or lieutenant-
general, the commanders of the forts, the civil officers
of the crown, the officers and foldiers of the gar-
rifons, the alcaldes in office and of the hermandad,
and the town-clerk. During our ftay here, the gar-
rifons of the forts confifted of about 125 men, being
detachments from Panama; and thefe, though coming
from a place fo near, are affected to fuch a degree,
that in lefs than a month they are fo attenuated, as to
be unable to do any duty, till cuftom again reftores
them to their ftrength. None of thefe, or of the natives of the country, above the Mulatto clafs, ever
fettle here, thinking it a difgrace to live in it: a certain proof of its unhealthinefs, fince thofe to whom
it gave birth forfake it.

IN

In manners and cuftoms, the inhabitants of Porto Bello refemble thofe of Carthagena, except that the latter are more free and generous, thofe in the parts round Port Bello being accufed of avarice; a vice natural to all the inhabitants of thefe countries.

Provisions are fcarce at Porto Bello, and confequently dear, particularly during the time of the galleons and the fair, when there is a neceffity for a fupply from Carthagena and Panama. From the former are brought maize, rice, cafava, hogs, poultry, and roots; and from the latter, cattle. The only thing in plenty here is fifh, of which there is a great variety and very good. It alfo abounds in fugar-canes, fo that the chacaras, or farm-houfes, if they may be fo called, are built of them. They have alfo ingenios * for making fugar and molaffes, and, from the latter, brandy.

Fresh water pours down in ftreams from the mountains, fome running without the town, and others croffing it. Thefe waters are very light and digeftive, and, in thofe who are ufed to them, good to create an appetite. Qualities, which in other countries would be very valuable, are here pernicious. This country feems fo curfed by nature, that what is in itfelf good, becomes here deftructive. For, doubtlefs, this water is too fine and active for the ftomachs of the inhabitants; and thus produces dyfenteries, the laft ftage of all other diftempers, and which the patient very feldom furvives. Thefe rivulets, in their defcent from the mountains, form little refervoirs, or ponds, whofe coolnefs is increafed by the fhade of the trees, and in thefe all the inhabitants of the town bathe themfelves conftantly every day at eleven in the morning; and the Europeans fail not to follow an example fo pleafant and conducive to health.

* Ingenio fignifies the mill, ftill, and apparatus, for making fugar, rum, &c. A.

H 2   As

As thefe forefts almoft border on the houfes of the town, the tigers often make incurfions into the ftreets during the night, carrying off fowls, dogs, and other domeftic creatures; and fometimes even boys have fallen a prey to them; and it is certain, that ravenous beafts, which provide themfelves with food in this manner, are afterwards known to defpife what the forefts afford; and that, after tafting human flefh, they flight that of beafts *. Befides the fnares ufually laid for them, the Negroes and Mulattos, who fell wood in the forefts of the mountains, are very dexterous in encountering the tigers; and fome, even on account of the flender reward, feek them in their retreats. The arms in this combat, feemingly fo dangerous, are only a lance, of two or three yards in length, made of a very ftrong wood, with the point of the fame hardened in the fire; and a kind of cimeter, about three quarters of a yard in length. Thus armed, they ftay till the creature makes an affault on the left arm, which holds the lance, and is wrapped up in a fhort cloak of baize. Sometimes the tiger, aware of the danger, feems to decline the combat; but his antagonift provokes him with a flight touch of the lance, in order, while he is defending himfelf, to ftrike a fure blow; for, as foon as the creature feels the lance, he grafps it with one of his paws, and with the other ftrikes at the arm which holds it. Then it is that the perfon nimbly aims a blow with his cimeter, which he kept concealed with the other hand, and hamftrings the creature, which immediately draws back enraged, but returns to the

* This is an error. Beafts of prey in America are not fo fierce as in Africa and Afia; they never attack the human fpecies, but when forced by hunger, or provoked. It is affirmed by the natives, that if an European, with his Negro and dog, were to meet with two hungry beafts of prey, whether tigers or ounces, they would feize the dog and Negro, and leave the European. But the truth I never knew experienced. A.

charge;

charge; when, receiving another such stroke, he is totally deprived of his most dangerous weapons, and
rendered incapable of moving.   After which the person kills him at his leisure, and stripping off the skin,
cutting off the head, and the fore and hind feet, returns to the town, displaying these as the trophies of
his victory.

Among the great variety of animals in this country,
one of the most remarkable is the Perico ligero, or
nimble Peter, an ironical name given it on account of
its extreme sluggishness and sloth.   It resembles a
middling monkey, but of a wretched appearance, its
skin being of a greyish brown, all over corrugated,
and the legs and feet without hair.   He is so lumpish,
as not to stand in need of either chain or hutch, for
he never stirs till compelled by hunger.   When he
moves, every effort is attended with such a plaintive,
and at the same time so disagreeable a cry, as at once
produces pity and disgust; and this even on the
slightest motion of the head, legs, or feet; proceeding
probably from a general contraction of the muscles
and nerves of his body, which puts him to extreme
pain when he endeavours to move them.   In this difagreeable cry consists his whole defence; for, it being
natural to him to fly at the first hostile approach of
any beast, he makes at every motion such howlings
as are even insupportable to his pursuer, who soon
quits him, and even flies beyond the hearing of his
horrid noise.   Nor is it only during the time he is in
motion that he utters these cries; he repeats them
while he rests himself, continuing a long time motionless before he takes another march.   The food of this
creature is generally wild fruits; when he can find
none on the ground, he looks out for a tree well loaded, which, with a great deal of pains, he climbs;
and, to save himself such another toilsome ascent,
plucks off all the fruit, throwing them on the ground;
and to avoid the pain of descending, forms himself

H 3   into

into a ball, and drops from the branches. At the foot of this tree he continues till all the fruits are confumed, never ftirring till hunger forces him to feek again for food.

SERPENTS are here as numerous and deadly as at Carthagena ; and toads * innumerable, fwarming not only in the damp and marfhy places, as in other countries, but even in the ftreets, courts of great houfes, and all open places in general. The great numbers of them, and their appearance after the leaft fhower, have induced fome to imagine, that every drop of water becomes a toad ; and though they allege, as a proof, the extraordinary increafe of them on the fmalleft fhower, their opinion does not feem to me well founded. It is evident, that thefe reptiles abound both in the forefts and neighbouring rivers, and even in the town itfelf; and produce a prodigious quantity of animalcula, from whence according to the beft naturalifts, thefe reptiles are formed. Thefe animalcula either rife in the vapours, which form the rain, and falling together with it on the ground, which is extremely heated by the rays of the fun, or being already depofited in it by the toads, grow, and become animated, in no lefs numbers than were formerly feen in Europe. But fome of them which appear after rains being fo large as to meafure fix inches in length, they cannot be imagined the effect of an inftantaneous production ; I am therefore inclined to think, from my own obfervations, that this part of the country, being remarkably moift, is very well adapted to nourifh the breed of thofe creatures, which love watery places ; and therefore avoid thofe parts of the ground expofed to the rays of the fun, feeking others where the earth is foft, and there form themfelves cavities in the ground, to enjoy the moif-

* Called by the natives ferpos : they appear every dewy evening in as great numbers as after a fhower. I never heard of the opinion the author fpeaks of. A.

ture;

ture; and as the furface over them is generally dry, the toads are not perceived; but no fooner does it begin to rain, than they leave their retreats, to come at the water, which is their fupreme delight; and thus fill the ftreets and open places. Hence the vulgar opinion had its rife, that the drops of rain were, tranf-formed into toads. When it has rained in the night, the ftreets and fquares in the morning feem paved with thefe reptiles; fo that you cannot ftep without tread-ing on them, which fometimes is productive of trou-blefome bites; for, befides their poifon, they are large enough for their teeth to be feverely felt. Some we have already obferved to be fix inches long, and this is their general meafure; and there are fuch numbers of them, that nothing can be imagined more difmal than their croakings, during the night, in all parts of the town, woods, and caverns of the moun-tains.

CHAP. VI.

Of the Trade of Porto Bello.

THE town of Porto Bello, fo thinly inhabited, by reafon of its noxious air, the fcarcity of provifions, and the barrennefs of its foil, becomes, at the time of the galleons, one of the moft populous places in all South America. Its fituation on the ifthmus betwixt the fouth and north fea, the goodnefs of its harbour, and its fmall diftance from Panama, have given it the preference for the rendezvous of the joint com merce of Spain and Peru, at its fair.

On advice being received at Carthagena, that the Peru fleet had unloaded at Panama, the galleons make the beft of their way to Porto Bello, in order to avoid the diftempers which have their fource from idlenefs. The concourfe of people, on this occafion, is fuch,

H 4   as

as to raife the rent of lodging to an exceffive degree; a middling chamber, with a clofet, lets, during
the fair, for a thoufand crowns, and fome large houfes
for four, five, or fix thoufand.

THE fhips are no fooner moored in the harbour,
than the firft work is, to erect, in the fquare, a tent
made of the fhip's fails, for receiving its cargo: at
which the proprietors of the goods are prefent, in
order to find their bales, by the marks which diftinguifh them. Thefe bales are drawn on fledges, to
their refpective places, by the crew of every fhip, and
the money given them is proportionally divided.

WHILST the feamen and European traders are thus
employed, the land is covered with droves of mules
from Panama, each drove confifting of above an hundred, loaded with chefts of gold and filver, on account
of the merchants of Peru. Some unload them at the
exchange, others in the middle of the fquare; yet,
amidft the hurry and confufion of fuch crowds, no
theft, lofs, or difturbance, is ever known. He who
has feen this place during the tiempo muerto, or dead
time, folitary, poor, and a perpetual filence reigning
every where; the harbour quite empty, and every
place wearing a melancholy afpect, muft be filled with
aftonifhment at the fudden change, to fee the buftling
multitudes, every houfe crowded. the fquare and ftreets
encumbered with bales and chefts of gold and filver
of all kinds; the harbour full of fhips and veffels,
fome bringing by the way of Rio de Chape the goods
of Peru, as cacao, quinquina or jefuits bark, Vicuna
wool, and bezoar ftones; others coming from Carthagena, loaded with provifions: and thus a fpot, at all
other times detefted for its deleterious qualities, becomes the ftaple of the riches of the old and new
world, and the fcene of one of the moft confiderable
branches of commerce in the whole earth.

THE fhips being unloaded, and the merchants of
Peru, together with the prefident of Panama, arrived,
                                                    the

the fair comes under deliberation. And for this purpose the deputies of the several parties repair on board
the commodore of the galleons, where, in presence of
the commodore, and the president of Panama; the
former, as patron of the Europeans, and the latter, of
the Peruvians; the prices of the several kinds of mer-
chandises are settled; and all preliminaries being ad-
justed in three or four meetings, the contracts are
signed, and made public, that every one may conform
himself to them in the sale of his effects. Thus all
fraud is precluded. The purchases and sales, as likewise
the exchanges of money, are transacted by brokers,
both from Spain and Peru. After this, every one begins to dispose of his goods; the Spanish brokers em-
barking their chests of money, and those of Peru sending away the goods they have purchased, in vessels
called chatas and bongos, up the river Chagre. And
thus the fair of Porto Bello ends.

FORMERLY this fair was limited to no particular
time; but as a long stay, in such a sickly place, extremely affected the health of the traders, his catholic
majesty transmitted an order, that the fair should not
last above forty days, reckoning from that in which
the ships came to an anchor in the harbour; and that,
if in this space of time the merchants could not agree
in their rates, those of Spain should be allowed to carry
their goods up the country to Peru; and accordingly
the commodore of the galleons has orders to re-
embark them, and return to Carthagena; but otherwise, by virtue of a compact between the merchants
of both kingdoms, and ratified by the king, no Spanish
trader is to send his goods, on his own account, beyond Porto Bello: and, on the contrary, those of Peru
cannot send remittances to Spain, for purchasing goods
there.

WHILST the English were permitted to send an annual ship, called Navio de Permisso, she used to bring
to the fair a large cargo on her own account, never
failing

failing firſt to touch at Jamaica, ſo that her loading alone was more than half of all thoſe brought by the galleons; for, beſides that her burden ſo far exceeded five hundred Spaniſh tons, that it was even more than nine hundred, ſhe had no proviſions, water, or other things, which fill a great part of the hold; ſhe indeed took them in at Jamaica, from whence ſhe was attended by five or ſix ſmaller veſſels, loaded with goods, which, when arrived near Porto Bello, were put on board her, and the proviſions removed into the tenders; by which artifice the ſingle ſhip was made to carry more than five or ſix of the largeſt galleons. This nation having a free trade, and ſelling cheaper than the Spaniards, that indulgence was of infinite detriment to the commerce of Spain.

In the dead time, all the trade ſtirring here conſiſts in proviſions from Carthagena; and cacao and quin-quina, down the river Chagre: the former is carried in ſmall veſſels to Vera Cruz, and the quinquina either depoſited in warehouſes, or put on board ſhips, which, with permiſſion, come from Spain to Nicaragua and Honduras; theſe ſhips alſo take in cacao Some ſmall veſſels likewiſe come from the iſlands of Cuba, La Trinidad, and St. Domingo, with cacao and rum.

Whilst the aſſiento of Negroes ſubſiſted either with the French or Engliſh, one of their principal factories was ſettled here; and was of conſiderable advantage to its commerce, as being the channel by which not only Panama was ſupplied with Negroes, but from whence they were ſent all over the kingdom of Peru: on which account the agents of the aſſiento were allowed to bring with them ſuch a quantity of proviſions as was thought neceſſary, both for their own uſe, and their ſlaves of both ſexes.

BOOK

BOOK III.

Voyage from Porto Bello to Panama.

CHAP. I.

Voyage up the Chagre, and Journey from Cruces to Panama by Land.

AS it had always been our fixed defign to ftay no longer than abfolutely neceffary in any place, till we had anfwered the great end of our commiffion, our ardour to enter upon it, together with a defire of quitting this dangerous climate, induced us to make the utmoft difpatch. In order to this, we fent advice from Porto Bello, to Don Dionyfio Martinez de la Vega, prefident of Panama, of our arrival, the motives of our voyage, and other circumftances, together with his majefty's orders relating to the affiftance to be given us by all his officers; adding our requefts, that he would be pleafed to fend one or two of thofe veffels ufed on the Chagre, to bring us to Panama, it being impracticable for us to travel thither by land, as fome of the inftruments were too large for the narrow craggy roads in many parts, and others of a nature not to be carried on mules. This gentleman, who has always fhewn a remarkable zeal for every thing dignified with his majefty's name, was not in the leaft wanting on this occafion; and his polite reply, which fully anfwered our moft fanguine hopes, was

followed

followed by two veffels, difpatched to Porto Bello. Immediately on their arrival, we put on board the inftruments and baggage, belonging both to the French gentlemen and ourfelves; and on the 22d of December 1735, departed from Porto Bello.

The land wind being contrary to us, we rowed out of Porto Bello harbour; but the brifas fetting in at nine in the morning, both veffels got under fail; and a frefh gale brought us, at four in the evening of the fame day, to the mouth of the river Chagre, where we landed at the cuftom houfe; and the next day we began to row up the river.

On the 24th, we endeavoured to proceed in the fame manner; but the force of our oars being too weak to ftem the current, we were obliged to fet the veffels along with poles. At a quarter after one in the afternoon, we meafured the velocity of the current, and found it ten toifes and one foot in forty feconds and a half. In this flow toilfome manner we proceeded till the 27th, at eleven in the morning, when we arrived at Cruces, the landing-place, about five leagues from Panama. As we advanced up the river, we found a great increafe in the velocity of the current, which on the 25th was ten toifes in twenty-fix feconds and a half: on the 26th, at the place where we anchored for that night, ten toifes in fourteen feconds and a half; and on the 27th, at the town of Cruces, the fame fpace in fixteen feconds. Confequently the greateft velocity of the water is two hundred and eighty-three toifes, or about a league, in an hour.

This river, which was formerly called Lagartos, from the number of alligators in it, though now better known by that of Chagre, has its fource in the mountains near Cruces. Its mouth, which is in the north fea, in 9° 18′ 40″ N. latitude, and 295° 6′ longitude, from the meridian of Teneriffe, was difcovered by Lopez de Olano. Diego de Alvites difcovered that
part

part of it where Cruces is fituated; but the firft Spaniard who failed down it, to reconnoitre it to its mouth, was captain Hernando de la Serna, in the year 1527. Its entrance is defended by a fort, fituated on a fteep rock on the eaft fide near the fea fhore. This fort is called San Lorenzo de Chagres, has a commandant and a lieutenant, both appointed by his majefty, and the garrifon is draughted from Panama.

ABOUT eight toifes from the above fort, is a town of the fame name. The houfes are principally of reeds, and the inhabitants Negroes, Mulattos, and Meftizos. They are a brave and active people, and, on occafion, take up arms to the number of triple the ufual garrifon of the fort.

OPPOSITE, on a low and level ground, ftands. the royal cuftom-houfe, where an account is taken of all goods going up the Chagre. Here the breadth of the river is about 120 toifes, but grows narrower gradually as you approach its fource. At Cruces, the place where it begins to be navigable, it is only twenty toifes broad; the neareft diftance between this town and the mouth is twenty-one miles, and the bearing N. W. 7° 24′ wefterly; but the diftance meafured along the feveral windings of the river, is no lefs than forty-three miles.

IT breeds a great number of caymanes or alligators : creatures often feen on its banks, which are impaffable, both on account of the clofenefs of the trees, and the bufhes, which cover the ground, as it were, with thorns. Some of thefe trees, efpecially the cedar, are ufed in making the canoes or banjas employed on the river. Many of them being undermined by the water, are thrown down by the fwellings of the river; but the prodigious magnitude of the trunk, and their large and extenfive branches, hinder them from being carried away by the current; fo that they remain near their original fituation, to the

great

great inconvenience and even danger of the veffels; for, the greater part of them being under water, a veffel, by ftriking fuddenly on them, is frequently overfet. Another obftruction to the navigation of this river is the races, or fwift currents, over the fhallows, where thofe veffels, though built for that purpofe, cannot proceed for want of a fufficient quantity of water; fo that they are obliged to be lightened, till they have paffed the fhallow.

The barks employed on this river are of two kinds, the chatas and bongos, called in Peru, bonques. The firft are compofed of feveral pieces of timber, like barks, and of great breadth, that they may draw but little water; they carry fix or feven hundred quintals. The bongos are formed out of one piece of wood; and it is furprifing to think there fhould be trees of fuch prodigious bulk, fome being eleven Paris feet broad, and carrying conveniently four or five hundred quintals. Both forts have a cabin at the ftern, for the conveniency of the paffengers, a kind of awning fupported with a wooden ftancheon reaching to the head, and a partition in the middle, which is alfo continued the whole length of the veffel; and over the whole, when the veffel is loaded, are laid hides, that the goods may not be damaged by the violence of the rains, which are very frequent here. Each of thefe require, befides the pilot, at leaft eighteen or twenty robuft Negroes; for, without fuch a number, they would not be able, in going up, to make any way againft the current.

All the forefts and woods near this river are full of wild beafts, efpecially different kinds of monkies. They are of various colours, as black, brown, reddifh, and ftriated; there is alfo the fame diverfity in their fize; fome being a yard long, others about half a yard, and others fcarce one third. The flefh of all thefe different kinds is highly valued by the Negroes, efpecially that of the red; but, however delicate the

meat

meat may be, the fight of them is, I think, enough
to make the appetite abhor them; for, when dead,
they are fcalded in order to take off the hair, whence
the fkin is contracted by the heat, and when thoroughly cleaned, looks perfectly white, and very
greatly refembles a child of about two or three years
of age, when crying. This refemblance is fhocking
to humanity, yet the fcarcity of other food in many
parts of America renders the flefh of thefe creatures
valuable; and not only the Negroes, but the Creoles
and Europeans themfelves, make no fcruple of eating it.

Nothing, in my opinion, can excel the profpects
which the rivers of this country exhibit. The moft
fertile imagination of a painter can never equal the
magnificence of the rural landfcapes here drawn by
the pencil of Nature. The groves which fhade the
plains, and extend their branches to the river; the
various dimenfions of the trees, which cover the eminences; the texture of their leaves; the figure of
their fruits, and the various colours they exhibit, form
a moft delightful fcene, which is greatly heightened
by the infinite variety of creatures with which it is diverfified. The different fpecies of monkies, fkipping
in troops from tree to tree, hanging from the branches,
and in other places fix, eight, or more of them linked
together, in order to pafs a river, and the dams with
their young on their fhoulders, throwing themfelves
into odd poftures, making a thoufand grimaces, will
perhaps appear fictitious to thofe who have not
actually feen it. But if the birds are confidered our
reafon for admiration will be greatly augmented:
for, befides thofe already mentioned (Book I. chap.
vii.), and which, from their abundance, feem to
have had their origin on the banks of this river, here
are a great variety of others, alfo eatable; as the
wild and royal peacock, the turtle-dove, and the
heron. Of the latter there are four or five fpecies;

fome entirely white; others of the fame colour, except the neck and fome parts of the body, which are red; others black, only the neck, tips of the wings and the belly white; and fome, with other mixture of colours; and all differing in fize   The fpecies firft mentioned are the leaft; the white mixed with black the largeft and moft palatable. The flefh of peacocks, phea-fants, and other kinds, is very delicate *. The trees along the banks of this river are furprifingly loaded with fruit; but the pine-apples, for beauty, fize, flavour, and fragrancy, excel thofe of all other countries, and are highly efteemed in all parts of America.

On our arrival at Cruces, we went on fhore, and were entertained by the alcalde of the town, whofe houfe was that of the cuftoms, where an account is taken of all goods brought up the river   Having, with all poffible difpatch, got every thing ready for our journey to Panama, on the 29th, at half an hour after eleven in the morning, we fet out, and reached that city by three quarters after fix in the evening. We made it our firft bufinefs to wait on the prefident, a mark of refpect due, not only to his dignity, but alfo for the many civilities he had fhewn us. This worthy gentleman received us all, and particularly the fo-reigners, in the moft cordial and endearing manner. He alfo recommended to all the king's officers, and other perfons of diftinction in the city, not to be want-ing in any good office, or mark of efteem : a beha-viour which fhewed at once the weight of the royal orders, and his zeal to execute his fovereign's plea-fure.

Some indifpenfable preparations, which were to be made for the profecution of our journey, detained us longer at Panama than we expected. We however

---

* The fifhy tafte, which moft of the fowls in this country have, is an exception to their delicacy as food. A.

employed

employed our time to the beſt advantage, making ſeveral obſervations, particularly on the latitude and the
pendulum; but the proximity of Jupiter at that time
to the Sun hindered us from ſettling the longitude.
I alſo employed myſelf in taking a plan of the place,
with all its fortifications, and adjacent coaſt. At
length, all things being in readineſs, we embarked
without any farther loſs of time.

CHAP. II.

Deſcription of the City of Panama.

PANAMA is built on an iſthmus of the ſame
name, the coaſt of which is waſhed by the South
Sea. From the obſervations we made here, we found
the latitude of this city to be 8° 57' 48"½ north. With
regard to its longitude, there are various opinions;
none of the aſtronomers having been able, from obſervations made on the ſpot, to aſcertain it; ſo that it
is ſtill doubtful whether it lies on the eaſt or weſt ſide
of the meridian of Porto Bello. The French geo-
graphers will have it to lie on the eaſt ſide, and accordingly have placed it ſo in their maps; but, in
thoſe of the Spaniards, it is on the weſt: and I conceive the latter, from their frequent journies from one
place to the other, may be concluded to have a more
intimate knowledge of their reſpective ſituations;
whereas the former, being ſtrangers in a great meaſure to thoſe places, have not the opportunity of
making ſo frequent obſervations. I allow indeed,
that, among the Spaniards who make this little
journey, the number is very ſmall of thoſe, who have
either capacity or inclination for forming a well-
grounded judgment of the road they travel; but there
have been alſo many expert pilots, and other perſons

of curiofity, who have employed their attention on it; and from their report, the fituation of the city has been determined. This opinion is in fome meafure confirmed by our courfe, the direction of which on the river, from its mouth to the town of Cruces, was eaft 6° 15′ foutherly; and the diftance being 21 miles, the difference between the two meridians is 20 minutes, the diftance Chagre is fituated to the weft of Cruces. We muft alfo confider the diftance between Porto Bello and Chagre. During the firft two hours and a half, we failed a league and a half an hour; when, the land breeze fpringing up, we failed two leagues an hour, for feven hours; which in all makes 18 leagues; and the whole courfe having been very nearly weft, the difference of longitude muft have been 44 miles; or 41, allowing for what might have been wanting of a due weft courfe; and from this again fubtracting the 20 minutes which Cruces lies to the eaft of Chagre, the refult is, that Cruces is fituated twenty-one minutes to the weftward of Porto Bello. To this laft refult muft be added the diftance of meridians between Cruces and Panama, the bearing of which is near S. W. and N. E.; and reckoning that we travelled, on account of the roughnefs and cragginefs of the road, only three quarters of a league an hour, during the feven hours, the whole is 14 miles, and the difference of meridians 10 minutes and a half. Confequently Panama is fituated about 30 minutes weft of Porto Bello; and the Spanifh artifts are nearer the truth than the French.

THE firft difcovery of Panama, the Spaniards owe to Tello de Guzman, who landed here in 1515; but found only fome fifhermen's huts, this being a very proper place for their bufinefs, and from thence the Indians called it Panama, which fignifies a place abounding in fifh. Before this, namely in the year 1513, Bafco Nunez de Balboa difcovered the South Sea, and took legal poffeffion of it in the names of the

kings

kings of Caſtile. The diſcovery of Panama was, in
the year 1518, followed by the ſettlement of a colony
there, under Pedrarias Davila, governor of Caſtilla
del Oro, the name by which this Terra Firma was
then called. And in 1521, his catholic majeſty, the
emperor Charles V. conſtituted it a city, with the
proper privileges.

It was this city's misfortune, in the year 1670, to
be ſacked and burnt by John Morgan, an Engliſh
adventurer. He had before taken Porto Bello and
Maracaybo; and, retiring to the iſlands, he every
where publiſhed his deſign of going to Panama; upon
which many of the pirates, who then infeſted thoſe
ſeas, joined him. He firſt ſailed for Chagre, where
he landed ſome of his men, and at the ſame time
battered the caſtle with his ſhips; but his ſucceſs
was owing to a very extraordinary accident. His
ſtrength was conſiderably diminiſhed by the great
numbers killed and wounded by the fort, and he
began to think it adviſable to retreat; when an
arrow, ſhot from the bow of an Indian, lodged in the
eye of one of Morgan's companions. The perſon
wounded, rendered deſperate by the pain, with a remarkable firmneſs and preſence of mind drew the
arrow from the wound, and wrapping one of its ends
in cotton, or tow, put it into his muſket, which was
ready loaded, and diſcharged it into the fort, where
the roofs of the houſes were of ſtraw, and the ſides of
wood, according to the cuſtom of that country. The
arrow fell on one of the roofs, and immediately ſet it
on fire, which was not at firſt obſerved by the be-
ſieged, who were buſy in defending the place; but
the ſmoke and flames ſoon informed them of the total
deſtruction of the fort, and of the magazine of
powder, which the flames muſt ſoon reach. This
unexpected accident filled them with terror and con-
fuſion; the courage of the ſoldiers degenerated into
tumult and diſobedience; and, every one being eager

to fave himfelf, the works were foon abandoned, in order to efcape the double danger of being either burnt or blown up The commandant, however, determined to do all in his power, ftill defended the fort, with fixteen or twenty foldiers, being all that were left him, till, covered with wounds, he fell a victim to his loyalty. The pirates, encouraged by this accident, pufhed their attack with the utmoft vigour; and the few people were obliged to furrender the place, which the violence of the flames foon laid in afhes. Having furmounted this difficulty, the greater part of them proceeded up the river in boats and launches, leaving the fhips at anchor, for the defence of their new conqueft. The detachment having landed at Cruces, marched towards Panama, and, on the Sabana, a fpacious plain before the city, they had feveral fkirmifhes, in which Morgan always gained the advantage; fo that he made himfelf mafter of the city, but found it almoft forfaken; the inhabitants, on feeing their men defeated, having retired into the woods. He now plundered it at his leifure; and, after ftaying fome days, agreed, for a large ranfom, to evacuate it without damaging the buildings; but, after the payment of the money, the city was fet on fire, by accident, as they gave out, and as the hiftory of his adventures relates; but it is much more probable that it was done by defign. To pretend it was owing to accident, feemed to them the beft palliative for their violating the treaty.

This misfortune rendering it abfolutely neceffary to rebuild the city, it was removed to its prefent fituation, which is about a league and a half from the former, and much more convenient. It has a wall of free-ftone, and is defended by a large garrifon of regulars; whence detachments are fent to do duty at Darien, Porto Bello, and Chagre. Near the city, on the north-weft, is a mountain called Ancon, whofe
perpendicular

perpendicular height, by a geometrical menſuration, we found to be 101 toiſes.

The houſes in general, when we viſited this city, were of wood, having but one ſtory, and a tiled roof, but large; and from their diſpoſition, and the ſymmetry of their windows, made a handſome appearance. A few were of ſtone. Without the walls is an open ſuburb, larger than the city itſelf, and the houſes of the ſame materials and conſtruction as thoſe within, except ſuch as border on the country, moſt of which are thatched with ſtraw; and among them ſome bujios, or huts. The ſtreets, both of the city and ſuburb, are ſtraight, broad, and for the moſt part paved.

Though the greater part of the houſes were formerly of wood, fires were rarely known at Panama, the nature of the timber being ſuch, that if any fire is laid on the floor, or placed againſt a wall, it is productive of no other conſequence than that of making a hole, without kindling into a flame; and the fire itſelf extinguiſhed by the aſhes. But, notwithſtanding this excellent quality in the wood, in the year 1737, the city was almoſt entirely conſumed, the goodneſs of the timber being unable to ſecure it from the ravages of the flames; indeed, by the concurrence of another cauſe the timber was then rendered more combuſtible. The fire began in a cellar, where, among other goods, there were great quantities of pitch, tar, naphtha, and brandy; theſe inflammable ſubſtances rendered this ſingular kind of wood a more eaſy prey to the devouring flames. In this conflagration the ſuburb owed its ſafety to its diſtance from the city, which is 1200 toiſes. Since this misfortune, it has been again rebuilt; and the greater part of the houſes are now of ſtone, all ſorts of materials for buildings of this kind being here in the greateſt plenty.

In this city is a tribunal or royal audience, in which the governor of Panama prefides; and to this employment is annexed the captainfhip general of Terra Firma, which is generally conferred on an officer of diftinction, though his common title is that of prefident of Panama.   It has alfo a cathedral, and a chapter confifting of the bifhop, and a number of prebendaries; an aujutamiento, or corporation, compofed of alcaldes and regidores; three officers of revenue, under an accomptant, treafurer, and agent; and a court of inquifition appointed by the tribunal of inquifition at Carthagena.   The cathedral, and alfo the convents, are of ftone; indeed, before the conflagration, feveral of the latter were of wood; but that terrible misfortune fhewed them the neceffity of ufing more folid materials.   The convents are thofe of the Dominicans, Francifcans, Auguftines, and Fathers of Mercy; a college of Jefuits, a nunnery of the order of St. Clara, and an hofpital of St. Juan de Dios.   The flender revenues will not admit of their being very numerous; and accordingly the ornaments of the churches are neither remarkably rich, nor contemptible.

The decorations of private houfes are elegant, but not coftly; and though there are here no perfons of fuch monftrous fortunes as in fome cities of America, it is not deftitute of wealthy inhabitants, and all have a fufficiency; fo that, if it cannot be claffed among opulent cities, it is certainly above poverty.

The harbour of this city is formed in its road, by the fhelter of feveral iflands, particularly Ifla de Naos, de Perico, and Flamencos: the anchoring-place is before the fecond, and thence called Perico.   The fhips here lie very fafe; and their diftance from the city is about two and a half, or three leagues.

The tides are regular; and, according to an obfervation we made on the day of the conjunction, it was high-water at three in the evening.   The water
rifes

rifes and falls confiderably; fo that the fhore, lying
on a gentle flope, is, at low water, left dry to a great
diftance.   And here we may obferve the great difference of the tides in the north and fouth feas, being directly oppofite: what in the ports on the north fea is accounted irregular, is regular in the fouth; and when
in the former it ceafes to increafe or decreafe, in the
latter it both rifes and falls, extending over the flats,
and widening the channels, as the proper effect of
the flux and reflux.   This particular is fo general, as
to be obferved in all the ports of the South Sea; for
even at Manta, which is almoft under the equinoctial,
the fea regularly ebbs and flows nearly fix hours; and
the effects of thefe two motions are fufficiently vifible
along the fhores.   The fame happens in the river of
Guayaquil, where the quantity of its waters does not
interrupt the regular fucceffion of the tides.   The like
phenomena are feen at Paita, Guanchaco, Callao,
and the other harbours; with this difference, that the
water rifes and falls more in fome places than in
others; fo that we cannot here verify the well-grounded
opinion entertained by failors, namely, that between
the tropics the tides are irregular, both in the difpro-
portion of the time of flood to that of the ebb, and
alfo in the quantity of water rifing or falling by each of
thefe motions; the contrary happening here.   This
phenomenon is not eafily accounted for; all that can
be faid is, that the ifthmus, or narrow neck of land,
feparating the two feas, confines their waters, whereby
each is fubject to different laws.

The variation of the magnetic needle, in this road,
is $7°\ 39'$ eafterly.   Both the road and whole coaft
abound in a great variety of excellent fifh, among
which are two kinds of oyfters, one fmaller than the
other; but the fmalleft are much the beft.

At the bottom of the fea are a great number of
pearls; and the oyfters, in which they are found, are
remarkably delicious.   This fifhery is of great ad-
I 4   vantage

vantage to the inhabitants of all the iflands in this bay.

The harbour of Perico is the rendezvous of the Peru fleet, during the time of the fair; and is never without barks loaded with provifions from the ports of Peru, and a great number of coafting veffels going from thence to Choco, and parts on the weftern coaft of that kingdom.

The winds are the fame as along the whole coaft; the tides or currents are ftronger near the iflands than at a diftance from them; but no general rule can be given as to their courfe, that depending on the place where the fhip is, with regard to the channels which they form. They alfo vary in the fame place according to the winds. Let it therefore fuffice that we have fhewn there are tides on this coaft that, on any occafion, this notice may be applied to ufe.

CHAP. III.

Of the Climate and Inhabitants of Panama.

MANY countries of America have fuch a refemblance, in refpect to the inhabitants and cuftoms, that they appear the fame. This is equally obfervable in the climate, when no difference is occafioned by the accidental difpofition of the ground, or quality of the foil. But, this fubject having been already fufficiently handled, a rational curiofity will require us only to mention thofe particulars in which they differ. Thus, after faying that the inhabitants of this city refemble thofe of Carthagena with regard to their conftitution, I muft add, that there is fome difference in their difpofition, thofe of Panama being more parfimonious, more defigning and infidious, and ftopping at nothing when profit is in view, the polefar both of Europeans and Creoles; and it is difficult

ficult to determine which fet the firft example. The fame felfifhnefs and parfimony reigns equally among the women, fome Spanifh ladies excepted, who have accompanied their hufbands, appointed auditors, or to fome other employments; thefe ftill retaining the qualities they imbibed from education.

THE women of Panama begin to imitate the drefs of thofe of Peru, which, when they go abroad, confifts only of a gown and petticoat, nearly refembling thofe worn in Spain; but at home, on vifits, and fome particular ceremonies, their fhift is their only clothing, from the waift upwards. The fleeves are very long and broad, and quite open in the lower part or near the hand; and thefe, like the bofom, are decorated with very fine lace, the chief pride of the ladies of Panama. They wear girdles, and five or fix chaplets or rows of beads about their necks, fome fet in gold, fome of coral mixed with fmall pieces of gold, and others lefs coftly; but all of different fizes, in order to make the greater fhow; and befides thefe, one, two, or more gold chains, having fome relics dependant from them. Round their arms they wear bracelets of gold and tombac; alfo ftrings of pearls, corals, and bugles. Their petticoat reaches only from their waift to the calf of their legs; and from thence to a little above their ancle, hangs, from their under petticoat, a broad lace. The Meftiza, or Negro women, or the coloured women as they are called here, are diftinguifhed in their drefs from thofe of Spain, only by the gown and petticoat; the particular privilege of the latter, and which alfo gives them the title of Signora; though many of them have little to boaft of, either with regard to rank or wealth *.

IF I omitted in Carthagena the following obfervation, it was in order to referve it for this place; name-

* Thefe cuftoms are general throughout all the northern parts of South America. A.

ly,

ly, that in Carthagena, Porto Bello, and Panama, the inhabitants have a very fingular pronunciation; and as fome nations have a haughty accent, fome a politenefs in their manner of expreffion, and others fpeak in a very quick manner; fo here their pronunciation has a faintnefs and languor, which is very difagreeable, till we are reconciled to it by cuftom. And what is ftill more particular, each of thefe three cities has a different accent in this languor; befides particular fyllables peculiar to each, and no lefs different than they are from the manner of fpeaking ufed in Spain. This may, in fome meafure, flow from an ill habit of body, weakened by the exceffive heat of the climate; but I believe it is principally owing to cuftom.

The only difference between the climate of Carthagena and this is, that fummer begins later, and ends fooner, as, the longer the brifas delay their return, the fooner they are over. From many thermometrical obfervations made on feveral days without any fenfible difference betwixt them at the fame hours, on the 5th and 6th of January 1736, at fix in the morning, they found the liquor at 1020½, at noon 1023½, and at three in the afternoon at 1025. But, at the fame time, it muft be obferved, that the brifas now began to blow, and, confequently, it was not the time of the greateft heats; thefe prevailing in the months of Auguft, September, and October.

Though this climate would naturally be fuppofed to produce the fame plants with others in the fame latitude, it is very different. Nor does this feem to proceed from any defect in the foil, but from the fondnefs of the inhabitants for trade, and their total neglect of agriculture, as too laborious. But, be the real caufe of it what it will, this is certain, that even in the parts contiguous to the city, the land is left entirely to nature; nor does the leaft veftige remain of its being formerly cultivated. From hence proceeds a fcarcity of all things, and, confequently, they are fold at a

high

high price. Here are no pulfe or pot-herbs of any kind; and that this is not owing to the fterility of the earth, we had an evident proof in a fmall garden, belonging to a Gallician, where all things of this kind were produced in great plenty. By this means Panama is under a neceffity of being fupplied with every thing, either from the coaft of Peru, or places in its own jurifdiction.

CHAP. IV.

Of the ufual Food of the Inhabitants of Panama.

THE very want of provifions caufes the tables at Panama to be better furnifhed; and it may be truly faid, that this city fubfifts wholly by commerce, whatever is confumed in it coming from other places. The fhips of Peru are continually employed in exporting goods from that country, and the coafting barks in bringing the products of the feveral places in its jurifdiction and that of Váraguas. So that Panama is plentifully furnifhed with the beft of wheat, maize, poultry, and cattle. Whether it be owing to the fuperior goodnefs of their food, the temperament of the climate, or to fome other caufe to me unknown, it is certain that the inhabitants of this city are not fo meagre and pale as thofe who live at Carthagena and Porto Bello.

THEIR common food is a creature called guaṇa. It is amphibious, living equally on the land and in the water. It refembles a lizard in fhape, but is fomething larger, being generally above a yard in length: fome are confiderably bigger, others lefs. It is of a yellowifh green colour, but of a brighter yellow on the belly than on the back, where the green predominates. It has four legs like a lizard; but its claws are much longer in proportion; they are joined by a
web,

web, which covers them, and is of the same form as those of geese, except that the talons at the end of the toes are much longer, and project entirely out of the web or membrane. Its skin is covered with a thin scale adhering to it, which renders it rough and hard; and, from the crown of its head to the beginning of its tail, which is generally about half a yard, runs a line of vertical scales, each scale being from one to two lines in breadth, and three or four in length, separated so as to represent a kind of saw. But from the end of the neck to the root of the tail, the scales gradually lessen, so as, at the latter part, to be scarce visible. Its belly is, in largeness, very disproportionable to its body; and its teeth separated, and very sharp pointed. On the water it rather walks than swims, being supported by the webs of its feet; and on that element its swiftness is such, as to be out of sight in an instant; whereas on the land, though far from moving heavily, its celerity is greatly less. When pregnant, its belly swells to an enormous size; and indeed they often lay sixty eggs at a time, each of which is as large as those of a pigeon. These are reckoned a great dainty, not only at Panama, but in other parts where this creature is found. These eggs are all inclosed in a long, fine membrane, and form a kind of string. The flesh of this animal is exceedingly white, and universally admired by all ranks. I tasted both the flesh and the eggs, but the latter are viscid in the mouth, and of a very disagreeable taste : when dressed, their colour is the same with that of the yolk of a hen's egg. The taste of the flesh is something better; but, though sweet, has a nauseous smell. The inhabitants, however, compared it to that of chicken; though I could not perceive the least similarity *.

These

---

* The flesh of the guana is whiter than chicken, and more pleasing to most palates, except as to the dryness of it. The common sauce to it is lime juice, seasoned with Chian pepper, which

sauce

Thefe people, who, by being accuftomed to fee them, forget the natural horror attending the fight of an alligator, delight in this food, to which the Europeans at firft can hardly reconcile themfelves.

Hᴇʀᴇ are two fingularities attributed to nature, and firmly believed by the inhabitants; one in the plant called yerva del gallo; the other the double-headed fnake called la cabeca.

Iᴛ is conftantly afferted in this city, that its neighbourhood produces a fnake having a head at each extremity; and that from the bite of each a poifon is conveyed equal in activity to that of the coral, or rattle fnake; we could not have the fatisfaction of feeing one of this ftrange fpecies, though we ufed all the means in our power to gratify our curiofity: according to report, its ufual length is about half a yard, in figure perfectly refembling an earth-worm. Its diameter is about fix or eight lines, and its head different from that of other fnakes; being of the fame dimenfions with its body. It is however very probable that the creature has only one head, and, from its refembling a tail, has been imagined to have two *. The motion of it is very flow, and its colour variegated with fpots of a paler tinct.

Tʜᴇ herb called del gallo, or cocks-herb, is so highly valued here, that they affirm, if an incifion be made round the neck of that fowl, provided the vertebra be not injured, on the application of this herb, the wound immediately heals. Whatever confiruction we put upon this pretended cure, it can only be confidered as a mere vulgar notion; and I mention it here with no other intention, than to fatisfy the world that we were not ignorant of it.

fauce the natives eat with their fifh, flefh, and fowl. If the guana were to be had in England, I doubt not but it would be ranked among the greateft dainties. A.

* This conjecture is very right. H.

Dᴜʀɪɴɢ

DURING our ſtay at Panama we were very urgent with thoſe who related this ſtory to procure us ſome of the herb, that we might make the experiment; but in this we were as unfortunate as in the article of the two-headed ſnake, none being to be had. I have, however, ſince been told, by perſons ſettled in Panama, that it was very common; a ſufficient proof, in my opinion, that the ſtory has no foundation; for, if it was ſo eaſy to be had, and of ſuch ſurpriſing virtue, what reaſon could they have for refuſing to convince us by ocular demonſtration? It may have a ſtyptic virtue, when none of the principal blood-veſſels are injured; but that it can join them after being cut, together with the nerves and tendons when totally ſevered, no perſon of any knowledge or judgment will ever be brought to believe. And if its effects are ſo remarkably happy on poultry, it is ſurely natural to think it ſhould have the ſame on any other animal; and, conſequently, on the human ſpecies. If this were the caſe, it would be of infinite value; and no ſoldier, eſpecially, ſhould be without it, as a few ounces of this grand reſtorative would immediately cure the moſt terrible wounds.

CHAP. V.

Of the Trade and Commerce of Panama.

FROM what has been ſaid relating to the commerce of Porto Bello in the time of the galleons, an idea may be formed of that of Panama on the ſame occaſion; this city being the firſt where the treaſure from Peru is landed, and likewiſe the ſtaple for the goods brought up the river Chagre. This commerce is of the greateſt advantage to the inhabitants, both with regard to letting their houſes, the freight of veſſels, the hire of mules and Negroes, who forming

themſelves

themselves into separate bodies, draw along from Cruces large bales, or any brittle and delicate wares; the roads here, though the distance is but short by crossing the chain of mountains called the Cordilleras, are in some parts so narrow, that a beast of burden can hardly pass along; and, consequently, an imminent danger would attend the employing of mules for this service.

This city, even during the absence of the armada, is never without a great number of strangers; it being the thoroughfare for all going to the ports of Peru, in the South Sea, as also for any coming from thence to Spain : to which must be added, the continual trade carried on by the Peruvian ships, which bring variety of goods, as meal of different sorts, wines, brandy from grapes, or brandy castilla, as it is called by all the Americans in these parts, sugar, tallow, leather, olives, oil, and the like. The ships from Guayaquil bring cacao, and quinquina or jesuits bark; which always meet with a quick exportation here, especially in time of peace. All goods, particularly those of Peru, are subject to great alterations in their prices, so that on many occasions the owners lose considerably, and sometimes their whole purchase : on the other hand, there are favourable opportunities, when they triple it, according to the plenty or scarcity of the commodity. The different forts of meal are in particular subject to this accident ; they soon becoming so extremely vitiated by the great heat, that there is an absolute necessity for throwing them overboard. The wines and brandies also, from the heat of the jars, contract a pitchy taste, and are soon unfit for use. The tallow melts, becomes full of maggots, and turns into a kind of earth ; the same may be observed of other goods. Hence, if the gain is sometimes great, the risk of the loss is proportional.

The coasting barks, which make frequent trips from the adjacent ports; supply the city with hogs, poultry,
hung-

hung-beef, hogs-lard, plantanes, roots, and other eatables; with all which, this city, by the induſtry of others, is abundantly ſupplied.

THE Peru and Guayaquil veſſels, unleſs at the time when the armada is here, return empty, except when they have an opportunity of taking Negroes on board; as, while the aſſiento ſubſiſts, there is at Panama a factory, or office, which correſponds with that at Porto Bello; and hither the Negroes are brought, as being, in ſome meaſure, the ſtaple for them, with regard to the kingdoms of Terra Firma and Peru.

THE preſident of Panama is inveſted with a power of licenſing every year one or two ſhips, which go to Sonſonate, el Realejo, and other ports in the province of Guatemala and New Spain, to fetch from thence tar, naphtha, and cordage, for the veſſels belonging to the Panama trade; they carry thither ſuch parts of the Peruvian goods as do not find a market at Panama; but few of the ſhips which have obtained this permiſſion return immediately; for the moſt profitable part of their trade conſiſting of indigo, they make the beſt of their way to Guayaquil, or other ports farther to the ſouthward. The dearneſs of proviſions in this city and its diſtrict, occaſioned by the large quantity required, and the great diſtance from whence they are brought, is amply compenſated by the multitude and value of the pearls found in the oyſters of its gulf; and particularly thoſe near the iſlands del Rey, Tabaga, and others, to the number of 43, forming a ſmall archipelago. The firſt to whom the Indians made this valuable diſcovery was Baſco Nunez de Balboa, who, in his paſſage this way, to make farther diſcoveries on the South Sea, was preſented with ſome by Tumaco, an Indian prince. At preſent they are found in ſuch plenty, that there are few perſons of ſubſtance near Panama, who do not employ all, or, at leaſt, part of their ſlaves in this fiſhery, the manner of which not being com-

2 monly

monly known, it will not be improper to defcribe it here.

The owners of the Negroes employ the moft proper perfons for this fifhery; which being performed at the bottom of the fea, they muft be expert fwimmers, and capable of holding their breath a long time. Thefe they fend to the iflands, where they have huts built for their lodgings, and boats which hold eight, ten, or twenty Negroes, under the command of an officer. In thefe boats they go to fuch parts as are known to produce pearls, and where the depth of water is not above ten, twelve, or fifteen fathom. Here they anchor; and the Negroes having a rope faftened round their bodies, and the other end to the fide of the boat, they take with them a fmall weight, to accelerate their finking, and plunge into the water. On reaching the bottom, they take up an oyfter, which they put under the left arm; the fecond they hold in their left hand, and the third in their right: with thefe three oyfters, and fometimes another in their mouth, they rife to breathe, and put them in a bag. When they have refted themfelves awhile, and recovered their breath, they dive a fecond time; and thus continue, till they have either completed their tafk, or their ftrength fails them. Every one of thefe Negro divers is obliged daily to deliver to his mafter a fixed number of pearls; fo that when they have got the requifite number of oyfters in their bag, they begin to open them, and deliver the pearls to the officer, till they have made up the number due to their mafter; and if the pearl be but formed, it is fufficient, without any regard to its being fmall or faulty. The remainder, however large or beautiful, are the Negro's own property, nor has the mafter the leaft claim to them; the flaves being allowed to fell them to whom they pleafe, though the mafter generally purchafes them at a very fmall price.

THESE Negroes cannot every day make up their number, as in many of the oyfters the pearl is not at all, or but imperfectly formed; or the oyfter is dead, whereby the pearl is fo damaged, as to be of no value; and as no allowance is made for fuch pearls, they muft make up their numbers with others.

BESIDES the toil of this fifhery, from the oyfters ftrongly adhering to the rocks, they are alfo in no fmall danger from fome kinds of fifh, which either feize the Negroes, or, by ftriking on them, crufh them by their weight againft the bottom. So that thefe creatures feem to know that men are robbing them of the moft valuable product of their element, and therefore make a vigorous defence againft their enemy. The fifhery on the whole coaft is obnoxious to the fame danger from thefe fifh; but they are much more frequent where fuch riches abound. The fharks and tintoreras, which are of an enormous fize, feed on the bodies of thefe unfortunate fifhermen; and the mantas, or quilts, either prefs them to death by wrapping their fins about them, or crufh them againft the rocks by their prodigious weight. The name manta has not been improperly given to this fifh, either with regard to its figure or property; for being broad and long like a quilt, it wraps its fins round a man, or any other animal that happens to come within its reach, and immediately fqueezes it to death. This fifh refembles a thornback in fhape, but is prodigioufly larger.

EVERY Negro, to defend himfelf againft thefe animals, carries with him a fharp knife, with which, if the fifh offers to affault him, he endeavours to ftrike it in a part where it has no power to hurt him; on which the fifh immediately flies. The officers keep a watchful eye on thefe voracious creatures, and, on difcovering them, fhake the ropes faftened to the Negroes' bodies, that they may be upon their guard; many, on the divers being in danger, have thrown themfelves
into

into the water, with the like weapon, and haften down to their defence: but too often all their dexterity and precaution is not fufficient to protect the diver from being devoured by thefe fifh, or lofing one of his legs or arms by their bite. Several ineffectual fchemes have been practifed, to prevent fuch melancholy accidents.

The pearls of thefe fifheries are generally of a good water, and fome very remarkable, both in their fhape and fize; but as there is a difference in both thefe properties, fo there is alfo a difference in their water and colour; fome being highly valuable, and others as remarkably defective. Some of thefe pearls, though indeed but few, are fent to Europe, the greater part being carried to Lima; where the demand for them is very great, being not only univerfally worn there by all perfons of rank, but alfo fent from thence into the inland parts of Peru.

Besides thefe pearls, the kingdom of Terra Firma was formerly equally remarkable for the fine gold produced by the mines in its territories; and which confequently proved a very confiderable addition to its riches. Part of thefe mines were in the province of Veraguas, others in that of Panama; but moft, alfo the richeft, and whofe metal was of the fineft quality, were in the province of Darien; and, on that account, the conftant object of the miners. But the Indians revolting, and making themfelves mafters of the whole province, there was a neceffity for abandoning thefe mines, by which means the greater part of them were loft; a few only remaining on the frontiers, which ftill yield a fmall quantity of gold. Their produce might indeed be increafed, did not the fear of the fickle nature of the Indians, and the fmall confidence that can be placed on their apparent friendfhip, deter the mafters of the mines from taking proper meafures for improving them.

Thought

Though the mines of Veraguas and Panama are not expofed to thefe dangers, yet they are not worked with more vigour than the others; and this for two reafons: the firft is, that, befides their being lefs rich in metal than the others, the gold they yield is not of fo good a quality as that of Darien: the fecond, and indeed the moft weighty, is, that as thefe feas, by their rich produce of pearls, offer a more certain, and at the fame time a more eafy profit, they apply themfelves to this fifhery preferably to the mines. Some, indeed, though but few, are worked, befides thofe above mentioned, on the frontiers of Darien.

Besides the advantage arifing to Panama from its commerce, as the revenue here is not equal to the difburfements, a very confiderable fum of money is annually remitted hither from Lima, for the payment of the troops, the officers of the audience, and others in employment under his majefty.

CHAP. VI.

Extent of the Audience of Panama, in the Kingdom of Terra Firma.

THE city of Panama is not only the capital of its particular province, but alfo of the whole kingdom of Terra Firma, which confifts of the three provinces of Panama, Darien, and Veraguas. The firft is the feat of every branch of the government, as being fituated between the other two; Darien lying on the eaft fide, and Veraguas on the welt.

The kingdom of Terra Firma begins northwards at the river of Darien, and ftretching along by Nombre de Dios, Bocas del Toro, Bahia del Almirante, is terminated weftward by the river de los Dorados in the north fea; and towards the fouth fea, beginning on the weftern part, it extends from Punta Gorda, in

Cofta

Cofta Rica, by Punta de Mariatos, Morro de Puercos, to the gulf of Darien; whence it continues fouthward along the coaft, by Puerto de Pinas, and Morro Quemado, to the bay of St. Bonaventura. Its length from eaft to weft is 180 leagues, but, if meafured along the coaft, it exceeds 230; and its breadth, from north to fouth, is the fame as that of the ifthmus, which includes the whole province of Panama, and part of that of Darien. The narroweft part of this ifthmus is from the rivers Darien and Chagre, on the north fea, to thofe of Pito and Camito on the fouth fea: and here the diftance, from fea to fea, is about 14 leagues. Afterwards it increafes in breadth towards Choco and Sitara; and the fame weftward in the province of Veraguas, forming an interval of forty leagues from fea to fea.

Along this ifthmus run thofe famous chains of lofty mountains, called the Andes, which, beginning at fuch a prodigious diftance as the Terra Magellanica, traverfe the kingdom of Chili, the province of Buenos Ayres, and thence through the provinces of Peru and Quito; and from the latter, contract themfelves, as it were, for a paffage through this narrow ifthmus. Afterwards, again widening, they continue their courfe through the provinces and kingdoms of Nicaragua, Guatemala, Cofta Rica, St. Miguel, Mexico, Guajaca, la Puebla, and others; with feveral arms or ramifications, for ftrengthening, as it were, the fouthern with the northern parts of America.

In order to give the reader a comprehenfive idea of this kingdom, I fhall fpeak particularly of each of its three provinces, beginning with that of Panama as the principal. Moft of its towns and villages are fituated in fmall plains along the fhore, the reft of the country being covered with enormous and craggy mountains, uninhabited on account of their fterility.

In this province are three cities, one town, a few forts, villages, and country feats; the names of

K 3   which,

which, with the tribes of the inhabitants, are here fub-joined.

The cities are Panama, Porto Bello, and Santiago de Nata de los Cavelleros. The fituation of the latter was firft difcovered, in the year 1515, by captain Alonzo Perez de la Rua, at which time Nata was prince of this diftrict. Gafpar de Efpinofa was firft commiffioned to people it, under the title of a town. It was afterwards taken and burnt by the Indians, but he rebuilt it, and called it a city. It is large, but the chief houfes are only of earth, or unburnt bricks, and the others of mud walls. Its inhabitants are a mixture of Spaniards and Indians.

The town called los Santos is a modern fettlement of Spaniards, who before lived at the city of Nata, but, with a view of augmenting their fortune by improving the ground, left the city; and the inhabitants of the town are at prefent more in number than thofe of Nata. Its environs were firft difcovered by Rodrigo Valenzuela, and at that time contained an Indian town, governed by a prince called Guazan: the origin of the town fufficiently fhews it is peopled by Spaniards and Indians.

The number of villages in this province is very confiderable, and of different kinds.

1. Nuestra Senora de Pacora. to which we give the preference, is inhabited by Mulattos and their defcendants.

2. San Chriftoval de Chepo owes its name to the caciques, or princes, Chepo and Chepauri, and was difcovered in 1515, by Tello de Guzman. Befides Indians, here is a company of foot, belonging to the garrifon of Panama, moft of whom are fettled here with their families.

Several Rancherias, or affemblages of Indian huts, are under the jurifdiction of a village. Thefe Rancherias are fituated to the fouthward, in the fmall chafms or breaches of the mountains.

In

Iɴ the favannahs of the river Mamoni are feveral fuch affemblages of huts, and within the fame jurifdiction; namely,

On the river de la Campana.
In the breach of Curcuti.
On the banks and at the mouth of the river Canas.
On the river del Platanar.
On the river de Pinganti.
On the river de Bayano.
In the breach de Terralbe.
In that of Platanar.
In that of Calobre.
In that of Pugibay.
In that of Marcelo.
On the river de Mange.

Under the jurifdiction of the fame village are alfo the following Rancherias, fituated to the northwards.

On the river del Playon.
On the fmaller river de la Conception.
On the river de Guanacati.
On the river del Caco, or Mandinga.
On the river de Sarati.

3. The village of San Juan, fituated on the road between Panama and Porto Bello, is inhabited by Mulattos and their defcendants.

4. The village of Nueftra Sinor de la Confolation, a Negro fettlement.

5. The village de la Santiffima Trinidad de Chame, difcovered by captain Gonzalo de Badajoz, and called Chame from its prince at that time, is inhabited by Spaniards and Indians.

6. The village of St. Ifidro de Quinones, difcovered by the fame officer, and then governed by its prince Totronagua: its prefent inhabitants Spaniards and Indians.

7. The village of St. Francifco de Paula, in the Cordillera; alfo inhabited by Spaniards and Indians.

K 4   8. The

8. The village of St. Juan de Pononome, fo called from the name of its cacique; its inhabitants are Indians, who ftill retain the bow and arrow, at which they are very dexterous, and of an intrepid bravery.

9. The village of Santa Maria is fituated in a tract of land difcovered by Gonzalo de Badajoz. The name of its laft prince was Efcolia; it is at prefent wholly inhabited by Spaniards.

10. The village of Santo Domingo de Parita, the laft word being the name of its prince. It was formerly inhabited wholly by Indians, but at prefent there are many Spaniards among them.

11. Taboga, Taboguilla, and other iflands, near which the pearl fifheries are carried on, were difcovered by the order of Pedro Arias Davila, the firft governor and captain-general of the kingdom of Terra Firma. In thefe iflands are houfes belonging to Spaniards, and huts for the Negro divers.

12. The iflands del Rey were difcovered by Gafper de Morales and captain Francifco Pizarro. In thefe iflands fome Spaniards have houfes, befides great numbers of Negro divers.

Second Province of Terra Firma.

THE fecond province of this kingdom is that of Veraguas, of which the city of Santiago is the capital. The firft who difcovered this coaft was admiral Chriftopher Columbus, in 1503. To the river now called Veragua, he gave the name of Verdes-aguas, on account of the green colour of its water; or, according to others, becaufe the Indians called it by that name in their language. But, however that be, it is from this river that the province derives its name. In 1508, the captains Gafpar de Efpinofa, and Diego de Alvirez, renewed the difcovery by land; but being repulfed by prince Urraca, were obliged to content themfelves with a fettlement in the neighbourhood:
and

and even here the Spaniards were not able to maintain their ground againſt the frequent incurſions of the Indians; ſo that finding the abſolute neceſſity of a ſtronger ſettlement, they built the city of Santiago de Veraguas on the ſpot where it now ſtands.

BESIDES this city the province contains two others, and ſeveral villages.

THE city of Santiago al Angel was founded in 1521 by Benedict Hurtado, governor of Panama: it has been twice deſtroyed and rebuilt: the inhabitants partly Spaniards, partly Mulattos.

THE city of Nueſtra Senora de los Remedios de Pueblo-Nuevo; the inhabitants the ſame as thoſe of the former.

1. The villages in this province are San Franciſco de la Montana, inhabited by Indians uſing bows and arrows.

2. San Miguel de la Haya, inhabited by different ſorts of people.

3. San Marcelo de Leonmeſa de Tabarana, inhabited by Indians.

4. San Raphael de Guaymi, by Indians.

5. San Philipe del Guaymi, by Indians.

6. San Martin de los Caſtos, by Indians.

7. San Auguſtin de Ulate, by Indians.

8. San Joſeph de Bugava, by Indians.

9. and 10. La Piedad, and San Miguel, by Indians.

11. San Pedro, and San Pablo de los Platanares, by Indians.

12. San Pedro Noloſco, by Indians.

13. San Carlos, by Indians.

Third Province of Terra Firma.

THE third province of Terra Firma is that of Darien, where the greater part of the inhabitants are wandering Indians, living without any religion, and in the moſt ſhocking barbariſm, which was indeed the

motive

motive of their revolt. In 1716 there was here a considerable number of villages, Rancherias, and Doctrinas *, whose inhabitants had sworn allegiance to the
king of Spain, and therefore under the governors of
Panama; though, at present, very few are remaining.
Those remaining in the above-mentioned year, were,

1. The village and staple for the mines of Santa
Cruz de Cana, a very considerable settlement of Spaniards and Indians.

2. The village de la Conception de Sabalo, inhabited
like the preceding, but less populous.

3. The village of San Miguel de Tayequa; inhabitants the same.

4. The village of San Domingo de Balsas, inhabitants like the others, being Spaniards and Indians.

5. Spanish village, in the territory of Santa Marica.

6. The Doctrina San Geronymo de Yabira, a word
in the Indian language signifying Doncel, i.e. a virgin;
and for this reason the river near it is called Rio Doncel, or Virgin river.

7. San Enrique de Capeti, or the sleepy.

8. Santa Cruz de Pucro. In the Indian language
Pucro signifies a sort of light wood, which, at Guayaquil, is called Balsa.

9. The Doctrina de San Juan de Tacaracuna, and
Matarnati; the names of two of the mountains of the
Andes, contiguous to the community.

10. The Indian village of San Joseph de Zete-Gaati,
is not a Doctrina. Zete-Gaati is the name of a kind
of willow growing in the neighbourhood.

Rancherias and Hamlets in the southern Parts.

The hamlet of Nuestra Senora del Rosario de Rio
Congo.

* A name, given by the Jesuits, to Indian communities, which
they have gathered together and civilized.

Other

Other hamlets on the rivers Zabalos, Balfas, and Uron.

On the river Tapanacul.

On the river Pucro.

On the banks and at the mouth of the river Paya.

At los Paparos, or the peafants.

On the river Tuquefa.

On the river Tupifa.

On the river Yabifa.

And at Chepigana.

Rancherias and Hamlets in the northern Parts.

On the river Queno.

On the Seraque.

On the Sutagunti.

On the Moreti.

On the Agrafenequa.

On the Ocabajanti.

On the Uraba.

ALL thefe Doctrinas and communities were formerly of Indians, and not inconfiderable, fome of the latter confifting of 400 perfons; but their general number was between 150 and 200; from whence we may form an idea of the populoufnefs of thefe Doctrinas. But, to fave the trouble of computing the feveral inhabited places in this kingdom, as I thought proper to infert their names, I fhall conclude with a concife lift of all thefe places, which will affift the reader in forming fome idea of this country.

Recapitulation of all the inhabited Places in the Kingdom of Terra Firma.

Four fortreffes.

Six cities.

One town of Spaniards and Indians.

Thirty-

Thirty-five villages { Eleven of Spaniards and Indians.
Two of Mulattos and Negroes.
Twenty-two of Indians, moſt of
  them Doctrinas.

Thirty-two Rancherias or hamlets, each containing ſeveral cottages ſcattered among the breaches, along the ſides of rivers and ſavannahs.

Forty-three iſlands, where the pearl-fiſhery is carried on, ſome of them in the bay of Panama, ſome near the coaſt of that city, and others ſouth of Veraguas.

BOOK

BOOK IV.

Voyage from Perico Harbour to Guayaquil.

CHAP. I.

Voyage from Perico to the City of Guayaquil.

OUR tents and other neceffaries being ready, we all embarked on board the St. Chriftopher, captain Don Juan Manuel Morel; and the next day, being the 22d of February 1736, we fet fail; but having little wind, and that variable, it was the 26th at funfet before we loft fight of the land, the laft we faw being Punta de Mala.

By remarks repeatedly made till we loft fight of this laft point, and which agreed with obfervations, but differed from thofe by account, we found the fetting of the current to be S. W. 5° wefterly; which obfervation correfponded with the accounts given us by able pilots, who affured us it continued to 3 or 4 degrees of latitude; and, according to their farther information, we corrected our daily account at one mile and one fixth per hour; and found their information to be well founded. But it is neceffary to obferve, that, till our fhip was off Punta de Mala, there was no vifible current; and that, whilft we continued failing in the gulf of Panama, the latitude by account agreed with the obferved.

FROM

FROM the time we set sail, till Punta de Mala bore from us N. W. 6° 30′ westerly, we continued to steer S. S. W. 1° 30′ and 8° 30′ westerly: the winds variable with calms.

AFTER passing Punta de Mala, we steered S. between 8° westerly and 2° 30′ easterly, till six in the evening of the first of March 1736, when we discovered the land contiguous to St. Matthews bay. Upon which we stood to the S. W. to avoid a ledge of rocks, which runs three leagues into the sea, and also the currents, which set towards it, and Gorgona bay.

THIS ledge of rocks was discovered in 1594, by a ship's striking on it.

FROM St. Matthew's bay, we, for some hours, steered S. W. 6° 15′ westerly; and the next day S. E. ¼ southerly; which, being the third day, at one in the afternoon, brought us in sight of Cape St. Francis, bearing N. ¼ easterly.

ACCORDING to the reckoning of Don George Juan, the difference of meridians between Panama and Cape St. Francis was 0° 36′; which nearly agrees with the map of this coast. It must, however, be supposed that the distance between each knot on the log-line was 47 feet by 5¼ royal inches, which is equal to 50½ English feet: this confirms what we have already observed, book I. chap. i. and proves the justness of our observations on the currents.

HAVING weathered this cape, we steered W. 3° southerly; S. W. 3° westerly; and on the 6th and 7th S. 7° easterly, and S. E. 6° easterly; till on the 7th, at 8 in the morning, we again made Cape St. Francis, bearing N. 5° easterly, and Cape Passado S.; after which we coasted along shore, observing the most remarkable parts, till the 9th, when, at half an hour after three in the evening, we came to an anchor in Manta bay, in eleven fathom water, the bottom mud
mixed

mixed with fand: Cape St. Lorenzo bearing W.S.W. and Monte Chrifto S. S. E. 6° eafterly.

Two reafons induced us to anchor here: the firft was, that as part of the intention of our original voyage was to meafure fome degrees of the equator, befides thofe of the meridian; and having been informed, at Panama, of the fituation of this coaft, we were defirous of viewing it, in order to know whether, by forming our firft bafe on its plains, the feries of triangles could be continued to the mountains contiguous to Quito. the fecond, the want of water and provifions; for the feafon being pretty far advanced, we had flattered our-felves, while at Panama, with falling in with the brifas, and by that means of foon reaching Guayaquil; and had therefore taken in provifions only for fuch a fhort voyage.

In order to fatisfy ourfelves with regard to our firft and principal view, we all went on fhore on the 10th in the evening to the village of Monte Chrifto, about two leagues and a half from the coaft. But we foon found any geometrical operations to be impracti-cable there, the country being every where extremely mountainous, and almoft covered with prodigious trees, an infurmountable obftruction to any fuch de-fign. This being farther confirmed to us by the Indian inhabitants, we determined to purfue our voyage to Guayaquil, and thence to Quito. Ac-cordingly on the 11th we returned to the coaft of Manta, where, whilft the fhip was taking in water and provifion, we employed ourfelves in making ob-fervations; by which we found the latitude of this place to be 0° 56′ 5½″ fouth. But Meffrs. Bouguer and de la Condamine, reflecting that our ftay at Guayaquil would be confiderable before the feafon would permit the mules to come from Guaranda to carry us to the mountains, and defirous of making the beft ufe of their time, determined to ftay here, in order to make further obfervations on the longitude

tude and latitude, that they might afcertain the place where the equator cuts this coaft, examine the length of the pendulum, and make other obfervations equally important. Accordingly proper inftruments were left with them.

On the 13th of the fame month of March, our vef-fel put to fea, keeping along the coaft, and paffed the next day within the ifland de la Plata. The 15th we began to lofe fight both of Cape St. Lorenzo, and alfo of the ifland; at one in the afternoon we fteered S. S. E. till the 17th, when we difcovered Cape Blanco, the fouth point of the bay of Guayaquil. From Cape Blanco we coafted along the bay, till, about noon on the 18th, coming to the mouth of the river Tumbez, we anchored about half a league from the land; the river's mouth bearing E. 5 degrees northerly, and the ifland of Santa Clara, commonly called Amortajado, or Muerto, from its refembling the figure of a human corpfe, N. 4 deg. eafterly, in fourteen fathom water, and a muddy bottom.

Some particular affairs of the captain of the fhip obliged us to remain here till the 20th, when, at fix in the morning, we weighed; and at half an hour after fix in the evening, the ftrength of the current on the ebb obliged us to come to an anchor. Thus we continued anchoring every ebb, and failing during the flood. And here we found that the current always fets out of the bay, though with much lefs velocity on the flood than on the ebb; for we obferved that the tide never altered its direction in 19 hours and a half. The caufe of this phenomenon is fuppofed to be, the pro-digious quantity of water difcharged into it by the rivers. On the 23d, having come to an anchor off Punta de Arenas in the ifland of Puna, we fent on fhore for a pilot to carry in our fhip; for, though the diftance was only feven leagues, the great number of fhallows in this fhort paffage rendered a precau-tion of this kind prudent, if not abfolutely necef-

fary.

fary.   And on the 24th, at feven in the morning, we fafely anchored in Puna harbour; Cape Centinela bearing S. S. W. 2° 30′ wefterly, and Cape Maria Mandinga W. N. W. 1° 15′ wefterly, diftant one quarter of a league.

FROM Punta de Mala to St. Matthew's bay, we had the wind firft at N. and N. W.; afterwards it fhifted to the N. E. and during the laft day veered to the E. N. E.; but when we came in fight of this bay, changed again to N. being preceded by rains, which continued till our arrival at Manta, the winds having fhifted to the S. E. fouth, and S. W. and weft, but with fome variations from all thofe points.

I HAVE already mentioned that at St. Matthew's bay, it was not only the opinion of the pilots relating to the currents which fet towards Gorgona, but alfo our own experience, that induced us to alter our courfe, which was neceffary, in order to continue our voyage.   All the reft of the coaft, from Cape St. Francis to Manta, they fet to the N. and this prevented us from getting to windward, and obliged us to tack, as the wind was contrary.

IN our paffage from Manta to Cape Blanco, the winds were not lefs favourable, continuing as before, except a few gales at N. W. and N. N. E. till we made the above cape.   The currents here alfo fet to the northward; and from Cape Blanco to Puna harbour, to feaward, that is, towards the weft; but, as we have before obferved, a greater velocity on the ebb than on the flood.

BEING very defirous of obferving an eclipfe of the moon, which was to happen on the 26th of March, and our time for preparing for it being but fhort, we concluded to ftay at a little village fituated in this harbour; but finding thefe houfes, which were entirely built of canes, too weak to fupport the pendulum, we determined to make the beft of our way to Guayaquil; and accordingly, at half an hour after

VOL. I.   L   eleven

eleven at night, we left the fhip at anchor, and went to the city in a boat; and, at five in the evening of the 25th, by the vigour of our rowers, we arrived at Guayaquil, notwithftanding the ftrength of the tide againft us. Here we immediately applied ourfelves to fettle the pendulum; but our diligence was entirely fruftrated, the air being fo filled with vapours, that nothing was to be feen.

IT may not be amifs here to infert the variations we obferved in different parts of the South Sea, in the fame order with thofe obferved from Cadiz to Carthagena.

A Table of Variations obferved in feveral Parts of the South Sea, the Longitude reckoned from the Meridian of Panama.

Latitudes. deg. min.		Longitude. deg. min.		Variation. deg. min.	
8	17 N.	359	55	8	45 E.
7	49	359	42	7	34
7	30	359	31	7	49
7	02	359	18	7	59
3	55	358	21	7	34
0	56	358	43	7	20
0	36	359	06	8	29
0	20	358	40	7	25
0	15	358	56	7	30
0	22 S.	359	50	8	17
0	51 Monte Chrifto bearing S. E. ¼ foutherly.			8	00

ISLAND de la Plata, bearing S. 15° 45′ wefterly, and Monte Chrifto E. S. E.   7°   46′

2   18 S.   8   00

Cape Blanco S. S. W.   3   30 W.

Punto de Mero eaft 7° northerly, } 8   ioo
diftant 3 leagues

On the coaft of Sumber, of which the } 8° 11′
latitude by obfervation was 3° 14′

We fhould, for feveral days, have been without knowing certainly the latitude, an objeft of the laft importance in any voyage, had not Mr. Godin had the precaution to take with him a Hadley's quadrant. This ingenious gentleman having been pitched upon for the voyage to America, undertook a journey to London, purely to purchafe feveral inftruments, and among others bought that already mentioned; and which proved of the greateft ufe to us, in finding the latitude during this paffage; a point the more difficult and neceffary, on account of feveral perplexing circumftances; the courfe being fometimes north, fometimes fouth, and the currents fetting in the fame direftion. Affifted by this inftrument, we were enabled to take the meridian altitude of the fun, whilft, from the denfity of the vapours which filled the atmofphere, the fhadow could not be defined on the ufual inftruments.

CHAP. II.

Account of the Voyage from Perico to Puna.

THE brifas, by their return, as we before obferved, occafion an alteration in the weather of Panama, by introducing the fummer, as they alfo do in the paffage from Perico to Puna; or, more properly, to Cape Blanco: for, after the brifas have begun to blow at Panama, they gradually increafe and fpread, in oppofition to the fouth winds, till, overcoming them, they are fettled: but their periods are not always equal, either on the land or in the ocean. Generally the brifas do not reach beyond the equator, or are fo faint, as often to be interrupted by calms, or other weak and unfettled winds. Sometimes, indeed,

L 2   deed,

deed, they have an extraordinary ftrength, being felt
even to the ifland of Plata. But their greateft force
is gradually increafed as we approach nearer to Panama. Thefe winds, which blow from between the
N. and N. E. clear the atmofphere, free the coaft
from fogs, and are not attended with tempefts of
rain; but frequently fo fqually, efpecially between
Cape Francifco and the Bay of Panama, that, without
particular care and the utmoft difpatch in putting
the fhip in a proper condition, they are often dangerous.

AT the period of the brifas, the fures or fouth
winds begin to blow; and, when fettled, are more
violent than the former. But they do not, as many
have imagined, blow always precifely from the fouth;
for they fhift from the S. E. even to the S. W. and
their diftance from the S. is obferved to be greateft
at particular times. When they incline to the S. E.
which is the land fide, they are accompanied with
violent, but happily fhort, tempefts of wind and rain.
The fhips which trade from the coafts of Peru and
Guayaquil to Panama, generally fail during the fures,
in order to take the benefit of the north wind at their
return; and, by that means, their voyages are eafily
and expeditionfly performed. Sometimes, indeed,
they fail with other winds, though they are generally
longer at fea, in order to reach Paita; but often this.
diligence, or rather avarice, is fo-far difappointed, that
they are obliged to put in at Tumaco, Acames,
Manta, or Punta de Santa Elena, for provifions and
water.

THESE are the principal winds in this paffage; and
whatever changes may fometimes happen, they are
not of any continuance, the fettled wind foon recovering its place.

THE currents, in thefe parts, are not fo regular as
the winds; for, during the brifas, the waters run
from Morro de Puercos S. W. and W. to the
height

height of Malpelo; and from thence E. and E. S. E. to Cape St. Francis, inclining fomething towards Gorgona. From Cape St. Francis their direction is S. and S. W. which continues for 30 or 40 leagues feawards, the ftrength of them being proportionate to that of the brifas.

Dᴜʀɪɴɢ the feafon of the fures, or fouth winds, the currents run N. and N. W. from Punta de Santa Elena, as far as Cape St. Francis, extending thirty or forty leagues feawards; from hence they run with a great velocity eaft, as far as the meridian of Malpelo; and from Morro de Puercos S. E. along the coaft, though at fome diftance from it, and tending partly to the bay of Gorgona. But from the meridian of Malpelo to Morro de Puercos, they run with great violence N. W. and W. Alfo in the paffage from Cape Blanco to Cape Santa Elena, a violent current runs weft from the river of Guayaquil, during its fwellings; but when the river is low, the current fets into Puna bay: the time of the former is during the brifas, and the latter in the feafon of the fures.

Aᴛ all times, in leaving Perico to fail to Guayaquil, or the coaft of Peru, care muft be taken to keep at a proper diftance from the ifland of Gorgona, many inftances having happened of fhips being loft either by this negligence, or, more frequently, by calms. It is alfo equally neceffary to be careful of the ifland of Malpelo; but the latter is of the two the leaft dangerous, as the greateft detriment is only a longer delay of the voyage.

Iғ a fhip happens to come in fight of the ifland of Gorgona, it will be found very difficult to get clear of it by fteering either S. S. W. or even N. fo that the fureft method is to return towards Panama along the coaft, the currents there changing their direction; at the fame time taking care not to keep at a great diftance from it, to avoid being again carried away by the current, which fets S. E.

Tʜᴇ

THE land all along the coaft from Panama to Santa Elena is of a middling height, except in fome parts, where we difcern mountains at a vaft diftance, and very high; being part of the Cordillera. Monte Chrifto is the land mark of Manta, being a high mountain, and having a village of the fame name at its foot.

In the bays along this coaft, and particularly at the mouths of rivers, it is dangerous to keep clofe to the fhore, there being many fhallows not known even to the pilots of the country. In the bay of Manta, there is one at the diftance of three or four leagues from the fhore, on which feveral fhips have ftruck; but the water is here fo fmooth, that all the damage they fuftained was, their being obliged to be immediately careened, in order to ftop the leaks occafioned by the accident.

In all this paffage a rough fea is feldom met with; for, if it be fometimes agitated by fqualls and fhort tempefts, it foon fubfides after the ftorm is over. Whilft the fouth winds prevail, fogs are very frequent, and fometimes fo thick as totally to preclude all fight of the coaft. This we ourfelves partly experienced in our paffage; whereas, during the brifas, it is quite the contrary; the air is ferene, and the coaft fo clear as to be approached with confidence and fafety.

CHAP. III.

Of our Stay at Guayaquil, and the Meafures taken for our Journey to the Mountains.

THE fhip St. Chriftopher, which we left at Puna, followed us fo foon, that on the 26th in the evening fhe came to an anchor before the city; the next day all our baggage and inftruments were landed, and we began our obfervations for determining

ing the fituation of Guayaquil, with regard to its latitude and longitude. The defire of fucceeding rendered us very attentive to obferve an immerfion of the fatellites of Jupiter, to make amends for our difappointment of the eclipfe of the Moon; but we were in this equally unfortunate; the denfity of the vapours which filled the atmofphere rendered our defign abortive; but, the days being more favourable than the nights for aftronomical obfervations, we took feveral meridian altitudes of the Sun, and never neglected any opportunities that offered, during the nights, of doing the fame with regard to fome particular ftars.

On our arrival at Guayaquil, the corregidor of that city, whofe great civility, together with that of all the king's officers and other perfons of diftinction, deferves our acknowledgments, fent notice of it to the corregidor of Guaranda, that he might order carriages to the port of Caracol, for conveying us to the mountains. The paffage thither was then indeed impracticable; it being in this country the end of winter, at which time the roads are extremely bad, and the rivers fwelled fo as not to be forded without the greateft rifk, and too wide for the bridges of this country.

The corregidor of Guaranda was then at Quito on fome bufinefs of his office; but the prefident and governor of that province, Don Dionyfio de Alcedoy Herrera, ordered him to return to his jurifdiction without delay, for providing every thing neceffary for our journey; fending, at the fame time, circular orders to all the other corregidors, through whofe jurifdictions we were to pafs to Quito, enjoining them not to be wanting in any kind of good office in their power. Every thing being thus happily difpofed, and advice arriving that the mules were on their way to Caracol, where they arrived the 6th of May, we were no lefs expeditious to embark on the river, which is the ufual paffage. There is indeed a road by land; but at all times extremely difficult and dangerous, on account of the many bays and large rivers which muft be paffed; fo that no perfon travels this road but in fummer, and then only fuch as have no baggage, and are, befides, well acquainted with the country and the ferries.

CHAP. IV.

Defcription of Guayaquil.

THOUGH there is no certainty with regard to the time when Guayaquil was founded, it is univerfally allowed to be the fecond city of Spanifh origin, both in its own province and the kingdom of Peru; it appearing, from ancient records preferved in its archives, that it was the next city founded after San Miguel de Piura; and the foundation laid of Los Reyes, Remac, or Lima, being in 1534, or, according to others, in 1535, the building of Guayaquil may be fixed between thofe two years; but the profperity it attained under its governor Belalcazar was of no long continuance, being, after feveral furious attacks, entirely deftroyed by the neighbouring Indians. It was, however, in 1537, rebuilt by captian Francifco de Orellana. The firft fituation of Guayaquil was in the bay of Charapoto, a little to the northward of the place where the village of Monte Chrifto now ftands; from whence it was removed to the prefent fpot, which is on the weft bank of the river of Guayaquil, in 2° 11′ 21″ of fouth latitude, as appeared from our obfervations. Its longitude was not determined by any accurate obfervations: but, by computing it from thofe made at Quito, it is 297° 17′ reckoning from the meridian of Teneriffe. On its removal by Orellana, from its firft fituation, it was built on the declivity of a mountain called Cerillo Verde, and is now termed Ciudad Vieja, or the old town.

town. Its inhabitants being afterwards ſtraitened by
the mountain on one ſide, and by ravines or hollows
made by floods of rain on the other, formed a deſign,
without entirely abandoning the place, to build the
principal part of the city at the diſtance of five or ſix
hundred toiſes; which was accordingly begun in
1693; and for preſerving a communication with the
old part, a bridge of timber was erected, of about
three hundred toiſes in length, by which means the
inconveniences of the ravines are avoided, and, the
intervals being filled with ſmall houſes, the old and
new towns are now united.

This city is of conſiderable extent, taking up, along
the bank of the river from the lower part of the old
town to the upper part of the new, near half a league;
but the breadth is not at all proportional, every perſon being fond of having a houſe near the river, both
for the amuſements it affords, and for the benefit of
refreſhing winds, which, in winter, are the more
eagerly coveted as they are very rare.

All the houſes of both towns are built of wood,
and many of them covered with tiles; though the
greater part of thoſe in the old town are only
thatched; but in order to prevent the ſpreading of
fires, by which this city has ſeverely ſuffered on ſeveral occaſions, ſuch covering is now prohibited. Moſt
of theſe conflagrations owed their riſe to the malevo-
lence of the Negroes, who, in order to revenge ſome
puniſhments inflicted on them by their maſters, took
the opportunity, during the night, of throwing fire
on the thatch, and by that means not only ruined
thoſe who were the immediate objects of their revenge,
but alſo the greater part of the inhabitants of the
city.

Though the houſes are wholly built of wood, they
are generally large and beautiful; have all one ſtory
and an entreſole; the back part of the ground floor
ſerves for warehouſes; in the front are ſhops of all
kinds,

kinds, and generally before them fpacious porticoes, which in winter are the only parts where you can walk, the ftreets being utterly impaffable.

As a further precaution againft fire, which they have fo much reafon to dread, the kitchens ftand twelve or fifteen paces from the houfes, with which they communicate by means of a long open gallery, refembling a bridge; but fo lightly built, that, on the leaft appearance of fire in the kitchen, it is demolifhed in an inftant; by which means the houfe is preferved. Perfons of rank and fortune live in the upper apartments, and the entrefoles are let to ftrangers who come to trade, or pafs through the city with their goods.

The ground on which the new city is built, and the favannahs in its neighbourhood, are not to be travelled over either on foot or horfeback during the winter; for, befides being a fpongy chalk, it is every where fo level, that there is no declivity for carrying off the water; and therefore, on the firft rain, it becomes one general flough. So that, from the time of the rains fetting in till the end of winter, it is necef-fary to lay in the parts not covered by the above-mentioned piazzas, very large planks for croffing over them; but thefe foon become flippery, and oc-cafion frequent falls into the chalky flough. The return of fummer, however, foon exhales the water, and renders the ground fufficiently dry for travelling. In this refpect the old town has the advantage, being built on a gravelly foil, which is never impaffable.

This city is defended by three forts, two on the river near the city, and the third behind it, guarding the entrance of a ravine. Thefe are all built after the modern method of fortification; but, before they were erected, it had only a platform, which is ftill remain-ing in the old town. Thefe forts are built of large pieces of very hard wood, forming a variety of pal-lifades, and the wood is particularly proper for this
country,

country, and the ufe it is here applied to; retaining
its folidity either under the water or in the mud.
Before thefe fortifications were erected, the city was
taken by European corfairs, in the years 1686 and
1709; but the fuccefs of the latter was owing to the
villany of a Mulatto, who, in order to revenge himfelf on fome particular perfons in the city, conducted
the enemy through a by-way, where they were not
expected; fo that the inhabitants, being furprifed, were
not prepared for defence.

ALL the churches and convents are of wood, except
that of St. Domingo, ftill ftanding in the old town,
which is of ftone; the great folidity of the ground in
that part being fufficient for fupporting buildings of
this kind. The convents in the new city, befides
the parochial church, are an Auguftine and a Francifcan, with a college of Jefuits; the members of
them not very numerous, on account of the fmallnefs
of the revenues. Here is alfo an hofpital, but without any other endowment than the fhell of the building. The city and its jurifdiction are under a corregidor, nominated by the king, who holds his office during five years. Notwithftanding he is fubordinate to
the prefident and audience of Quito, he appoints the
deputies in the feveral departments of his jurifdiction;
and, for the police and civil government, Guayaquil
has ordinary alcaldes and regidores. The revenue is
managed here by a treafurer and an accomptant, who
receive the tributes of the Indians, the duties on imports and exports, and the taxes on commodities,
which are either confumed there, or carried through it.

THE ecclefiaftical government is lodged in the
bifhop of Quito's vicar, who is generally alfo the
prieft of the town.

CHAP

CHAP. V.

Of the Inhabitants, Cuſtoms, and Riches of Guayaquil.

GUAYAQUIL contains, in proportion to its dimenſions, as many inhabitants as any city in all America; the continual reſort of ſtrangers, drawn thither by commerce, contributing very greatly to increaſe the number, generally computed at twenty thouſand. A great part of its eminent families are Europeans, who have married there; beſides which, and ſubſtantial Creoles, the other inhabitants are of different caſts, as in the cities already deſcribed.

THE inhabitants capable of bearing arms, are divided into companies of militia, according to their rank and caſt; ſo that on occaſion they may be ready to defend their country and property. One of theſe conſiſting entirely of Europeans, and called the foreign company, is the moſt numerous, and makes the moſt ſplendid appearance among the whole militia. Without conſidering their wealth or ſtation, they appear in arms, and pay a proper obedience to their officers, who are choſen by themſelves, from their own body, being generally ſuch as have ſerved in Europe, and, conſequently, more expert in military affairs. The corregidor is the commander in chief; having under him a colonel and major, for diſciplining the other companies.

THOUGH the heat here is equal to that of Panama, or Carthagena, yet the climate diſtinguiſhes itſelf in the colour of the human ſpecies; and if a certain author has ſtyled it the equinoctial Low Countries, in alluſion to the reſemblance it bears to the Netherlands of Europe, it may, with equal propriety, bear that appellation from this ſingularity, namely, that all the natives, except thoſe born from a mixture of blood,

are

are frefh-coloured, and fo finely featured, as juftly to
be ftyled the handfomeft, both in the province of
Quito, and even in all Peru.   Two things are here
the more remarkable, as being contrary to common
obfervation; one, that, notwithftanding the heat of
the climate, its natives are not tawny; the other,
that, though the Spaniards have not naturally fo fair
a complexion as the northern nations, their children
born here of Spanifh women are very fair; nor has
this phenomenon hitherto been fufficiently explain-
ed.   To attribute it to the effluvia exhaling from the
contiguous river, appears to me little fatisfactory;
other cities having the fame advantageous fituation,
without producing any improvement in the com-
plexions of the inhabitants; whereas here fair perfons are the moft common, and the children have
univerfally light hair and eyebrows, and very beautiful faces.

To thefe perfonal advantages beftowed by nature
in a diftinguifhed manner on the inhabitants, it has
added the no lefs pleafing charms of elegance and
politenefs; fo that feveral Europeans, who intended
only a fhort ftay here, have married and fettled;
nor were their marriages owing to the immenfe fortunes of their ladies, as in fome other cities of this
country, the inhabitants not being at all famous for
their riches.

The drefs of the women at Guayaquil nearly refembles that at Panama, except only when they either
pay or receive a vifit; inftead of the pollera, they
wear a faldellin, which is not longer than the pollera,
but, being open before, and croffing one fide over
the other, is adorned in the moft profufe manner.   It
is furbeloed with a richer ftuff, near half a yard in
depth, and bordered with fine laces, gold or fringe,
or ribands, difpofed with an air which renders the
drefs extremely rich and becoming.   When they go
abroad without a veil, they wear a light brown co-
                                    loured

loured mantelet, bordered with broad ſtrips of black velvet, but without laces or any other decorations. Beſides necklaces and bracelets, they wear roſaries, of the ſame degree of richneſs as at Panama; and not only load their ears with brilliant pendants, but add tufts of black ſilk, about the ſize of a filberd, and ſo full of jewels as to make a very ſplendid appearance.

FROM the commerce of this city, a ſtranger would imagine it richer than it actually is. This is partly owing to the two dreadful ·pillages it has ſuffered, and partly to fires, by both which it has been totally ruined. And though the houſes here, as already obſerved, are only of wood, the whole charge of which is the cutting and bringing it to the city; yet the expenſe of a houſe of any figure amounts to 15 or 20,000 dollars, workmen's wages being very high, and iron remarkably dear. Europeans, who have raiſed any thing of a fortune here, when they have no immoveable goods to detain them, retire to Lima, or ſome other city of Peru, where they may improve their ſtocks with greater ſecurity.

CHAP. VI.

Of the Temperature of the Air, and the different Seaſons at Guayaquil; its Inconveniencies and Diſtempers.

IN Guayaquil, the winter ſets in during December, ſometimes at the beginning, ſometimes in the middle, and ſometimes not till the end of the month, and laſts till April or May. During this ſeaſon, the elements, the inſects, and vermin, ſeem to have joined in a league to incommode the human ſpecies. Its extreme heat appeared from ſome thermometrical experiments; for, on the 3d of April, when its intenſeneſs had
begun

begun to abate, at fix in the morning the liquor flood at 1022; at noon at 1025; and at three in the afternoon at 1027; which fhews the heat in the middle of winter to be greater than at Carthagena. The rains alfo continue day and night, accompanied with frequent and dreadful tempefts of thunder and lightning; fo that every thing feems to confpire to diftrefs the inhabitants. The river, and all thofe which join it, overflow their banks, and lay under water the whole country. The long calm renders the refrefhing winds very defirable; and the innumerable fwarms of infects and vermin infeft both the air and ground in an intolerable manner.

The fnakes, poifonous vipers, fcorpions, and fcolopendræ, in this feafon find methods of getting into the houfes, to the deftruction of many of the inhabitants. And though they are not actually free from them all the reft of the year, yet at this time they are far more numerous, and alfo more active; fo that it is abfolutely neceffary to examine carefully the beds, fome of thefe animals having been known to find their way into them : and both as a fafeguard againft the danger, and to avoid the tortures of the mofchitos and other infects, all perfons, even the Negro flaves and Indians, have toldos or canopies over their beds. Thofe ufed by the lower clafs of people are made of tucuyo, or cotton, wove in the mountains : others ufe white linen laced, according to the temper or ability of the owner.

Though all thefe hot and moift countries fwarm with an infinite variety of volatile infects, yet the inhabitants are no where fo greatly incommoded as at Guayaquil, it being impoffible to keep a candle burning, except in a lantern, above three or four minutes, numberlefs infects flying into its flame and extinguifhing it. Any perfon therefore being obliged to be near a light, is foon driven from his poft by the infinite numbers which fill his eyes, ears, and noftrils. Thefe

infects

infects were almoft infupportable to us, during the
fhort clear intervals of fome nights, which we fpent
in making obfervations on the heavenly bodies.
Their ftings were attended with great tortures; and
more than once obliged us to abandon our obfervations, being unable either to fee or breathe for their
multitudes *.

ANOTHER terrible inconvenience attending the
houfes here, are the numbers of pericotes, or rats;
every building being fo infefted with them, that, when
night comes on, they quit their holes, and make fuch
a noife in running along the ceiling, and in clamber-
ing up and down the fides of the rooms and canopies
of the beds, as to difturb perfons not accuftomed to
them. They are fo little afraid of the human fpecies,
that, if a candle be fet down without being in a lantern,
they immediately carry it off; but, as this might be
attended with the moft melancholy confequences, care
is taken, that their impudence is feldom put to this
trial, though they are remarkably vigilant in taking
advantage of the leaft neglect. All thefe inconveniences, which feem infupportable to ftrangers, and
alone fufficient to render fuch a country uninhabited,
little affect the natives, as having been ufed to them
from their infancy: they are more affected with
cold on the mountains, which the Europeans fcarce
feel, or, at leaft, think very moderate, than with all
thefe difagreeable particulars.

THE leaft troublefome feafon is the fummer, as then
both the number and activity of thefe vermin are di-
minifhed; it being a miftake in fome authors to fay
they abound moft in that feafon. The heat is then
abated, by the fetting in of the· S. W. and W.S.W.
breezes, called here chandui, as coming over a moun-

* This account is too hyperbolical. They are, however, troublefome enough, and almoft infupportable, throughout all South America, except in the plains and deferts. A.

tain of that name. These begin conftantly at noon, and continue to refrefh the earth till five or fix in the following morning. The fky is always ferene and bright, the gentleft fhowers being rarely known. Provifions are in greater plenty, and thofe produced in the country of a very agreeable tafte, if ufed while frefh. Fruits are more common, efpecially melons and water-melons, which are brought in large balzas * to the city. But the capital advantage is the remarkable falubrity of the air in that feafon.

During the winter, tertian fevers are very common, and are here particularly painful and dangerous, owing partly to neglect, and partly to an averfion to the ufe of the bark, being prepoffeffed with a notion, that on account of its hot quality it can have no good effect in that climate; fo that, blinded with this prejudice, without ever confulting phyficians, who would undeceive them, they fuffer the diftemper to prey upon them, till they are often reduced to an irrecoverable ftate. The natives of the mountains, who are enured to a cold air, cannot endure that of Guayaquil, it having a natural tendency to debilitate them; and by an intemperate ufe of its delicious fruits they throw themfelves into thofe fevers, which are as common to them in one feafon as another.

Besides this difeafe, which is the moft general, fince the year 1740 the black vomit has alfo made its appearance, the galleons of the South Sea having, on account of the war, touched here in order to fecure the treafure among the provinces of the Cordillora. At that time great numbers died on board the fhips, together with many foreigners, but very few of the natives. In faying that the galleons brought

---

* Called by the natives jungadas: they are rafts made by pinning or tying feveral bodies of fmall trees together; the author defcribes them particularly in the next chapter. A.

this diftemper to Guayaquil, I follow the general opinion, as it was before that epocha unknown there.

THE natives are very fubject to cataracts, and other diftempers of the eye, which often caufe a total blind-nefs. Though thefe diftempers are not general, yet they are much more common than in other parts; and I am inclined to think it proceeds from the aqueous exhalations during the winter, when the whole country is overflowed with water, and which, from the chalky texture of the foil, muft be vifcid in the higheft degree; and, penetrating the external tunic, not only foul the cryftalline humour, but alfo cover the pupil, from whence cataracts, and other diforders of the eyes, have their origin.

CHAP. VII.

Provifions, and Manner of Living at Guayaquil.

HERE, as at Carthagena, nature and neceffity have introduced feveral kinds of bread, made from different grains and roots, to fupply the want of wheat. The moft ufual here is the criollo, or natural bread, being unripe plantanes, cut into flices, roafted, and ferved up as bread. But this is not entirely owing to neceffity, as feveral kinds of meal might eafily be brought from the neighbouring mountains in fuf-ficient quantities to fupply all the inhabitants of the city; though only a fmall fhare of it would fall to the lot of the poor, on account of the price, which vaftly exceeds that of the plantanes. However this be, the latter are defervedly preferred to wheat bread, which is fo badly made, that even the Europeans refufe to eat it, and accuftom themfelves to the criollo, which is far from being unpalatable.

MOST of the other provifions, except beef, fruits, and roots, are imported from the provinces of the
Cordel-

Cordelleros and Peru. It would naturally be expected, that the feveral branches of this river, which
abounds in fifh, would caufe a great plenty of them
in the city; but it is quite otherwife, and the fmall
quantity caught near it is far from being good, and
fo bony, that none but the inhabitants can eat them
without danger. Their badnefs in the neighbourhood
of the city is probably owing to the brackifh water;
but fome leagues above the city, the river affords a
great fupply of what is very excellent. In fuch hot
climates, however, they cannot be kept without falt;
and it is feldom the fifhermen venture to carry any to
the city, left, after all their labour, they fhould be
obliged to throw them away.

THE coafts and neighbouring ports abound in very
delicious fifh, fome of which are carried to the city,
as keeping better than the fpecies in the river; and
thefe, together with feveral of the teftaceous kind,
conftitute a confiderable part of the food of the inhabitants of Guayaquil. In the falt creek are taken
very large and fine lobfters, of which they make delicious ragouts: and from Jambeli creek, on the coaft
of Tumbez, are brought great quantities of oyfters,
which, in every refpect, furpafs thofe of all the coafts
from Panama to Peru, where there is alfo a great demand for them.

THE fame caufe which drives from that part of the
river near the city the fineft fifh, fome to the falt and
others to the frefh waters, according to their refpective
natures, renders good water very fcarce at Guayaquil,
efoecially in fummer; none being to be had at a lefs
diftance than four or five leagues up the river, according to the height of its waters. Many balzas are
therefore employed in fetching water, and felling it to
the inhabitants. During the winter, this trade is
partly at a ftand, as, by the increafe of the rivers, the
water at Guayaquil is rendered fit for ufe.

M 2

INSTEAD

INSTEAD of lard, as at Carthagena and other places, they commonly ufe, in drefling their food at Guayaquil, beef fuet. But whether the climate will not permit the beafts to acquire a proper degree of fatnefs, whether the fuet itfelf be not good, or whether they are carelefs in feparating it from the tallow; the fmell and tafte of both are much the fame, which render their difhes extremely naufeous to ftrangers; and, what is little better, they feafon all of them with Guiney pepper, which, though fmall, is fo very ftrong, that the fmell of it, when whole, fufficiently declares its furprifing activity; fo that perfons, not accuftomed to it, fuffer either way. If they eat, their mouths feem in a flame; if they forbear, they muft endure hunger, till they have overcome their averfion to this feafoning; after which they think the Guiney pepper the fineft ingredient in the world for giving a relifh to their food.

THE inhabitants of Guayaquil affect great fplendour in their formal entertainments; but the method of them is not very agreeable to an European gueft. The firft courfe confifts of different kinds of fweetmeats, the fecond of high-feafoned ragouts; and thus they continue to ferve up an alternate fucceffion of fweet and high-feafoned difhes. The common drink on thefe occafions is grape brandy, there called Aqua ardiente de Caftilla, cordials and wine: of all which they drink freely during the entertainment, heightening the pleafure by the variety; but the Europeans generally prefer wine.

THE cuftom of drinking punch has lately increafed confiderably in this city; and, when drunk in moderation, is found to agree very well with the conftitutions of this climate. Accordingly it has obtained greatly among perfons of diftinction, who generally drink a glafs of it at eleven, and again in the evening; thus allaying their thirft, and at the fame time correcting the water, which, befides the difagreeable tafte com-

municated

municated to it by heat, promotes an exceffive perfpiration : and this cuftom is fo prevailing, that even
the ladies punctually obferve it; and the quantity
both of acid and fpirit being but fmall, it becomes
equally wholefome and refrefhing.

CHAP. VIII.

Extent of the Jurifdiction of Guayaquil.

THE moft northern part of the jurifdiction of
Guayaquil begins at Cape Paffado, fo called
from its lying 21′ fouth of the equinoctial; and about
half a degree north of the bay of Manta.   From this
cape it continues all along the coaft, including the ifle
of Puna, to the town of Machala on the coaft of
Tumbez, where it is terminated by the jurifdiction of
Piura.   From thence it runs away caftward, and is
bounded by that of Cuenca ; and then, turning northwards along the weftern fkirts of the Andes, it terminates on thofe of Bamba and Chimbo.   Its length
from N. to S. is about 60 leagues, and its breadth
from E. to W. 40 or 45 ; reckoning from the point
of Santa Elena to the parts called Ojibar.   Its whole
country, like that in the neighbourhood of the city, is
one continued plain, and in winter univerfally overflowed.   It is divided into feven lieutenancies or departments, for each of which the corregidor appoints
a lieutenant or deputy, who, however, muft be confirmed by the audience of Quito.   Thefe departments
are, Puerto Viejo, Punta de Santa Elena, Puna,
Yaguache, Babahoyo, Baba, and Daule.

THE lieutenancy de San Gregorio de Puerto Viejo
is bounded northward by the government of Atacames,
and fouthward by the lieutenancy of Santa Elena.
Its capital of the fame name, though fmall, thinly
peopled, and poor, enjoys the privileges of a city,

and includes the towns of Monte Chrifto, Picoafa, Charapoto, and Xipijapa. Thefe have their particular priefts, who are likewife the fpiritual directors of all the fmaller villages in this diftrict.

THE town of Monte Chrifto flood at firft in the bay of Manta, and was called by that name. It had then a confiderable commerce by means of veffels paffing from Panama to the ports of Peru; but, having been pillaged and deftroyed by fome foreign adventurers, who infefted thofe feas, the inhabitants removed it to the foot of Monte Chrifto, where it now ftands, and from whence it has its name.

SOME tobacco is planted in this jurifdiction, but is not much efteemed; and the reft of its products, as wax, cotton, and pita, are barely fufficient to fupport its inhabitants, though they are far from being numerous; occafioned by the general poverty which reigns through all its towns and villages. The kinds of timber natural to fuch hot and moift countries grow here in prodigious quantities.

FORMERLY along the coaft, and in the bay belonging to this lieutenancy, was a confiderable pearl fifhery; but it has been totally difcontinued for fome years; occafioned partly from the dangers the divers were expofed to, from the mantas and tintoreas already defcribed; and partly from the poverty of the inhabitants of this country, who being in general Indians and Cafts, want ability to purchafe Negroes for this occupation. The bay has probably its name from the great number of mantas in thofe parts, efpecially as the common employment of the inhabitants is the taking of that fifh, which they falt, and carry into the inland provinces. The Europeans cannot help admiring their dexterity in this kind of fifhery, which they carry on in the following manner: they throw into the water a log of wood, fuch as they ufe in making a balza, being about five or fix yards in length, and near a foot in diameter, and fufficient to

bear

bear the weight affigned it, which is a net lying acrofs one end of it, while an Indian ftands in an erect pofition on the other; and, by help of a canalete or oar, puts off to fea, to the diftance of half a league or more, where he fhoots his net.   Another Indian, who follows him on a fimilar log, takes hold of the rope faftened to one end of the net; and when the whole is extended, they both move towards the land, where their partners wait to draw the net afhore.   And here one cannot help obferving with aftonifhment the dexterity and agility of the Indians, in maintaining an equilibrium on round logs, where, by the continual agitations of the fea, they muft be always changing their pofition, and making different motions with their body; and what ftill heightens the difficulty is, that he is obliged, at the fame time, to mind both his oar and the net, in drawing it towards the land.   They are indeed excellent fwimmers; fo that if they happen (which is very feldom) to flip off, they are foon on the log again, and in their former pofture; at leaft, they are in no danger of being fhipwrecked.

I shall place Punta de Santa Elena as the fecond lieutenancy, becaufe it joins to the fouth part of the former.   It extends all along the weftern coaft from the ifles of Plata and Salango, to the fame Punta de Santa Elena; from thence it ftretches along the north coaft, formed by the bay of Guayaquil; comprehending in this extent the towns of Punta, Chongon, Morro, Colonche, and Chandui.   At Chongon and Morro two priefts refide, to whofe parifhes the others belong.   The lieutenant, invefted with the civil government, refides in the town of Punta, two leagues from the port, where there are indeed warehoufes, or rather fheds, for receiving falt and other goods, but no dwelling-houfes.

The port of Punta has fo many falt-works, that it fupplies the whole province of Quito and jurifdiction

of

of Guayaquil. The falt is not the fineft, but remarkably compact, and anfwers very well the principal intention, that of falting flefh.

On the coaft belonging to this lieutenancy is found that exquifite purple, fo highly efteeméd among the ancients; but the fifh from which it was taken, having been either unknown or forgotten, many moderns have imagined the fpecies to be extinct. This colour, however, is found in a fpecies of fhell-fifh growing on rocks wafhed by the fea. They are fomething larger than a nut, and are replete with a juice, probably the blood, which, when expreffed, is the true purple; for if a thread of cotton, or any thing of a fimilar kind, be dipt in this liquor, it becomes of a moft vivid colour, which repeated wafhings are, fo far from obliterating, that they rather improve it; nor does it fade by wearing. The jurifdiction of the port of Nicoya, in the province of Guatemala, alfo affords this fpecies of turbines, the juice of which is alfo ufed in dying cotton threads, and, in feveral parts, for ribands, laces, and other ornaments. Stuffs died with this purple are alfo highly valued. This precious juice is extracted by different methods. Some take the fifh out of its fhell, and laying it on the back of their hand, prefs it with a knife from the head to the tail, feparating that part of the body into which the compreffion has forced the juice, and throw away the reft. In this manner they proceed till they have provided themfelves with a fufficient quantity. Then they draw the threads through the liquor, which is the whole procefs. But the purple tinge does not immediately appear, the juice being at firft of a milky colour; it then changes to green; and, lastly, into this celebrated purple. Others purfue a different method in extracting the colour; for they neither kill the fifh, nor take it entirely out of its fhell; but fqueeze it fo hard as to exprefs a juice, with which they die the thread, and afterwards replace the fifh on the rock

whence

whence it was taken. Some time after, it undergoes a second operation; but without yielding so much juice as at first; and at the third or fourth very little, by which means the fish is exhausted beyond recovery. In 1744, being in the lieutenancy of Santa Elena, I had the satisfaction to see this liquor extracted according to the first process, and some threads died with it. This purple is far from being so common as some authors have imagined; for, though the fish increases, yet so large a quantity is necessary to die a few ounces of thread, that little of it is seen; and indeed its great price is partly owing to its scarcity. Another circumstance worthy observation, and which increases or diminishes the value, is the difference of weight and colour of the cotton died with it, according to the different hours of the day. I could not find any satisfactory account of this property at Punta de Santa Elena, where the inhabitants, being less curious, have not carried their speculations so far as to be acquainted with this remarkable singularity; whereas at Nicoya it is so well known, that the dealers in it, both buyers and sellers, are exactly acquainted with the times of its increase or decrease, so that one of the first preliminaries to a contract is, to settle the time when it shall be weighed. From this alteration of the weight of the purple thread at Nicoya it may be inferred, that the same happens at Punta de Santa Elena; the turbines at both places being exactly of the same species, and without the least visible difference in colour. Another very remarkable particular relating to its tinct, and which I have heard from persons of undoubted veracity, is, that the colour of a thread of flax is very different from that of a thread of cotton. It would therefore be proper to make repeated experiments, on threads of silk, flax, and wool.

Some, by saying that the fish, from whence this die is extracted, breeds in a shell, by which either the

flat

flat or acaracolada or spiral may be understood; it may not be improper to remark, that it is the last species, and accordingly the cotton tinged with this juice is called Caraçolillo. This department also abounds in fruits, cattle of all kinds, wax, and fish; so that the inhabitants have very profitable motives for industry; accordingly it is very populous, and though it does not abound in towns, the number of inhabitants far exceeds that of the preceding government, and the harbour of Punta is much frequented by vessels, that is, by such as trade between Panama and the ports of Peru, in order to purchase different kinds of provisions, as calves, kids, fowl, and other kinds, of which there is here a great plenty. Vessels belonging to the merchants of Guayaquil of two hundred tons, load here with salt; a trade, which from the cheapness of that commodity turns to a very good account.

The next lieutenancy southward is Puna, an island in the mouth of Guayaquil river. It extends N. E. and S. W. between six and seven leagues, and is of a quadrilateral figure. According to an ancient tradition, its inhabitants were once between twelve and fourteen thousand; but, at present, it has only one small town, situated at the head of its harbour in the north-east part; and the few inhabitants consist chiefly of Casts, and some Spaniards, but very few Indians. To this lieutenancy has been annexed the town of Machala, on the coast of Tumbez, together with that of Naranjal, the landing-place of the river of the same name, called also the Suya; near which is a road leading to the jurisdictions of Cuenca and Alausi. But neither of these towns is in a more flourishing condition than that on the island. In the latter reside both the lieutenant and priest, to whom the others are subject, both in civil and ecclesiastical affairs; Puna not only being the principal town, but great ships, by reason of the depth of its harbour, load there,

which

which cannot be done at Guayaquil, on account of the fand in its river; while others come there to load with wood.

The jurifdictions of Machala and Manaranjol produce great quantities of cacao, and that of the former is efteemed the beft in all Guayaquil. In its neighbourhood, as in the ifland of Puna, are great numbers of mangles, or mangrove-trees, whofe interwoven branches and thick trunks cover all thofe plains; which, lying low, are frequently overflowed. As this tree is little known in Europe, it muft not be paffed over without a fhort defcription.

The mangrove is fo far different from other trees, that it requires a foil daily overflowed by the fea. Accordingly, when the water is ebbed away from the fpots where the mangroves thrive, they exhale very difagreeable effluvia from their muddy furface. This tree no fooner appears above the ground, than it divides itfelf into very knotty and diftorted branches; and from each knot germinates a multitude of others, increafing fo as to form, when grown up, an impenetrable thicket. Nor is it poffible to difcern the fhoots belonging to the principal branches; for, befides this entangled labyrinth, thofe of the fifth or fixth production are equal in magnitude to thofe of the firft, which is generally of an inch and a half or two inches in diameter; and all fo flexible, that the only method of fevering them is by fome edged tool. Though they extend themfelves nearly horizontally, yet the trunk and principal branches increafe both in height and thicknefs. Its leaves are very fmall, in proportion to the branches, not being above an inch and a half or two inches in length, oval, thick, and of a pale green. The ufual height of the principal ftems of the mangrove is eighteen or twenty yards, ten or twelve inches in diameter, and covered with a thin, rough bark. But its wood is fo folid and heavy, that it finks in water, and, when ufed in fhips or v 	effels, veſſels, is found very durable, being not ſubject either to ſplit or rot *.

The Indians here pay their annual tribute in the wood of the mangrove, which is uſed occaſionally in ſuch works as its nature is beſt adapted to.

The lieutenancy of Yaguache is at the mouth of the river of the ſame name, which falls into that of Guayaquil on the ſouth ſide; and has its riſe from the ſkirts of the Cordillera, ſouth of the river Bamba. Its juriſdiction contains three towns; the principal, that where the cuſtom-houſe is erected, is San Jacinto de Yaguache; the two others are Nouſa and Antonche. To theſe belong two prieſts, one reſiding at Yaguache, and the other at Nouſa. Though theſe towns are but thinly inhabited, the farms and country have great numbers, particularly of the poorer ſort.

The chief production of Yaguache is wood, and a little cacao: but cattle and cotton are the principal objects of their attention.

Babahoyo, a name ſufficiently known in all theſe countries, it being the ſeat of the grand cuſtom-houſe for every thing going into the Cordillera, or coming from thence, has a very large juriſdiction, in which, beſides the principal town, are thoſe of Ujiba, Caracol, Quilea, and Mangaches; the two laſt border on the Cordillera, and are a conſiderable diſtance from Ujiba, where the prieſt reſides during the winter, removing in the ſummer to Babahoyo; which, beſides its ſettled inhabitants, has always a great number of traders from other parts

* The mangrove ſhoots out collateral branches, which bend down, take root, and put out others which do the ſame, ſo that one tree in a few years covers a large ſpace of ground. Thoſe ſtems that are within the reach of high water mark are generally covered with a ſmall kind of ovſter, called mangrove oyſters, which are eaten by the natives. The bark of the tree is uſed to tan leather, in which it ſucceeds very well, but gives the leather a much higher colour than oak bark. A.

The

The country of this jurifdiction, being level and low, on the firft fwellings of the rivers Caluma, Ujiba, and Caracol, is overflowed to a prodigious diftance, though at different depths, particularly at Babahoyo, where the waters rife to the firft ftory of the houfes ; fo that during the winter it is entirely forfaken.

The country of this jurifdiction, as well as that of Baba, contiguous to it, abounds in fuch numbers of cacao plantations, that many are neglected, and their fruit left to the monkies and other animals, which are thus happily provided for by the fpontaneous fertility of the ground, without any affiftance from agriculture. It alfo produces cotton, rice, Guiney pepper, and a great variety of fruits. It has likewife large droves of black cattle, horfes, and mules, which, during the time the country is under water, are kept in the mountains ; but, as foon as the lands are dry, are driven down to fatten on the gamalotes, a plant of fuch luxuriance, as to cover entirely the ground ; its height exceeds two yards and a half. It alfo grows fo thick, as to preclude all paffage, even along the paths made by the traders.

The blade of the gamalote refembles that of barley, but longer, broader, thicker, and rougher. The green is deep, but lively, and the ftalk diverfified with knots, from which the leaves, which are ftrong, and fomething above two lines in diameter, have their origin. When the gamalote is at its full growth, the height of water, during the floods, by rifing above its top, preffes it down, and rots it ; fo that, when the waters ebb away, the earth feems covered with it ; but at the firft impreffion of the fun it fhoots again, and, in a few days, abounds in the fame plenty as before. One thing remarkable in it is, that, though it proves fo nourifhing to the cattle of this diftrict, it is very noxious to thofe from the Cordillera, as has been often experienced.

Baba

BABA is one of the largeſt lieutenancies of Guayaquil, reaching to the ſkirts of the Cordillera, or the
mountains of Anga Marca, belonging to the juriſdiction of Latacunga, or, according to the Indian pronunciation, Llatacunga. Beſides the principal town
of the ſame name, it has others annexed to it, ſo far as
to be under one prieſt, who, with the corregidor's
lieutenant, reſides continually at Baba. Formerly,
the river of the ſame name ran cloſe by this town; but
Don En Vinces having cut a canal for watering the
cacao plantations on his eſtate, the river inclining
more to this courſe than its former, it was found im-
poſſible to ſtop it; ſo that, leaving its original channel, it has ever ſince continued to run in a courſe ſome
diſtance from the town. The other two places are
San Lorenzo and Palenque, both at a great diſtance
from the capital, and near the Cordillera; ſo that their
Indian inhabitants are but little civilized.

THE cacao tree, which, as I have already obſerved,
abounds in this diſtrict, inſtead of being only four or
five, according to ſome authors, who poſſibly ſaw it
when very young, is generally not leſs than eighteen
or twenty feet high. It begins from the ground to divide itſelf into four or five ſtems, according to the
vigour of the root, from whence they all proceed.
They are generally between four and ſeven inches in
diameter; but their firſt growth is in an oblique direction, ſo that the branches are all expanded and ſeparated from one another. The length of the leaf is
between four and ſix inches, and its breadth three or
four. It is very ſmooth, ſoft, and terminates in a
point, like that of the China orange tree, but with
ſome difference in colour, the former being of a dull
green, and has nothing of the gloſs obſervable on the
latter: nor is the tree ſo full of leaves as that of the
orange. From the ſtem, as well as the branches,
grow the pods which contain the cacao. The firſt
appearance is a white bloſſom, not very large, whoſe
piſtil

piſtil contains the embryo of the pod, which grows to
the length of ſix or ſeven inches, and four or five in
breadth, reſembling a cucumber in ſhape ; and ſtriated
in a longitudinal direction, but deeper than the cucumber.   The pods are not preciſely of the above dimenſions, nor are they always proportionate to the
ſtem or branch, to which they adhere in the form of
excreſcences, ſome being much ſmaller ; and it is not
extraordinary to ſee one of the leaſt ſize on the principal trunk, and one prodigiouſly large near the extremity of a ſlender branch.   But it is obſerved,
that, when two grow in contact, one of them attracts
all the nutritive juice, and thrives on the decay of the
other.

THE colour of the pod, while growing, is green,
nearly reſembling that of the leaf ; but, when arrived
at its full perfection, it gradually changes to a yellow.
The ſhell which covers it is thin, ſmooth, and clear.
When the fruit is arrived at its full growth, it is gathered ; and being cut into ſlices, its pulp appears
white and juicy, with ſmall ſeeds regularly arranged,
and at that time of no greater conſiſtence than the reſt
of the pulp, but whiter, and contained by a very fine
delicate membrane, full of liquor, reſembling milk,
but tranſparent, and ſomething viſcid ; at this time it
may be eaten like any other fruit.   Its taſte is a
ſweetiſh acid ; but in this country is thought to be
promotive of fevers.   The yellowneſs of the pod indicates that the cacao begins to feed on its ſubſtance,
to acquire a greater conſiſtence, and that the ſeeds
begin to fill ; the colour gradually fading till they are
fully completed, when the dark brown colour of the
ſhell, into which the yellow has deviated, indicates
that it is a proper time to gather it.   The thickneſs
of the ſhell is now about two lines, and each ſeed
found incloſed in one of the compartments formed
by the tranſverſe membranes of the pod.   After gathering the fruit, it is opened, and the ſeeds taken
out

out and laid on fkins kept for that purpofe, or more generally on vijahua leaves, and left in the air to dry. When fully dried, they are put into leather bags, fent to market, and fold by the carga or load, which is equal to eighty-one pounds; but the price is far from fixed, being fometimes fold for fix or eight rials *per* carga, though lefs than the charge of gathering; but the general price is between three and four dollars, and, at the time of the armadas, when the demand is very large, rifes in proportion.

THIS tree produces its fruit twice a year, and in the fame plenty and goodnefs. The quantity gathered throughout the whole jurifdiction of Guayaquil amounts at leaft to 50,000 cargas.

THE cacao trees delight fo exceffively in water, that the ground where they are planted muft be reduced to a mire; and if not carefully fupplied with water, they die. They muft alfo be planted in the fhade, or at leaft defended from the perpendicular rays of the fun. Accordingly, they are always placed near other larger trees, under the fhelter of which they grow and flourifh. No foil can be better adapted to the nature of thefe trees than that of Guayaquil, as it favours them in both refpects; in the former, as confifting wholly of favannahs or wide plains overflowed in winter, and in fummer plentifully watered by canals; and with regard to the latter, it abounds in other trees, which afford them the requifite fhelter.

ALL the care neceffary in the culture of this tree confifts in clearing the ground from the weeds and fhrubs abounding in fo wet a foil: and this is fo neceffary, that, if neglected, in a few years thefe vegetables will deftroy the cacao plantations, by robbing the foil of all its nourifhment.

THE laft lieutenancy to be defcribed, is that of Daule. The principal town is of the fame name, and wafhed by the river, to which it owes its appellation. It contains many fpacious houfes belonging to the inhabitants

habitants of Guayaquil. It is alfo the refidence of a lieutenant and a parifh prieft having under their infpection the two towns of Santa Lucia and Valfar. Here are a great number of plantations of tobacco and fugar-canes, cacao, and cotton; together with large orchards of fruit-trees, and extenfive corn-fields.

The river Daule, which, like that of Baba, difcharges itfelf into Guayaquil river, is very large, and on both a great trade is carried on with that city. By the former, it receives the great plenty and variety of fummer fruits, and a confiderable part of the plantanes, which conftitute the bread ufed there during the whole year. Though great quahtities of tobacco grow in other parts of the jurifdiction of Guayaquil, yet none equals that of Daule.

The bufinefs of grazing is followed in all thefe lieutenancies; but more or lefs, in proportion to their extent, the nature of the foil, and the conveniency of driving the cattle to the mountains, beyond the reach of the inundations.

CHAP. IX.

Defcription of the River of Guayaquil, and of the Veffels trading on it.

THE river of Guayaquil being the channel of the commerce of that place, it will be proper to give fome account of it, in order to affift the reader in forming an idea of the trade carried on in that city.

The diftance of the navigable part of this river, from the city to the cuftom-houfe at Babahoyo, the place where the goods are landed, is, by thofe who have long frequented it, commonly divided into reaches, of which there are twenty, its courfe being wholly ferpentine; but to Caracol, the landing-place in winter, there are twenty-four reaches, the longeft of which are the three neareft the city; and thefe may be about two leagues and a half in length, but the others not above one. Whence it may be inferred, on an average, that the diftance, meafured on the furface of the river, between Guayaquil and the cuftom-houfe of Babahoyo, is twenty-four leagues and a half, and to Caracol twenty-eight and a half. The time requifite to perform this paffage is very different, according to the feafon, and nature of the veffel. During the winter, a chata generally takes up eight days in going from Guayaquil to Caracol, being againft the current of the river; whereas two days are fufficient to perform the paffage downwards. In fummer a light canoe goes up in three tides, and returns in little more than two; the fame may be faid of other veffels, the paffage downwards being always performed in much lefs time than the other, on account of the natural current of the river, in the reaches near the cuftom-houfe, where the ftrongeft flood only ftops the water from running downwards.

THE diftance from Guayaquil to Ifla Verde, fituated at the mouth of the river in Puna bay, is by pilots computed at about fix leagues, and divided, like the other part, into reaches; and from Ifla Verde to Puna three leagues: fo that the whole diftance from Caracol, the moft inland part up the river, to that of Puna, is thirty-feven leagues and a half. Between Ifla Verde and Puna it widens fo prodigioufly, that the horizon towards the north and fouth is bounded by the fky, except in fome few parts northwards, where the plantations of mangroves are perceived.

THE mouth of the river at the Ifla Verde is about a league in breadth, and even fomething broader at Guayaquil, above which it contracts itfelf as it advances nearer the mountains, and forms other creeks, the mouth of one of which, called Eftero de Santay,

faces

faces the city; another, termed Lagartos, is near the custom-house at Babahoyo. These are the largest, and at the same time extend to such a distance from the principal river, as to form very considerable islands.

The tides, as we have before observed, in summer-time reach up to the custom-house, checking the velocity of the waters, and consequently causing them to swell; but, in winter, the current being stronger and more rapid, this increase of the water is visible only in the reaches near Guayaquil; and in three or four different times of the year the great velocity of the current renders the tides imperceptible: the first of this season happens about Christmas.

The principal cause of the swellings of this river arises from the torrents rushing down from the Cordillera into it For though rain is frequent here, great part of the water is received by its lakes, or stagnates on the plains: so that the increase of the river is entirely owing to the torrents from the mountains.

One particular inconvenience of these floods is, their shifting the banks of sand lying between the city and Isla Verde; so that no ships of any considerable burden can go up with safety, without continually sounding with the lead, unless care has been taken to mark the banks since their last change.

The borders of this river, like those of Yaguache, Baba, and Daule, as well as those of the creeks and canals, are decorated with country-seats, and cottages of poor people of all casts, having here both the convenience of fishing and agriculture; and the. intermediate spaces filled with such a variety of thickets, that art would find it difficult to imitate the delightful landscape here exhibited by nature.

The principal and most common materials used in buildings on these rivers, are canes, whose dimensions and other particulars shall be taken notice of in their place. These also form the inward parts, as walls, floors, and rails of the stairs; the larger houses differ

only

only in fome of the principal pieces, which are of wood. Their method of building is, to fix in earth, eight, ten, or twelve pieces of wood, more or lefs, according to the dimenfions of the houfe, forked at the top, and of a proper length, all the apartments being on the firft ftory, without any ground floor. Beams are then laid acrofs on thefe forks, at the diftance of four or five yards from the ground. On thefe beams canes are laid in fuch a manner as to form a kind of rafters, and over thefe boards of the fame canes a foot and a half in breadth, which form as firm and handfome a flooring as if of wood. The partitions of the feveral apartments are of the fame materials, but the outer walls are generally latticed, for the free admiffion of the air. The principal beams of the roof of large houfes are of timber, the rafters of cane, with fmaller, in a tranfverfe direction, and over thefe vijahua leaves *. Thus a houfe is built at very little expenfe, though containing all the necef-fary conveniences. With regard to the poorer fort, every one's own labour fuffices to procure him a habitation. He goes up a creek in a fmall canoe, and from the firft wood cuts down as many canes, vijahuas, and bejucos †, as he wants, and, bringing the whole to the fhore, he makes a balza or float, on which he loads his other materials, and falls down the river to the place where he intends to erect his cottage. After which, he begins his work, faftening with bejucos thofe parts which are ufually nailed; and, in a few days, finifhes it in the completeft manner. Some of thefe cottages are almoft equal in dimenfions to thofe of timber.

The lower part, both of thefe houfes, as well as thofe in the greater part of the jurifdiction of Gua-yaquil (which are of the fame form), are expofed to

---

* This leaf is three or four feet long, and about one broad. A.
† A long pliant twig, ufed as a cord by the natives; defcribed B. V. Ch. I. A.

all winds, being entirely open, without having any wall, or fence, except the posts or stancheons by which the building is supported. For whatever cost was expended on the ground floor, it would be wholly useless in the winter, when all the country is turned to mud. Such houses, however, as stand beyond the reach of inundations, have ground floors, walled and finished like the other apartments, and serve as warehouses for goods; but those within the inundations are built, as it were, in the air, the water having a free passage under them. All the inhabitants have their canoes for passing from one house to another, and are so dexterous in the management of these skiffs, that a little girl ventures alone in a boat so small and slight, that any one less skilful would overset in stepping into it, and without fear crosses rapid currents, which an expert sailor, not accustomed to them, would find very difficult.

The continual rains in winter, and the slightness of the materials with which these houses are built, render it necessary to repair them during the summer; but those of the poorer sort, which are low, must be every year rebuilt, especially those parts which consist of cane, bejuco, and vijahua, while the principal stancheons, which form the foundation, still continue serviceable, and able to receive the new materials.

From the houses I proceed to give an account of the vessels, which (omitting the Chatas and canoes as common) are called Balzas, i. e. rafts. The name sufficiently explains their construction, but not the method of managing them, which these Indians strangers to the arts and sciences, have learned from necessity.

These Balzas, called by the Indians Jungadas*, are composed of five, seven, or nine beams of a sort of wood, which, though known here only by the name

---

* They are the same that are called Catamorans in the East Indies. A.

of

of Balza, the Indians of Darien called Puero; and, in all appearance, is the ferula of the Latins, mentioned by Columella; Pliny takes notice of two species of it, the leffer by the Greeks called Nartechia, and the larger Narthea, which grows to a great height. Nebrija calls it in Spanifh Canna Beja, or Canna Heja. Don George Juan, who faw it growing in Malta, found no other difference betwixt it and the Balza or Puero, only the Canna Beja, called ferula by the Maltefe, is much fmaller. The balza is a whitifh, foft wood, and fo very light, that a boy can eafily carry a log of three or four yards in length, and a foot in diameter. Yet, of this wood are formed the Janjades or Balzas, reprefented in Plate IV. Over part of it is a ftrong tilt, formed of reeds. Inftead of a maft, the fail is hoifted on two poles or fheers of mangrove wood, and thofe which carry a forefail have two other poles erected in the fame manner.

BALZAS are not only ufed on rivers, but fmall voyages are made at fea in them, and fometimes they go as far as Paita. Their dimenfions being different, they are alfo applied to different ufes; fome of them being fifhing Balzas; fome carry all kinds of goods from the cuftom-houfe to Guayaquil, and from thence to Puna, the Salto de Tumbez, and Paita; and others, of a more curious and elegant conftruction, ferve for removing families to their eftates and country-houfes, having the fame convenience as on fhore, not being the leaft agitated on the river; and that they have fufficient room for accommodations, may be inferred from the length of the beams, which are twelve or thirteen toifes, and about two feet or more in diameter: fo that the nine beams of which they confift, form a breadth of between twenty and twenty-four Paris feet; and proportional in thofe of feven, or any other number of beams.

THESE beams are faftened or lafhed together by bejucos, and fo fecurely, that with the crofs-pieces at
each

each end, which are alſo laſhed with all poſſible
ſtrength, they reſiſt the rapidity of the currents in
their voyages to the coaſt of Tumbez and Paita.
The Indians are ſo ſkilful in ſecuring them, that they
never looſen, notwithſtanding the continual agitation;
though by their neglect in examining the condition of
the bejucos, whether they are not rotten or worn, ſo
as to require others, there are ſome melancholy in-
ſtances of Balzas, which, in bad weather, have ſeparated, and, by that means, the cargo loſt, and the
paſſengers drowned. With regard to the Indians,
they never fail of getting on one of the beams, which
is ſufficient for them to make their way to the next
port. One or two unfortunate accidents of this kind
happened even while we were in the juriſdiction of
Quito, purely from the ſavage careleſſneſs of the
Indians.

THE thickeſt beam of thoſe which compoſe the
Balza, is placed ſo as to project beyond the other in
its after-part; and to this are laſhed the firſt beams on
each ſide, and thus, ſucceſſively, till the whole are ſe-
cured; that in the middle being the principal piece,
and thence the number of beams is always odd. The
larger ſort of Balzas generally carry between four
and five hundred quintals, without being damaged
by the proximity of the water; for the waves of the
ſea never run over the Balza; neither does the water
ſplaſh up between the beams, the Balza always following the motion of the water.

HITHERTO we have only mentioned the conſtruction and the uſes they are applied to ; but the greateſt
ſingularity of this floating vehicle is, that it ſails,
tacks, and works as well in contrary winds, as ſhips
with a keel, and makes very little lee-way. This advantage it derives from another method of ſteering
than by a rudder ; namely, by ſome boards, three or
four yards in length, and half a yard in breadth,
called Guaras, which are placed vertically, both in the

N 4   head

head and stern between the main beams, and by
thrusting some of these deep in the water, and raising
others, they bear away, luff up, tack; lie to, and
perform all the other motions of a regular ship: an
invention hitherto unknown to the most intelligent
nations of Europe, and of which even the Indians
know only the mechanism, their uncultivated minds
having never examined into the rationale of it. Had
this method of steering been sooner known in Europe,
it would have alleviated the distress of many a ship-
wreck, by saving numbers of lives; as in 1730, the
Genovesa, one of his majesty's frigates, being lost on
the Vibora, the ship's company made a raft; but com-
mitting themselves to the waves, without any means
of directing their course, they only added some melancholy minutes to the term of their existence. Such
affecting instances induced me to explain the reason
and foundation of this method of steering, in order
to render it of use in such calamitous junctures; and,
that I may perform it with the greater accuracy, I
shall make use of a short memoir, drawn up by Don
George Juan.

THE direction, says he, in which a ship moves
before the wind, is perpendicular to the sail, as Mess.
Renau, in the *Theorie de Manœuvres*, chap. ii. art. 1.
Bernouilli, cap. i. art. 4. *Pilot*, sect. ii. art. 13. have
demonstrated. And re-action being contrary and equal
to the action, the force with which the water opposes
the motion of the vessel, will be applied in a perpendicular direction to the sail, and continued from lee-
ward to windward, impelling with more force a greater
body than a smaller, in proportion to the superficies,
and the squares of the sines of the angle of incidence,
supposing their velocities equal. Whence it follows,
that a Guara being shoved down in the fore-part of
the vessel, must make her luff up; and by taking it
out, she will bear away or fall off. Likewise on a
Guara's being shoved down at the stern, she will bear
away;

away; and by taking it out of the water, the Balza will luff, or keep nearer to the wind. Such is the method used by the Indians in steering the Balzas; and sometimes they use five or six Guaras, to prevent the Balza from making lee-way; it being evident, that the more they are under water, the greater resistance the side of the vessel meets with; the Guaras performing the office of lee-boards, used in small vessels. The method of steering by these Guaras is so easy and simple, that when once the Balza is put in her proper course, one only is made use of, raising or lowering it as accidents require; and thus the Balza is always kept in her intended direction.

We have before observed, that this river and its creeks abound in fish, which for some time in the year afford employment for the Indians and Mulattos inhabiting its banks, and for which they prepare towards the end of summer, having then sown and reaped the produce of their little farms. All their preparatives consist in examining their Balzas, giving them the necessary repairs, and putting up a fresh tilt of vijahua leaves. This being finished, they take on board the necessary quantity of salt, harpoons, and darts. With regard to their provision, it consists only of maize, plantanes, and hung beef. Every thing being ready, they put on board the Balzas their canoes, their families, and the little furniture they are masters of. With regard to the cattle and horses, of which every one has a few, they are driven up to winter in the mountains.

The Indians now steer away to the mouth of some creek, where they expect to take a large quantity of fish, and stay there during the whole time of the fishery, unless they are disappointed in their expectations; in which case they steer away to another, till they have taken a sufficient quantity, when they return to their former habitations; but not without taking with them vijahua leaves, bejucos, and canes,

for

for making the neceffary repairs. When the communication is opened with the provinces of the Cordilleras, and the cattle begin to return into the plains, they carry their fifh to the cuftom houfe of Babahoyo, where they fell it; and, with the produce, purchafe baize, tucuyo, and other ftuffs, for clothing themfelves and families.

THEIR method of fifhing is thus: Having moored their Balza near the mouth of a creek, they take their canoes, with fome harpoons and fpears, and on fight of a fifh make towards it, till they arrive at a proper diftance, when they throw their fpear at it with fuch dexterity, that they feldom mifs; and if the place abounds in fifh, they load their canoes in three or four hours, when they return to their Balzas to falt and cure them. Sometimes, efpecially in places where the creeks form a kind of lake, they make ufe of a certain herb called Barbafco, which they chew, mix with fome bait, and fcatter about on the water. The juice of this herb is fo ftrong, that the fifh on eating a very little of it become inebriated, fo as to float on the furface of the water, when the Indians have no other trouble than to take them up. This juice is actually fatal to the fmaller fifh, and the larger do not recover for fome time; and even thefe, if they have eaten a confiderable quantity, perifh. It is natural to think, that fifh caught in this manner muft be prejudicial to health; but experience proves the contrary, and accordingly the moft timorous make no difficulty of eating them. Their next method of fifhing is with nets; when they form themfelves into companies, for the better management of them.

THE largeft fort of fifh caught here is called Bagre, fome of which are a yard and a half long; but flabby, and of an ill tafte, fo that they are never eaten frefh. The Robalo, a fort of large trout, is the moft palatable; but being only taken in the creeks a great
way

way above Guayaquil, the diftance will not admit
their being brought to that city.

THE increafe of fifh in this river is greatly hindered by the prodigious numbers of alligators, an
amphibious creature, living both in the rivers and the
adjacent plains, though it is not often known to go
far from the banks of the river. When tired with
fifhing, they leave the water, to bafk themfelves in the
fun, and then appear more like logs of half-rotten
wood thrown afhore by the current, than living creatures; but upon perceiving any veffel near them,
they immediately throw themfelves into the water.
Some are of fo monftrous a fize as to exceed five
yards in length. During the time they lie bafking on
the fhore, they keep their huge mouths wide open,
till filled with mofchitos, flies, and other infects,
when they fuddenly fhut their jaws and fwallow their
prey. Whatever may have been written with regard
to the fiercenefs and rapacity of this animal, I and all
our company know, from experience, they avoid a
man, and, on the approach of any one, immediately
plunge into the water. Their whole body is covered
with fcales impenetrable to a mufket-ball, unlefs it
happens to hit them in the belly near the fore legs;
the only part vulnerable.

THE alligator is an oviparous creature. The female makes a large hole in the fand near the brink
of a river, and there depofits her eggs; which are as
white as thofe of a hen, but much more folid. She
generally lays about a hundred, continuing in the
fame place till they are all depofited, which is about
a day or two. She then covers them with the fand;
and, the better to conceal them, rolls herfelf not only
over her precious depofitum, but to a confiderable
diftance. After this precaution, fhe returns to the
water till natural inftinct informs her that it is time
to deliver her young from their confinement; when
fhe comes to the fpot, followed by the male, and
tearing

tearing up the fand, begins breaking the eggs, but
fo carefully, that fcarce a fingle one is injured; and a
whole fwarm of little alligators are feen crawling
about. The female then takes them on her neck
and back, in order to remove them into the water;
but the watchful gallinazos make ufe of this opportunity to deprive her of fome; and even the male
alligator, which indeed comes for no other end, devours what he can, till the female has reached the
water with the few remaining; for all thofe which
either fall from her back, or do not fwim, fhe herfelf
eats; fo that of fuch a formidable brood, happily not
more than four or five efcape.

THE gallinazos, mentioned in our account of Carthagena, are the moft inveterate enemies of the alligators, or rather extremely fond of their eggs, in
finding which they make ufe of uncommon addrefs.
Thefe birds often make it their whole bufinefs to
watch the females during the fummer, the feafon
when they lay their eggs, the fands on the fides of the
river not being then covered with water. The gallinazo perches in fome tree, where it conceals itfelf
among the branches, and there filently watches the
female alligator, till fhe has laid her eggs and retires,
pleafed that fhe has concealed them beyond difcovery.
But fhe is no fooner under the water, than the gallinazo darts down on the repofitory, and, with its beak,
claws, and wings, tears up the fand, and devours the
eggs, leaving only the fhells. This banquet would
indeed richly reward its long patience, did not a multitude of gallinazos, from all parts, join the fortunate
difcoverer, and fhare in the fpoil. I have often been
entertained with this ftratagem of the gallinazos, in
paffing from Guayaquil to the cuftom-houfe of Babahoyo; and my curiofity once led me to take fome of
the eggs, which thofe who frequent this river, particularly the Mulattos, make no difficulty of eating,
when frefh. Here we muft remark the methods ufed
by

by Providence in diminifhing the number of thefe
deftruclive creatures, not only by the gallinazos, but
even by the males themfelves.   Indeed, neither the
river nor the neighbouring fields would otherwife be
fufficient to contain them ; for, notwithftanding the
ravages of thefe two infatiable enemies, their numbers
can hardly be imagined.

THESE alligators are the great deftroyers of the fifh
in this river, it being their moft fafe and general food;
nor are they wanting in addrefs to fatisfy their defires;
eight or ten, as it were by compact, draw up at the
mouth of a river or creek, whilft others go a confiderable diftance up the river, and chafe the fifh downwards, by which none of any bignefs efcape them.
The alligators, being unable to eat under water, on
feizing a fifh, raife their heads above the furface, and
by degrees draw the fifh from their jaws, and chew
it for deglutition.   After fatisfying their appetite, they
retire to reft on the banks of the river.

WHEN they cannot find fifh to appeafe their hunger,
they betake themfelves to the meadows bordering on
the banks, and devour calves and colts; and, in order
to be more fecure, take the opportunity of the night,
that they may furprife them in their fleep ; and it is
obferved, that thofe alligators which have once tafted
flefh, become fo fond of it, as never to take up with
fifh but in cafes of neceffity.   There are even too
many melancholy inftances of their devouring the
human fpecies, efpecially children, who, from the in-
attention natural to their age, have been without doors
after it is dark ; and though at no great diftance,
thefe voracious animals have dared to attack them,
and having once feized them, to make fure of their
prey againft that affiftance which the cries of the
victim never fail to bring, haften into the water,
where they immediately drown it, and then return to
the furface, and devour it at leifure.

THEIR

THEIR voracity has also been felt by the boatmen, whom, by inconsiderately sleeping with one of their arms or legs hanging over the side of the boat, these animals have seized, and drawn the whole body into the water. Alligators who have once feasted on human flesh, are known to be the most dangerous, and become, as it were, inflamed with an insatiable desire of repeating the same delicious repast. The inhabitants of those places where they abound, are very industrious in catching and destroying them. Their usual method is by a casonate, or piece of hard wood sharpened at both ends, and baited with the lungs of some animal. This casonate they fasten to a thong, the end of which is secured on the shore. The alligator, on seeing the lungs floating on the water, snaps at the bait, and thus both points of the wood enter his jaws, in such a manner that he can neither shut nor open his mouth. He is then dragged ashore, where he violently endeavours to rescue himself, while the Indians bait him like a bull, knowing that the greatest damage he can do, is to throw down such as, for want of care or agility, do not keep out of his reach.

THE form of this animal so nearly resembles that of the lagarto or lizard, that here they are commonly called by that name; but there is some difference in the shape of the head, which in this creature is long, and towards the extremity slender, gradually forming a snout like that of a hog, and, when in the river, is generally above the surface of the water; a sufficient demonstration, that the respiration of a grosser air is necessary to it. The mandibles of this creature have each a row of very strong and pointed teeth, to which some writers have attributed particular virtues; but all I can say to this is, that they are such as I and my companions, notwithstanding all our inquiries to attain a complete knowledge of every particular, could never hear any satisfactory account of.

CHAP.

CHAP. X.

Of the Commerce carried on by means of the City and River of Guayaquil, betwixt the Provinces of Peru and Terra Firma, and the Coast of New Spain.

THE commerce of Guayaquil may be divided into two parts; one reciprocal, being that of the products and manufactures of its jurisdiction; the other transitory, its port being the place where the goods from the provinces of Peru, Terra Firma, and Guatemala, consigned to the mountains, are landed; and on the other hand, those from the mountains, designed for the above-mentioned provinces, are brought hither and shipped for their respective ports. And as these two branches are very different, I shall first treat particularly of its reciprocal commerce.

THE cacao, one of its principal products, is chiefly exported to Panama, the ports of Sonsonate, el Realejo, and other ports of New Spain; and also to those of Peru, though the quantity sent to the latter is but small. It is something singular, that in this city and jurisdiction, where cacao grows in such plenty, little or no use should be made of it.

TIMBER, which may be esteemed the second article of its commerce, is chiefly sent to Callao; though a little is sold to the places between Guayaquil and that port. All the expense of it here is the charge of felling, carrying it to the next creek or river, and floating it down to Guayaquil; where, or at Puna, it is shipped for the ports it is consigned to.

THOUGH both these branches of trade are very advantageous to Guayaquil, as may easily be imagined, from the prodigious quantities exported; yet the trade of salt is not inferior to either, though the principal markets to which this is sent are only the inland towns

in

in the province of Quito. To thefe may be added cotton, rice, and fifh, both falted and dried; the two firft of which deferve to be mentioned, as they are exported both to the maritime and inland provinces.

The fourth and laft article of the commerce of this jurifdiction, is the trade in horned cattle, mules, and colts, of which great numbers are bred in the extenfive favannahs of this province. Thefe turn to good account in the provinces of the mountains, where there is not a fufficiency to anfwer the neceffary demands.

Besides thefe four capital articles, there are others, though fingly of little confequence, yet jointly are equal to any one of the former, as tobacco, wax, Guiney pepper, drugs, and lana de ceibo, by which great numbers of the lower clafs of people acquire a comfortable fubfiftence.

The lana de ceibo, or ceibo wool, is the product of a very high and tufted tree of that name. The trunk is ftraight, and covered with a fmooth bark; the leaf round, and of a middling fize. At the proper feafon the tree makes a very beautiful appearance, being covered with white bloffoms; and in each of thefe is formed a pod, which increafes to about an inch and a half or two inches in length, and one in thicknefs. In this pod the lana or wool is contained. When thoroughly ripe and dry, the pod opens, and the filamentous matter or wool gradually fpreads itfelf into a tuft refembling cotton, but of a reddifh caft. This wool is much more foft and delicate to the touch than cotton itfelf, and the filaments fo very tender and fine, that the natives here think it cannot be fpun; but I am perfuaded that this is entirely owing to their ignorance: and if a method be ever difcovered of fpinning it, its finenefs will entitle it rather to be called ceibo filk than wool. The only ufe they have hitherto applied it to, is to fill matraffes; and in this particular, it muft be allowed to have no equal,

equal, both with regard to its natural softness, and its rising so, when laid in the sun, as even to stretch the covering of the mattress; nor does it sink on being brought into the shade, unless accompanied with dampness, which immediately compresses it. This wool is here thought to be of an extreme cold quality, which is abundantly sufficient to hinder it from being generally used; though great numbers of persons of rank, and tenderly brought up, have never slept on any thing else, but without any injury to their health.

THE goods imported into this jurisdiction from Peru, in return for the above-mentioned commodities, are wine, brandy, oil, and dried fruits. From Quito it receives bays, tucuyos, flour, papas, bacon, hams, cheese, and other goods of that kind. From Panama, European goods purchased at the fairs. The chief commodities it receives from New Spain are iron, found in that country, but much inferior to that of Europe, being brittle and vitreous. It, however, serves for such uses where malleability is of no great importance, but is rarely used in building ships; also naphtha, and tar for the use of shipping. From the same coast, as well as from Peru, they have also cordage; though the last article, together with European iron, the owners of ships import on their own account; and therefore make no part of the commerce.

THE transitory commerce is in quantity much more considerable than that of the preceding, as it consists of the reciprocal exchange between the large kingdoms of Quito and Lima, of their respective commodities both natural and factitious. Lima sends the products of its vineyards and oliveyards; and Quito furnishes cloth, bays, tucuyos, serges, hats, stockings, and other woollen goods; but indigo being necessary for increasing the beauty of the colours, and none of it growing in the province of Quito, the merchants of Guayaquil import it from New Spain, and send it to the Quito manufacturers.

Summer is the proper season for carrying on these branches of commerce; because then the manufactures of the mountains can be brought down to Guayaquil, and the goods sent from other parts carried up to the mountainous parts. But the river of Guayaquil is never without vessels loading with goods of that jurisdiction, the sea here being always open. The profits resulting from this large and constant commerce could alone have preserved it from a total desertion, after being so frequently pillaged by pirates, and wasted by fire. And it is owing to the advantages resulting from this commerce, that we now behold it large, flourishing, and magnificent, as if it had enjoyed an uninterrupted prosperity from its very foundation.

BOOK V.

Journey from Guayaquil to the City of Quito.

────────────

CHAP. I.

Passage from Guayaquil to the Town of Caracol, and from thence to Quito.

ON receiving advice that the mules, provided by the corregidor of Guaranda, were on the road to Caracol, we immediately embarked at Guayaquil, on the 3d of May 1736, on board a large chata: but the usual impediment of the current, and several unfortunate accidents, rendered the passage so very long, that we did not land at Caracol before the 11th. The tortures we received on the river from the moschitos were beyond imagination. We had provided ourselves with guetres, and moschito cloths; but to very little purpose. The whole day we were in continual motion to keep them off; but at night our torments were excessive. Our gloves were indeed some defence to our hands, but our faces were entirely exposed, nor were our clothes a sufficient defence for the rest of our bodies; for their stings, penetrating through the cloth, caused a very painful and fiery itching. The most dismal night we spent in this passage was when we came to an anchor near a large and handsome house, but un-inhabited; for we had no sooner seated ourselves in

O 2   it,

it, than we were attacked on all sides with innumerable swarms of moschitos; so that we were so far from having any rest there, that it was impossible for a person, susceptible of feeling, to be one moment quiet. Those who had covered themselves with their moschito cloths, after taking the greatest care that none of these malignant insects were contained in them, found themselves in a moment so attacked on all sides, that they were obliged soon to return to the place they had quitted. Those who were in the house, hoping that they should find some relief in the open fields, ventured out, though in danger of suffering in a more terrible manner from the serpents; but were soon convinced of their mistake; it been impossible to determine which was the most supportable place, within the moschito cloth, without it, or in the open fields. In short, no expedient was of any use against their numbers. The smoke of the trees we burnt, to disperse these infernal insects, besides almost choking us, seemed rather to augment than diminish their multitudes. At day-break, we could not without concern look upon each other. Our faces were swelled, and our hands covered with painful tumours, which sufficiently indicated the condition of the other parts of our bodies exposed to the attacks of those insects. The following night we took up our quarters in a house inhabited, but not free from moschitos; though in much less numbers than before. On informing our host of the deplorable manner in which we had spent the preceding night, he gravely told us that the house we so greatly complained of had been forsaken on account of its being the purgatory of a soul. To which one of our company wittily answered, that it was much more natural to think that it was forsaken on account of its being a purgatory for the body.

The mules being arrived at Caracol, we set out on the 14th of May, and after travelling four leagues,

through

through savannahs, woods of plantain, and cacao-
trees, we arrived at the river Ojibar; and continued
our journey, during the whole day, along its banks,
fording it no less than nine times, though with no
small danger, from its rapidity, breadth, depth, and
rocky bottom; and, about three or four in the afternoon, we halted at a placed called Puerto de Mus-
chitos.

ALL the road from Caracol to the Ojibar is so deep
and boggy that the beasts at every step sunk almost
up to their bellies; but along the banks of that river
we found it much more firm and commodious.   The
name of the place where we were to take up our
lodging that night sufficiently indicates its nature.
The house had been for some time forsaken, like that
already mentioned on Guayaquil river, and become
a nest of moschitos of all kinds; so that it was impossible to determine which was the worst.   Some, to
avoid the torture of these insects, stripped themselves,
and went into the river, keeping only their heads
above water; but the face, being the only part exposed, was immediately covered with them; so that
those who had recourse to this expedient, were soon
forced to deliver up their whole bodies to these tor-
menting creatures.

ON the 15th we continued our journey through a
very thick forest, the end of which brought us once
more to the banks of the same river, which we again
forded four times, and rather with more danger than
at first.   About five, we halted on its banks, at a place
called Caluma, or the Indian post.   Here was no
house for lodging in, nor had we seen one during the
whole day's journey; but this inconvenience was in
some measure removed by the surprising dexterity of
our Indians, who, running into the woods, soon returned with branches of trees and vijahua leaves, with
which, in less than an hour, they erected several huts
large enough to contain our whole company; and so

well

well covered, that the rain, which came on very violently, did not penetrate them *.

THE thermometer at Caluma, on the 16th, at six in
the morning, was at 1016; and we were ourselves sensible that the air began to grow cool. At half an
hour after eight in the morning we began our journey, and at noon passed by a place called Mamarumi,
or mother of stone, where there is an inconceivably
beautiful cascade. The rock from which the water
precipitates itself is nearly perpendicular, and fifty
toises in height, and on both sides bordered with lofty
and spreading trees. The clearness of the water daz-
zles the sight, which is, however, charmed with its
lustre as it falls from the precipice; after which it
continues its course in a bed along a small descent,
and is crossed by the road. These cattaracts are by
the Indians called Paccha, and by the Spaniards of
the country Chorrera. From hence we continued
our journey; and after crossing the river twice on
bridges, but with equal danger as in fording it, we
arrived at two in the evening at a place called Tarigagua, where we rested in a large structure of timber,
covered with vijahua leaves, built for our reception.
Indeed we were no less fatigued with this day's journey
than with any of the preceding; some parts of it
being over dreadful precipices, and the road in others
so narrow, as hardly to afford a passage for the mules,
that it was impossible to avoid frequently striking
against the trees and rocks; few of us therefore reached Tarigagua without several bruises.

IT must not be thought strange that I should say
the bridges are equally dangerous with the fords; for
these structures, all of wood, and very long, shake in
passing them; besides, their breadth is not above three

* The natives, when they travel, erect new huts every night in
this manner, except they have the conveniency of tying their hammocks up in trees, by which means they save the trouble of a
watch and fire all night to keep off the wild beasts. A.

feet,

feet, and without any rail; so that one false step precipitates the mule into the torrent, where it is inevitably lost; accidents, according to the report of our guides, not uncommon. These bridges, by the rotting of the wood under water, are annually repaired towards winter, the only season when they are used; the rivers during the summer being fordable.

WHEN a person of distinction, as a president, a bishop, &c. is on a journey from Caracol or Babahoyo, the corregidor of Guaranda dispatches Indians for building cottages at the usual resting places, like that we found at Tarigagua; and these being left standing, serve afterwards for other passengers, till the rains destroy them. When these are thrown down, travellers must content themselves with the huts which their Indian guides build with wonderful dispatch.

AT Tarigagua, on the 17th, at six in the morning, the thermometer stood at $1014\frac{1}{4}$. And having been for some time accustomed to hot climates, we now sensibly felt the cold. It is remarkable, that we here often see instances of the effects of two opposite temperatures, in two persons happening to meet, one of them coming from Guayaquil, and the other from the mountains: the latter finding the heat so great that he is scarce able to bear any clothes, while the former wraps himself up in all the garments he can procure. The one is so delighted with the warmth of the water of the river, that he bathes in it; the other thinks it so cold, that he avoids being spattered by it. Nor is the case very different even in the same person, who, after a journey to the mountains, is returning to Guayaquil, or *vice versa*, provided the journey and return be made at the same season of the year. This sensible difference proceeds only from the change naturally felt at leaving a climate to which one has been accustomed, and coming into another of an opposite temperature; and thus two persons, one used to a

O 4 cold

cold climate, like that of the mountains, the other to a hot, like that of Guayaquil, must, at coming into an intermediate temperature, as at Tarigagua, feel an equal difference; one with regard to heat, and the other with regard to cold; which demonstrates that famous opinion—that the senses are subject to as many apparent alterations, as the sensations are various in those who feel them. For the impressions of objects are different, according to the different disposition of the senses; and the organs of two persons differently disposed are differently affected. At a quarter past nine in the morning we began to ascend the mountain of San Antonia, the foot of which is at Tarigagua; and, at one, came to a place called by the Indians Guamac, or Cross of Canes. Here is a small but inclining plain; and being told that it was half way up the acclivity, and our beasts requiring rest, we halted here.

THE ruggedness of the road from Tarigagua leading up this mountain is not easily described. It gave us more trouble and fatigue, besides the dangers we were every moment exposed to, than all we had experienced in our former journeys. In some parts the declivity is so great that the mules can scarce keep their footing, and in others the acclivity is equally difficult. In many places the road is so narrow that the mules have scarce room to set their feet; and in others a continued series of precipices. Besides, these roads, or rather paths, are full of holes, or camelones, near three quarters of a yard deep, in which the mules put their fore and hind feet; so that sometimes they draw their bellies and riders' legs along the ground. Indeed these holes serve as steps, without which the precipices would be in a great measure impracticable. But should the creature happen to put his foot between two of these holes, or not place it right, the rider falls, and, if on the side of the precipice, inevitably perishes. It may perhaps be said,

said, that it would be much safer to perform this part of the journey on foot: but how can any person be sure always of placing his feet directly on the eminences between the holes? and the least false step throws him up to the waist in a slimy mud, with which all the holes are full; and then he will find it very difficult either to proceed or return back.

These holes, or camelones, as they are called, render all this road very toilsome and dangerous, being as it were so many obstacles to the poor mules; though the danger is even greater in those parts where they are wanting. For as the tracks are extremely steep and slippery, from the soil, which is chalky and continually wet; so they would be quite impracticable, did not the Indians go before, and dig little trenches across the road, with small spades which they carry with them for this purpose; and thus both the difficulty and danger of these craggy paths are greatly lessened. This work is continual, every drove requiring a repetition of it; for in less than a night the rain utterly destroys all the trenches cut by several hands the preceding day. The trouble of having people going before to mend the road; the pains arising from the many falls and bruises; and the disagreeableness of seeing one's self entirely covered with dirt, and wet to the skin, might be the more cheerfully supported, were they not augmented by the sight of such frightful precipices, and deep abysses, as must fill the traveller's mind with terror. For, without the least exaggeration, it may be said, that in travelling this road, the most resolute tremble.

The manner of descending from these heights is not less difficult and dangerous. In order to understand this, it is necessary to observe, that in those parts of the mountains, the excessive steepness will not admit of the camelones being lasting; for the waters, by continually softening the earth, wash them away. On one side are steep eminences, and on the other frightful

ful abysses; and as they generally follow the direction of the mountain, the road, instead of lying in a level, forms two or three steep eminences and declivities, in the distance of two or three hundred yards; and these are the parts where no camelones can be lasting. The mules themselves are sensible of the caution requisite in these descents; for, coming to the top of an eminence, they stop, and having placed their fore feet close together, as in a posture of stopping themselves, they also put their hinder feet together, but a little forwards, as if going to lie down. In this attitude, having as it were taken a survey of the road, they slide down with the swiftness of a meteor. All the rider has to do is to keep himself fast in the saddle without checking his beast; for the least motion is sufficient to disorder the equilibrium of the mule, in which case they both unavoidably perish. The address of these creatures is here truly wonderful; for, in this rapid motion, when they seem to have lost all government of themselves, they follow exactly the different windings of the road, as if they had before accurately reconnoitred, and previously settled in their minds, the route they were to follow, and taken every precaution for their safety, amidst so many irregularities. There would indeed otherwise be no possibility of travelling over such places, where the safety of the rider depends on the experience and address of his beast.

But the longest practice of travelling these roads cannot entirely free them from a kind of dread or horror which appears when they arrive at the top of a steep declivity. For they stop without being checked by the rider; and if he inadvertently endeavours to spur them on, they continue immoveable; nor will they stir from the place till they have put themselves in the above-mentioned posture. Now it is that they seem to be actuated by reason; for they not only attentively view the road, but tremble and snort at the danger,

danger, which, if the rider be not accustomed to these emotions, cannot fail of filling him with terrible ideas. The Indians go before, and place themselves along the sides of the moutain, holding by the roots of trees, to animate the beasts with shouts, till they at once start down the declivity.

There are indeed some places where these declivities are not on the sides of precipices ; but the road is so narrow and hollow, and the sides nearly perpendicular, that the danger is almost equal to the former; for the track being extremely narrow, and the road scarce wide enough to admit the mule with its rider, if the former falls, the latter must be necessarily crushed; and for want of room to disengage himself, generally has a leg or an arm broken, if he escapes with life. It is really wonderful to consider these mules, after having overcome the first emotions of their fear, and are going to slide down the declivity, with what exactness they stretch out their fore-legs, that by preserving the equilibrium they may not fall on one side; yet at a proper distance make, with their body, that gentle inclination necessary to follow the several windings of the road ; and, lastly, their address in stopping themselves at the end of their impetuous career. Certainly the human species themselves could not show more prudence and conduct. Some mules, after being long used to these journeys, acquire a kind of reputation for their skill and safety, and accordingly are highly valued.

The worst seasons for these journeys, though difficult and dangerous at all times, are the beginnings of summer and winter; the rain then causing such dreadful torrents, that in some places the roads are covered with water ; and in others so damaged, that there is no possibility of passing, but by sending Indians before to mend them ; though after all their labour, which must be done in haste, and when those

people

people think them both safe and easy, they are such as an European stranger would willingly avoid.

BESIDES, the natural difficulty of all the roads among the mountains is increased by the neglect of them, which is greater than could easily be conceived. If a tree, for instance, happens to fall down across the road, and stop up the passage, no person will be at the pains to remove it ; and though all passing that way are put to no small difficulty by such an obstacle, it is suffered to continue ; neither the government, nor those who frequent the road, taking any care to have it drawn away. Some of these trees are indeed so large, that their diameter is not less than a yard and half, and consequently fill up the whole passage ; in which case, the Indians hew away part of the trunk, and assist the mules to leap over what remains ; but, in order to this, they must be unloaded ; and, after prodigious labour, they at last surmount the difficulty; though not without great loss of time, and damage to the goods : when, pleased with having got over the obstacle themselves, they leave the tree in the condition they found it ; so that those who follow are obliged to undergo the same fatigue and trouble. Thus the road, to the great detriment of trade, remains encumbered till time has destroyed the tree. Nor is it only the roads over San Antonio, and other mountains between Guayaquil and the Cordillera, that are thus neglected ; the case is general all over this country, especially where they lead over mountains, and through the forests.

ON the 18th, at six in the morning, the thermometer at Cruz de Canos was at 1010, and after travelling along a road no better than the day before, we arrived at a place at the end of the acclivity of the mountain, by the Indians called Pucara, which signifies a gate or narrow pass of a mountain ; it also signifies a fortified place, and possibly derived its name from its narrowness and the natural strength of its situation.
We

We now began to descend with more ease towards the province of Chimbo, though the road was not much better than the former. Here we were met by the corregidor of Guaranda or Chimbo, attended by the provincial alcalde, and the most eminent persons of the town. After complimenting us in the most cordial manner on our arrival, we proceeded together, and within a league of the town were met by the priest, a Dominican, accompanied by several of his order, and a great number of the inhabitants, who also left the town on the same friendly occasion; and, to heighten the ceremony, had brought with them a troop of cholos, or Indian boys.

Tʜᴇsᴇ cholos were dressed in blue, girded round their waists with sashes, on their heads a kind of turban, and in their hands they carried flags. This little corps was divided into two or three companies, and went before us dancing, and singing some words in their language, which, as we were told, expressed the pleasure they received from the sight of such persons arrived safe in their country. In this manner our cavalcade entered the town, on which all the bells in the place were rung, and every house resounded with the noise of trumpets, tabors, and pipes.

Oɴ expressing to the corregidor our surprise at this reception, as a compliment far above our rank, he informed us, that it was not at all singular, it being no more than what was commonly practised when persons of any appearance enter the town; and that there was no small emulation between the several towns, in paying these congratulations.

Aꜰᴛᴇʀ we had passed the mountains beyond Pacara, the whole country, within the reach of the eye, during a passage of two leagues, was a level and open plain, without trees or mountains, covered with fields of wheat, barley, maize, and other grain, whose verdure, different from that of the mountain, naturally gave us great pleasure, our sight for near a twelve-month

month having been conversant only with the products of hot and moist countries, very foreign to these, which nearly resemble those of Europe, and excited in our minds the pleasing idea of our native soil.

THE corregidor entertained us in his house at Guaranda till the 21st of the same month, when we continued our journey to Quito. The thermometer was for three days successively at 1004$\frac{1}{4}$.

ON the 22d, we began to cross the desert of Chimborazo, leaving the mountain of that name on the left, and travelling over different eminences and heights, most of which were of sand, the snow for a great distance forming, as it were, the sides of the mountain. At half an hour after five in the evening we arrived at a place called Rumi Machai, that is, a stony cave, an appellation derived from a vast cavity in a rock, and which is the only lodging travellers find here.

THIS day's journey was not without its trouble; for though we had nothing to fear from precipices, or dangerous passes, like those in the road to Guaranda, yet we suffered not a little from the cold of that desert, then increased by the violence of the wind. Soon after we had passed the large sandy plain, and being thus got over the severest part of the desert, we came to the ruins of an antient palace of Yncas, situated in a valley between two mountains; but these ruins are little more than the foundations of the walls.

ON the 23d, at three quarters after five in the morning, the thermometer was at 1000, or the freezing point, and accordingly we found the whole country covered with a hoar frost; and the hut in which we lay had ice on it. At nine in the morning we set out, still keeping along the side of Chimborazo. At two in the afternoon we arrived at Mocha, a small, mean place; but where we were obliged to pass the night.

ON the 24th, at six in the morning, the thermometer was at 1006; and at nine we set out for Hambato,

I   which

which we reached at one in the afternoon, after passing several torrents, breaches, or chasms of the mountain Carguairaso, another mountain covered with snow, a little north of Chimborazo. Among these chasms is one without water, the earth remaining dry to the depth of twelve feet. This chasm was caused by a violent earthquake, which shall be spoken of in its place.

On the 25th, the thermometer at Hambato, at half an hour after five in the morning, stood at 1010, and on the 26th, at six in the morning, at 1009¾. This day, having passed the river of Hombato, and afterwards that of St. Miguel, by help of a wooden bridge, we arrived at Latacunga.

On the 27th, at six in the morning, the thermometer was at 1007, when leaving Latacunga we reached in the evening the town of Mula-Halo, having in the way forded a river called Alaques.

On the 28th, the liquor of the thermometer was at the same height as at Latacunga and we proceeded on our journey, arriving in the evening at the mansion-house or villa called Chi Shinche. The first part of this day's journey was over a large plain, at the end of which we had the pleasure of passing by a structure that belonged to the Pagan Indians, being a palace of the Yncas. It is called Callo, and gave name to the plain. We afterwards came to an acclivity, at the top of which we entered on the plain of Tiopullo, not less in extent than the first; and at the bottom, towards the north, is the house where we were entertained that night.

On the 29th, the thermometer at six in the morning was at 1003¾. We set out the earlier, as this was to be our last journey. A road crossing several breaches and beaten tracks, brought us to a spacious plain called Tura-Bamba, that is, a muddy plain; at the other extremity of which stands the city of Quito, where we arrived at five in the evening. The president

sident of the province was Don Dionesio de Alzedo y Herrera, who, besides providing apartments for us in the palace of the Audencia, entertained us the first three days with great splendour, during which we were visited by the bishop, the auditors, the canons, the regidores, and all other persons of any distinction, who seemed to vie with each other in their civilities towards us.

In order to form an adequate idea of this country, it will not be amiss, after being so particular in describing the disagreeable parts, and the many dangers to which travellers are exposed, to add a description of the most remarkable productions of nature. The lands between the custom-house of Babahoyo, or Caracol, and Guaranda, are of two kinds: the first, which extends to Tarigagua, is entirely level; and the second, which begins at that part, wholly mountainous. But both, and even two leagues beyond Pucara, are full of thick forests of various kinds of large trees, differing in the foliage, the disposition of their branches, and the size of their trunks. The mountains, which form this chain of the Andes, are, on the west side, covered with woods; but on the east entirely bare. Among these mountains is the source of that river which, b ing increased on all sides by brooks, makes so grand an appearance between Caracol and Guayaquil, and proves so advantageous to the commerce of the country.

In the level part of this woody extent are a great number of animals and birds, of the same kind with those described in our account of Carthagena, except that to the last may be added wild peacocks, bustards, pheasants, and a few others, which are here in such abundance, that, did they not always rest on the tops of the trees, where, either from their enormous height, or being covered with leaves, they are secure, a traveller, with a good fowling-piece and ammunition, might at any time procure himself an elegant repast. But these

7

these forests are also terribly infested with snakes and monkeys, particularly a kind called marimondas, which are so very large, that, when standing on their hind legs, they are little less than six feet high. They are black, and, in every respect, very ugly; but easily tamed. None of the forests are without them; but they seem most common in those of Guayaquil.

Among the vegetable productions, I shall select three, which to me seemed worthy of a particular description; namely, the cana, vijahua, and the bejuco; as they are not only the materials of which the houses in the jurisdiction of Guayaquil are built, but also applied to various other uses.

The canas, or canes, are remarkable both for their length and thickness, and the water contained in their tubes. Their usual length is between six and eight toises; and though there is a difference in their size, the largest do not exceed six inches diameter. The wood or side of the tube is about six lines in diameter; so that, when the cana is opened, it forms a board near a foot and a half in breadth; and hence it will not appear strange, that houses should be built of such materials. From the time of their first appearance, till they attain their full perfection, when they are either cut down, or of themselves begin to dry, most of their tubes contain a quantity of water; but with this remarkable difference, that at full moon they are entirely, or very nearly, full; and with the decrease of the moon the water ebbs, till at the conjunction little or none is to be found. I have myself cut them at all seasons, so that I here advance nothing but what I know to be true from frequent experience. I have also observed that the water during its decrease appears turbid, but about the time of the full moon it is as clear as crystal. The Indians add another particular, that the water is not found in all the joints, one having water, and another not, alternately. All I can say to this singularity is, that on opening a joint which happens to be empty, the two contiguous ones have water; and this is commonly the case in almost all the canes. This water is said to be an excellent preservative against the ill consequence of any bruises; at least it is drunk as such by all who come from the mountains, where such accidents are unavoidable.

The canes being cut, they are left to dry, or, as they say here, to be cured; whence they acquire such a degree of strength, that they serve either for rafters, beams, flooring, or even masts for balzas. Ships which load with cacao are also cieled with them, to preserve the timbers from the great heat of that fruit. They are also used as poles for litters, and in an infinite number of other particulars.

The vijahua is a leaf generally five feet in length, and two and a half in breadth. They grow wild, and without any stem. The principal rib in the middle, is between four and five lines in breadth, but all the other parts of the leaf are perfectly soft and smooth: the under side is green, and the upper white, covered with a very fine white and viscid down. Besides the common use of it in covering houses, it also serves for packing up salt, fish, and other goods sent to the mountains; as it secures them from the rain. They are also, in these desert places, of singular use for running up huts on any exigency.

The bejucos are a kind of ligneous cordage, and of two kinds; one growing from the earth, and twining round trees; the other strike their roots into certain trees: and from thence derive their nourishment. Both kinds, after growing to a great height, incline again to the earth, on which they creep till they meet with another tree, to the top of which they climb as before, and then again renew their inclination towards the earth; and thus form a labyrinth of ligatures. Some are even seen extended from the top of one tree to another, like a cord. They are so remarkably
flexile,

flexile, that no bending or twisting can break them. But if not cut at the proper time, they grow of an unwieldy bigness. The slenderest of them are about four or five lines in diameter, but the most common size is between six and eight; though there are others much thicker, but of little or no use, on account of the hardness contracted in their long growth. The chief use of them is for lashing, tying, or fastening different things together; and, by twisting several of them in the nature of ropes, they make cables and hawsers for the balzas and small vessels; and are found by experience to last a long time in the water.

In these forests also grows a tree, called very properly Matapalo; i. e. kill timber. It is of itself a weak tree; but, growing near another of considerable bulk, and coming into contact with it, shoots above it, when, expanding its branches, it deprives its neighbour of the rays of the sun. Nor is this all; for, as this imbibes the juices of the earth, the other withers and dies. After which, it becomes lord of the soil, and increases to such a bulk, that very large canoes are made of it; for which its wood is, of all others, the best adapted, being very light and fibrous.

CHAP. II.

Difficulties attending our making the necessary Observations for measuring the Length of an Arch of the Meridian, and the Manner of our Living during the Operations.

ALL the progress made during one whole year, which we spent in coming to Quito, was the surmounting the difficulties of the passage, and at length reaching that country where we were to enter

on

on the principal part of our commission. Nor will
even this appear a small matter, if the great distance
and diversity of climates be considered. A few of
the first days after our arrival were spent in making
proper returns for the civilities we had received from
all persons of rank; after which, we began to deli-
berate on the best methods of performing our work;
and the rather, as M. Bouguer and de la Condamine
were now arrived. The former reached Quito on
the 10th of June, by the same road of Guaranda;
and the latter on the 4th of the same month, having
taken his route by the river of Emeralds, in the government of Atacames.

OUR first operation was, to measure a piece of
ground, which was to be the base of the whole work;
and this we finished during the remainder of the current year. But it proved a very difficult and fatiguing
operation, from the heat of the sun, and the winds and
rains, which continually incommoded us. The plain
made choice of for this base is situated 249 toises
lower than Quito, and four leagues to the N. E. of
that city. It is called the plain of Yaruqui, from a
village of that name near it. This plain was particularly chosen, as the best adapted to our operations;
for though there are several others in this district,
yet all of them lay at too great a distance from the
direction of our base. The quality, disposition, and
lower situation, all contribute to render it less cold
than Quito. Eastward it is defended by the lofty
Cordillera of Guamani and Pambamarca, and westward by that of Pichincha. The soil is entirely
sand; so that, besides the heat naturally resulting
from the direct rays of the sun, it is increased by the
rays being reverberated by the two Cordilleras:
hence it is also exposed to violent tempests of thunder, lightning, and rain; but, being quite open towards the north and south, such dreadful whirlwinds
form here, that the whole interval is filled with columns

lumns of sand, carried up by the rapidity and gyra-
tions of violent eddy winds, which sometimes produce fatal consequences: one melancholy instance
happened while we were there; an Indian, being
caught in the centre of one of these blasts, died on
the spot. It is not, indeed, at all strange, that the
quantity of sand in one of these columns should totally stop all respiration in any living creature, who
has the misfortune of being involved in it.

OUR daily labour was, to measure the length of
this plain in a horizontal direction, and, at the same
time, by means of a level, to correct the inequalities
of the ground; beginning early in the morning, and
continuing to pursue our task closely till evening,
unless interrupted by extreme bad weather; when we
retired to a tent always pitched for that purpose, as
well as for a retreat at noon, when the heat of the sun
became too great for us, after the fatigue of the
morning.

WE at first intended to have formed our base in
the plain of Cayambe, situated twelve leagues to the
north of Quito. Accordingly, the company first repaired to this plain, to view it more attentively. In
this place we lost M. Couplet, on the 17th of September 1736, after only two days illness. He was indeed
slightly indisposed when we set out from Quito ; but,
being of a strong constitution, his zeal for the service
would not permit him to be absent at our first essay.
On his arrival, however, his distemper rose to such
a height, that he had only two days to prepare for his
passage into eternity ; but we had the satisfaction to
see he performed his part with exemplary devotion.
This almost subitaneous death of a person in the flower
of his age, was the more alarming, as none of us could
discover the nature of his disease.

THE mensuration of the base was succeeded by observing the angles, both horizontal and vertical, of the
first triangles we intended to form ; but many of them

P 3   were

were not pursued, the form and disposition of the series being. afterwards altered to very great advantage. In order to this, M. Verguin, with some others, was sent to draw a geographical map of the parts south of Quito; whilst M. Bouguer did the same with regard to the northern parts; a task we found absolutely necessary, in order to determine the points where the signals should be placed, so as to form the most regular triangles, and whose sides should not be intercepted by higher mountains.

DURING these operations, M. de la Condamine went to Lima, in order to procure money on recommendatory letters of credit, which he had brought from France, for defraying the expences of the company, till remittances arrived; and Don George Juan followed him, in order to confer with the viceroy of Peru, for amicably determining some differences which had arisen with the new president.

THESE two gentlemen, having happily terminated their respective affairs, returned to Quito about the middle of June, when both M. Bouguer and those who surveyed the southern. parts had finished their plans. It was now determined to continue the series of triangles to the south of Quito; and the company accordingly divided themselves into two bodies, consisting of French and Spaniards, and each retired to the part assigned him; Don George Juan and M. Godin, who were at the head of one party, went to the mountain of Pambamarca; while M. Bouguer, de la Condamine, and myself, together with our assistants, climbed up to the highest summit of Pichincha. Both parties suffered not a little, both from the severity of the cold, and the impetuosity of the winds, which on these heights blew with incessant violence; and these difficulties were the more painful to us, as we had been little used to such sensations. Thus in the torrid zone, nearly under the equinoctial, where it is natural to suppose we had most to fear from the heat,

heat, our greatest pain was caused by the excessiveness of the cold, the intenseness of which may be conjectured from the following experiments made by the thermometer, carefully sheltered from the wind, on the top of Pichincha; the freezing point being at 1000.

On the 15th of August, 1737, at twelve at noon, the liquor was at the height of 1003. At four in the evening, at 1001¼. At six in the evening, at 998¼.

On the 16th of August, at six in the morning, at 997. At ten in the forenoon, at 1005. At twelve at noon, at 1008. At five in the evening, at 1001¼. At six in the evening, at 999½.

On the 17th, at three quarters after five in the morning, at 996. At nine in the morning, at 1001. At ¼ after twelve, at 1010. At ¼ after two in the afternoon, at 1012¼. At six in the evening, at 999. And at ten in the evening, at 998.

Our first scheme for shelter and lodging, in these uncomfortable regions, was, to pitch a field-tent for each company; but on Pichincha this could not be done, from the narrowness of the summit; and we were obliged to be contented with a hut, so small, that we could hardly all creep into it. Nor will this appear strange, if the reader considers the bad disposition and smallness of the place, it being one of the loftiest crags of a rocky mountain, one hundred toises above the highest part of the desert of Pichincha. Such was the situation of our mansion, which, like all the other adjacent parts, soon became covered with ice and snow. The ascent up this stupendous rock, from the base, or the place where the mules could come, to our habitation, was so craggy, as only to be climbed on foot, and to perform it, cost us four hours continual labour and pain, from the violent efforts of the body, and the subtility of the air; the latter being such, as to render respiration difficult. It was my misfortune, when I climbed something above half way, to be so

over

overcome, that I fell down, and remained a long time without sense or motion; and, as I was told, with all the appearances of death in my face. Nor was I able to proceed after coming to myself, but was obliged to return to the foot of the rock, where our servants and instruments remained. The next day I renewed the attempt of climbing the rock; though probably I should have had no better success than before, had not some Indians· assisted me in the most steep and difficult places.

The strange manner of living which we were reduced to, may not, perhaps, prove unentertaining to the reader; and therefore I shall, as a specimen of it, give a succinct account of what we suffered on Pichincha. For this desert, both with regard to the operations we performed there, and its inconveniences, differing very little from others, an idea may be very easily formed of the fatigues, hardships, and dangers, to which we were continually exposed. The principal difference between the several deserts, consisted in their greater or lesser distance from places where we could procure provisions; and in the inclemency of the weather, which was proportionate to the height of the mountains, and the season of the year when we visited them.

We generally kept within our hut. Indeed, we were obliged to do this, both on account of the intenseness of the cold, the violence of the wind, and our being continually involved in so thick a fog, that an object at six or eight paces was hardly discernible. When the fog cleared up, the clouds, by their gravity, moved nearer to the surface of the earth, and on all sides surrounded the mountain to a vast distance, representing the sea, with our rock like. an island in the centre of it. When this happened, we heard the horrid noises of the tempests, which then discharged themselves on Quito and the neighbouring country. We saw the lightnings issue from the
clouds,

clouds, and heard the thunders roll far beneath us; and whilst the lower parts were involved in tempests of thunder and rain, we enjoyed a delightful serenity; the wind was abated, the sky clear, and the enlivening rays of the sun moderated the severity of the cold. But our circumstances were very different when the clouds rose; their thickness rendered respiration difficult; the snow and hail fell continually, and the wind returned with all its violence; so that it was impossible entirely to overcome the fears of being, together with our hut, blown down the precipice on whose edge it was built, or of being buried under it by the daily accumulations of ice and snow.

The wind was often so violent in these regions, that its velocity dazzled the sight; whilst our fears were increased by the dreadful concussions of the precipice by the fall of enormous fragments of rocks. These crashes were the more alarming, as no other noises are heard in these deserts. And, during the night, our rest, which we so greatly wanted, was frequently disturbed by such sudden sounds. When the weather was any thing fair with us, and the clouds gathered about some of the other mountains which had a connection with our observations, so that we could not make all the use we desired of this interval of good weather, we left our hut, to exercise ourselves, in order to keep us warm. Sometimes we descended to some small distance, and at others amused ourselves with rolling large fragments of rocks down the precipice; and these many times required the joint strength of us all, though we often saw the same performed by the mere force of the wind. But we always took care in our excursions not to go so far, but that on the least appearance of the clouds gathering about our cottage, which often happened very suddenly, we could regain our shelter. The door of our hut was fastened with thongs of leather, and on the inside not the smallest crevice was left unstopped; besides which,

it

it was very compactly covered with straw. But, notwithstanding all our care, the wind penetrated through.
The days were often little better than the nights; and
all the light we enjoyed was that of a lamp or two,
which we kept burning, that we might distinguish
one another, and improve our time as much as possible in reading. Though our hut was small, and
crowded with inhabitants, besides the heat of the
lamps, yet the intenseness of the cold was such, that
every one of us was obliged to have a chafingdish of
coals. These precautions would have rendered the
rigour of the climate supportable, had not the imminent danger of perishing by being blown down the
precipice roused us, every time it snowed, to en-
counter the severity of the outward air, and sally out
with shovels, to free the roof of our hut from the
masses of snow which were gathering on it. Nor
would it, without this precaution, have been able to
support the weight. We were not, indeed, without
servants and Indians; but they were so benumbed
with the cold, that it was with great difficulty we
could get them out of a small tent, where they kept a
continual fire. So that all we could obtain from them
was, to take their turns in this labour; and even then
they went very unwillingly about it, and consequently
performed it slowly.

It may be easily conceived what we suffered from
the asperities of such a climate. Our feet were swelled,
and so tender, that we could not even bear the heat,
and walking was attended with extreme pain. Our
hands were covered with chilblains; our lips swelled
and chopped; so that every motion, in speaking or
the like, drew blood; consequently we were obliged
to a strict taciturnity, and but little disposed to laugh,
an extension of the lips producing fissures, very painful for two or three days together.

Our common food in this inhospitable region was
a little rice boiled with some flesh or fowl, which we

procured

procured from Quito; and, instead of fluid water, our pot was filled with ice: we had the same resource with regard to what we drank: and, while we were eating, every one was obliged to keep his plate over a chafingdish of coals, to prevent his provisions from freezing. The same was done with regard to the water. At first we imagined, that drinking strong liquors would diffuse a heat through the body, and consequently render it less sensible of the painful sharpness of the cold; but, to our surprise, we felt no manner of strength in them, nor were they any greater preservative against the cold than common water. For this reason, together with the apprehension that they might prove detrimental to our health; besides the danger of contracting an ill habit, we discontinued their use, having recourse to them but very seldom, and then sparingly. We frequently gave a small quantity to our Indians, together with part of the provisions which were continually sent us from Quito; besides a daily salary of four times as much as they usually earn.

But, notwithstanding all these encouragements, we found it impossible to keep the Indians together. On their first feeling the rigours of the climate, their thoughts were immediately turned on deserting us. The first instance we had of this kind was so unexpected, that, had not one of a better disposition than the rest staid with us, and acquainted us of their design, it might have proved of very bad consequence. The affair was this: there being on the top of the rock no room for pitching a tent for them, they used every evening to retire to a cave at the foot of the mountain, where, besides a natural diminution of the cold, they could keep a continual fire; and consequently enjoyed more comfortable quarters than their masters. Before they withdrew at night, they fastened on the outside the door of our hut, which was so low that it was impossible to go in or out without stooping;

and

and as every night the hail and snow which had fallen formed a wall against the door, it was the business of one or two to come up early and remove this obstruction, that, when we pleased, we might open the door. For though our Negro servants were lodged in a little tent, their hands and feet were so covered with chilblains, that they would rather have suffered themselves to have been killed than move. The Indians therefore came constantly up to dispatch this work betwixt nine and ten in the morning; but we had not been there abdve four or five days, when we were not a little alarmed to see ten, eleven, and twelve come, without any news of our labourers; when we were relieved by the honest servant mentioned above, who had withstood the seduction of his countrymen, and informed us of the desertion of the four others. After great difficulty, he opened a way for us to come out, when we all fell to clearing our habitation from the masses of snow. We then sent the Indian to the corregidor of Quito, with advice of our condition, who, with equal dispatch, sent others, threatening to chastise them severely, if they were wanting in their duty.

But the fear of punishment was not sufficient to induce them to support the rigour of our situation; for within two days we missed them. On this second desertion the corregidor, to prevent other inconveniences, sent four Indians under the care of an alcalde, and gave orders for their being relieved every fourth day.

Twenty-three tedious days we spent on this rock, viz. to the 6th of September, and even without any possibility of finishing our observations of the angles; for, when it was fair and clear weather with us, the others, on whose summits were erected the signals which formed the triangles for measuring the degrees of the meridian, were hid in clouds; and when (as we conjectured, for we could never plainly discern them) those

those were clear, Pichincha was involved in clouds. It was therefore necessary to erect our signals in a lower situation, and in a more favourable region. This, however, did not produce any change in our habitation till December, when, having finished the observations which particularly concerned Pichincha, we proceeded to others; but with no abatement either of inconveniencies, cold or fatigue, the places where we made all our observations being necessarily on the highest parts of the deserts; so that the only respite, in which we enjoyed some little ease, was during the short interval of passing from one to the other.

In all our stations subsequent to that on Pichincha, during our fatiguing mensuration of the degrees of the meridian, each company lodged in a field-tent, which, though small, we found less inconvenient than our Pichincha hut, though at the same time we had more trouble, being oftener obliged to clear it from the snow; as the weight of it would otherwise have demolished the tent. At first, indeed, we pitched it in the most sheltered places; but, on taking a resolution that the tents themselves should serve for signals, to prevent the inconvenience of those of wood, we removed them to a more exposed situation, where the impetuosity of the winds sometimes tore up the piquets, and blew them down. Then we were not a little pleased with our having brought supernumerary tents, and with our dexterity in pitching another instead of that which the wind had torn away. Indeed, without this precaution, we should have been in the utmost danger of perishing. In the desert of Asuay we particularly experienced the benefit of this expedient; three tents belonging to our company being obliged to be pitched one after another, till at last they all became unfit for use, and two stout poles were broken. In this terrible condition our only resource was to quit the post, which was next to the signal of Sinasaguan, and shelter ourselves in a breach or chasm. The two companies

panies were both at that time on this desert, so that the sufferings of both were equal. The Indians who attended us, not willing to bear the severity of the cold, and disgusted with the frequent labour of clearing the tent from the snow, at the first ravages of the wind, deserted us. Thus we were obliged to perform every thing ourselves, till others were sent us from a seat about three leagues distant at the bottom of the mountain.

While we were thus labouring under a variety of difficulties from the wind, snow, frost, and the cold, which we here found more severe than in any other part; forsaken by our Indians, little or no provisions, a scarcity of fuel, and, in a manner, destitute of shelter, the good priest of Cannar, a town situated at the foot of these Cordilleras, south west from the signal of Sinasaguan, about five leagues from it, and the road very difficult, was offering up his prayers for us; for he, and all the Spaniards of the town, from the blackness of the clouds, gave us over for lost; so that, after finishing our observations, when we passed through the town, they viewed us with astonishment, and received us with the most cordial signs of delight, adding their congratulations, as if we had, amidst the most threatening dangers, obtained a glorious victory. And, doubtless, our operations must appear to them a very extraordinary performance, if we consider the inexpressible horror with which they view those places where we had passed so many days.

It was at first determined to erect signals of wood in the form of a pyramid; but to render our stay in the piercing colds of these regions as short as possible, we abandoned that intention, of which there would have been no end; because, after remaining several days in the densest parts of the clouds, when a clear interval happened, the signals could not be distinguished: some the winds had blown down, and others had been carried away by the Indians who tended
their

their cattle on the sides of the mountains, for the sake of the timber and ropes. To remedy which, the only expedient was to make the very tents in which we lodged, serve for signals ; for the orders of the magistrates, and threatenings of the priests, were of little consequence in such a desert country, where it was almost impossible to discover the delinquents.

The deserts of the mountains of Pambamarca and Pichincha were the noviciates, in which we were inured to the severe life we led from the beginning of August 1737, to the end of July 1739. During which time, our company occupied thirty-five deserts; and that of Don George Juan, thirty-two; the particulars of which shall be enumerated, together with the names of all those on which we erected signals for forming the triangles ; in all which, the inconveniencies were the same, except that they became less sensible, in proportion as our bodies became inured to fatigue, and naturalized to the inclemencies of those regions; so that in time we were reconciled to a continual solitude, coarse provisions, and often a scarcity of these. The diversity of temperatures did not in the least affect us, when we descended from the intense cold of one of those deserts into the plains and valleys, where the heat, though but moderate, seemed excessive to those coming from such frozen regions. Lastly, without any concern, we encountered the dangers unavoidable among those steep precipices, and a great variety of others to which we were continually exposed. The little cabins of the Indians, and the stalls for cattle scattered up and down on the skirts of the mountains, and where we used to lodge in our passage from one desert to another, were to us spacious palaces; mean villages appeared like splendid cities, and the conversation of a priest; and two or three of his companions, charmed us like the banquet of Xenophon : the little markets held in those towns, when we happened to pass through them on
a Sunday,

a Sunday, seemed to us as if filled with all the variety of Seville fair.   Thus the least object became magnified, when we descended for two or three days from our exile, which, in some places, lasted fifty days successively; and it must be owned, that there were particular occasions when our sufferings were such, that nothing could have supported us under them, and animated us to persevere, but that honour and fidelity which jointly conspired to induce both companies, whatever should be the consequence, not to leave imperfect a work so long desired by all civilized nations, and so particularly countenanced by the two powerful monarchs our sovereigns.

It may not be amiss here to inform the reader of the different opinions conceived by the neighbouring inhabitants, with regard to our enterprise.   Some admired our resolution, others could not tell what construction to put upon our perseverance; and even those of the best parts and education among them were utterly at a loss what to think.   They made it their business to examine the Indians concerning the life we led, but the answers they received only tended to increase their doubts and astonishment.   They saw that those people, though naturally hardy, robust, and inured to fatigues, could not be prevailed upon, notwithstanding the encouragement of double pay, to continue any time with us.   The serenity in which we lived on those dreaded places was not unknown to them; and they saw with what tranquillity and constancy we passed from one scene of solitude and labour to another.   This to them appeared so strange, that they were at a loss what to attribute it to.   Some considered us as little better than lunatics; others more sagaciously imputed the whole to covetousness, and that we were certainly endeavouring to discover some rich minerals by particular methods of our own invention; others again suspected that we dealt in magic; but all were involved in a
labyrinth

labyrinth of confusion with regard to the nature of our design. And the more they reflected on it, the greater was their perplexity, being unable to discover any thing proportionate to the pains and hardships we underwent. And even when we informed them of the real motive of this expedition, which caused so much astonishment, their ignorance of its importance would not suffer them to give credit to what we said ; suspecting that we concealed, under the veil of an incomprehensible chimera, our real practices, of which, as I have already observed, they had no good opinion.

Among several pleasant adventures which this occasioned, I shall only mention two, both which are still fresh in my memory ; and may serve to illustrate the strange ideas these ignorant people formed of us. While we were at the signal of Vengotasin, erected on a desert at no great distance from the town of Latacunga, about a league from the place where we had pitched our field-tent was a cow-house, where we constantly passed the night.; for the ascent not being remarkably difficult, we could every morning, in fair weather, return soon enough to the tent to begin our observations. One morning, as we were passing to the signal, we saw at a distance three or four Indians, in appearance on their knees ; and we found indeed, on our approaching nearer, that this was their real posture ; we also observed that their hands were joined, and that they uttered words in their language with the greatest fervour and the most supplicant accent ; but, by the position of their eyes, it was evident that we were the persons whom they thus addressed. We several times made signs for them to rise, but they still kept their posture till we were got at a considerable distance. We had scarce begun to prepare our instruments within the tent, when we were alarmed with a repetition of the same supplicant vociferations. On going out to know the cause, we found the same Indians again on their knees before the tent; nor were we able, by all the signs we could make, to raise them from that posture. There fortunately happened at that time to be with us a servant who understood both the Indian and Spanish languages; and having directed him to ask these poor people what they wanted of us, we were informed, that the eldest of them was the father of the others, and that his ass being either strayed or stolen, he came to us, as persons who knew every thing, to intreat us to commiserate his great loss, and put him in a method of recovering his beast. This simplicity of the Indians afforded us no small entertainment; and though we did all we could, by means of our interpreter, to undeceive them, we found they were equally tenacious of this strange error as of genuflexion; and would still believe, that nothing was hid from us; till, having wearied themselves with these clamorous vociferations, and finding we took no notice of them, they retired, with all the marks of extreme sorrow that we would not condescend to inform them where they might find the ass; and with a firm persuasion that our refusal proceeded from ill-nature, and not from ignorance.

The other adventure I shall mention, happened to myself in particular, and not with simple and ignorant Indian peasants, but with one of the principal inhabitants of Cuença. While the whole company were on the mountain of Bueran, not far from the town of Cannar, I received a message from the priest of that place, informing me, that two Jesuits of my acquaintance were passing that way, and, if I was desirous of seeing them, I might find them at his house. As I was cheerfully descending the mountain to enjoy this pleasing invitation, I happened to be overtaken by a gentleman of Cuenca, who was going to take a view of his lands in that jurisdiction, and had observed me coming from our tent. He was, it seems, acquainted
ed

ed with my name, though he had never seen me;
but observing me dressed in the garb of the Mestizos,
and the lowest class of people, the only habit in
which we could perform our operations, he took me
for one of the servants, and began to examine me;
and I was determined not to undeceive him till he
had finished.   Among other things, he told me, that
neither he nor any body else would believe, that the
ascertaining the figure and magnitude of the earth, as
we pretended, could ever induce us to lead such a
dismal and uncouth life; that, however we might
deny it, we had doubtless discovered many rich minerals on those lofty deserts; adding, that persons in
his circumstances were not to be satisfied with fine
words.   Here I laboured to remove the prejudices he
entertained against our operations; but all I could
say only tended to confirm him in his notion; and, at
parting, he added, that, doubtless, by our profound
knowledge in the magic art, we might make much
greater discoveries than those who were ignorant of it.
These opinions were blended with others equally ab-
surd and ridiculous; but I found it impossible to undeceive him, and accordingly left him to enjoy his
own notions.

Our series of triangles in the south part being
finished, and a second base measured by each company to prove the truth of our work, we began our
astronomical observations; but, our instruments not
being perfectly adapted to that intention, we were
obliged, in the month of December of the same year,
to return to Quito, in order to construct another, on
whose accuracy we could safely rely; and this employed us till the first of August of the following
year 1740; when, without any further loss of time,
we again repaired to Cuença, and immediately began
our observations: but these, being very tedious, were
not finished before the end of September; the atmosphere of that country being very unfavourable to

astronomical observations. For, in the deserts, the clouds in which we were so frequently involved hindered us from discerning the other signals; and in the city, over which they spread a kind of perpetual pavilion, they hid the stars from us while they passed the meridian; but patience and resolution, inspired by the importance of our enterprise, having enabled us at last to perform our task on the south side of the equator, we prepared for our journey to the north of it, in order to make the astronomical observations at the other extremity of the arch of the meridian, and thus put the finishing hand to our work: but this was for some time retarded by an accident of importance which called us to Lima, as will be related in the second volume.

In December 1743, the reasons which detained us at Lima, Guayaquil, and in Chili, no longer subsisting, we returned to Quito in January 1744, when Don George Juan and I prolonged the arch of the meridian four triangles, by which it was extended to the place where M. Godin, in 1740, had made the second astronomical observation, and which he now repeated, and finished in the month of May 1744.

Mess. Bouguer and M. de la Condamine having at that time finished the several parts assigned to them, had left Quito, in order to return to France; the former by the way of Carthagena, and the latter by the river of the Amazons; but the rest of the company remained there some time; some for fear of being taken by the enemy, some for want of the means to defray the charges necessary in so long a journey, and others on account of their having contracted some obligations, and were unwilling to leave the country till they could discharge them. So that in the former only the natural desire of returning to their country prevailed, in order there to repose themselves after such a series of labours and hardships, by which the

2   health

health and vigour of all were in some measure impaired.

CHAP. III.

The Names of the Deserts and other Places where the Signals were erected for forming the Series of Triangles for measuring an Arch of the Meridian.

IN order to gratify the curiosity of the reader with regard to our operations, I shall mention, in separate articles, the places where each company made their observations, and the time they were obliged to remain there; omitting a detail of circumstances, many of which would be little more than a paraphrase on the subject of the preceding chapter. Nor shall I here include those stations used in the year 1736, after measuring the base of Yaruqui, both on its extremities and in the deserts of Pambamarca and Yllahalo; for the disposition of the triangles being afterwards altered, they were repeated. Therefore, considering them as not used at that time, I shall begin with those stations in which no such circumstances happened, and range them in the order they were occupied.

Deserts on which the signals were erected for the operations conducted by M. de la Condamine and myself.

I. Signal on the desert of Pichincha.

The signal was at first erected on the highest summit of Pichincha, but afterwards removed to another station at the foot of the pic; the top having been afterwards found not to be the most proper place. We began our observations on this mountain on the

Q 3   14th

14th of August 1737, but could not finish them before the beginning of December following.

II. THE signal on Oyambaro, the south extremity of the base of Yaruqui.

ON the 20th of December 1737, we removed to Oyambaro; and finished our observations necessary to be made there on the 29th of the same month.

III. SIGNAL on Caraburu, the northern extremity of the base of Yaruqui.

ON the 30th of December we passed to Caraburu, and continued there till the 24th of January 1738. This long stay was partly occasioned by the badness of the weather, and partly by the want of signals.

IV. SIGNAL on the desert of Pambamarca.

ON this desert of Pambamarca, where we had be-fore been in 1736, on finishing the measurement at Yaruqui, a second signal was erected here, and we went up the 26th of January, 1738, where we re-mained till the 8th of February; and though we had not here the difficulties of the ice and snow to strug-gle with, as on Pichincha and other subsequent sta-tions, yet we were extremely incommoded by the velocity of the winds, which were so violent that it was difficult to stand; and, notwithstanding the best shelter possible to be procured, we often found it very difficult to keep the instrument steady; which, of consequence, greatly increased the difficulty of making the observations with the necessary accuracy.

V. SIGNAL on the mountain of Tanlagua.

ON the 12th of February we ascended the moun-tain of Tanlagua; and having the next day finished our observations, returned. If this mountain be but
small

small in comparison of others in this Cordillera, and
thus saved us the many inconveniencies of a lofty
station, yet the steepness of its sides put us to no
small difficulty, there being no other possible method
of going up than by climbing; and the greatest care
is requisite in fixing the hands and feet close and firm;
nor is it possible to climb it in less than four hours.
The descent, as may naturally be concluded, is little
less hazardous, as you must sit and slide down much
the greater part of it; and this must be done gently,
lest, by celerity of motion, you tumble down the
precipice.

VI. Signal on the plain of Changalli.

On the 7th of March we removed to the signal of
Changalli, and finished the necessary observations on
the 20th. We spent the time here very comfortably.
The signal was erected on a plain, where neither the
air nor weather molested us; and being lodged in a
farm-house near the signal, and not far from the town
of Pintac, we had all the necessary conveniencies of
life, the want of which we often severely felt in the
deserts. These comforts did not, however, in the least
abate our diligence to avail ourselves of every instant
when the signals on the mountains were not concealed in clouds. But one circumstance which lengthen-
ed our stay was, that some of the signals were wanting, having been blown down by the wind; it was
therefore resolved, that for the future the field-tents
should serve for signals. And, accordingly, we afterwards constantly pursued this method.

VII. Signal on the derert of Pucaguaico, on the side of the mountain Catopaxi.

This mountain we ascended the 21st of March,
and on the 4th of April were obliged to return, after
in vain endeavouring to finish our observations. For,

Q 4   not

not to mention our own sufferings, the frost and snow, together with the winds, which blew so violently that they seemed endeavouring to tear up that dreadful volcano by its roots, rendered the making observations absolutely impracticable. Such is indeed the rigour of this climate, that the very beasts avoid it; nor could our mules be kept at the place where we, at first, ordered the Indians to take care of them; so that they were obliged to wander in search of a milder air, and sometimes to such a distance that we had often no small trouble in finding them.

At Pucaguaico we however saw the necessity of either erecting the signal further to the south, or setting up another in the intermediate space. Several consultations were held, to determine on the best method; but, as other things were necessary to be done before we came to a conclusion, the operations were suspended, and the interval spent in making observations on the velocity of sound, and other physical subjects. Every thing being ready for renewing our operations, we a second time ascended Pucaguaico on the 16th of August, and it was our good fortune by the 22d to have finished all our necessary operations.

VIII. Signal on the desert Corazon.

On the 12th of July, before we had finished our operations at the station of Pucaguaico, we ascended to the desert Corazon, where we staid till the 9th of August. This mountain is nearly of the same height with that of Pichincha; and its loftiest summit, like that of the former, a rock of considerable altitude. At the foot of this rock the signal was erected; and thus our station nearly resembled that of Pichincha. There was indeed this considerable difference, that our sufferings from the winds, frost, and snows, were considerably less.

IX. Sig-

IX. Signal on Papa-urco.

It had been determined that Papa-urco should be the place where the intermediate signal betwixt those of Pucaguaico and Vengotasin should be erected. This mountain, which is of a middling height, we ascended the 11th of August, and continued on it till the 16th, when we returned to Pucaguaico ; so that this easy mountain was a kind of resting-place between the two painful stations of Corazon and Pucaguaico.

X. Signal on the mountain of Milin;

Whose height is nearly the same with that of the Papa-urco. We ascended it on the 23d of August, and by the 29th had finished the necessary observations.

XI. Signal on the mountain Vengotasin.

The mountain of Vengotasin is not remarkably high, but our stay on it was longer than we at first imagined; for, after finishing our observations on the 4th of September, some difficulties which arose with regard to the position of the following signal towards the south detained us till the 18th. However, the town of Latacungo being contiguous to the skirts of this mountain, and having several farms in its neighbourhood, we were at no loss for many conveniencies of which we were destitute in several other stations.

XII. Signal on the mountain of Chalapu.

Our stay on this mountain was shorter than on any other in the whole series of triangles; for we continued only part of four days, going up the 20th and coming down the 23d. It is none of the highest mountains, and has in its neighbourhood the town of Hambato, and its skirts diversified with seats and
farms ;

farms; but the acclivity is so steep, that the safest way is to ascend it on foot.

XIII. Signal of Chichichoco.

The signal of Chichichoco was erected on the side of the mountain of that name, which is a branch of the famous snowy mountain of Carguairaso. Here we stayed only from the 24th to the 29th of September. Though the spot where we placed the signal was of a very inconsiderable height when compared with that of the other mountains, yet, from its proximity to Carguairaso, when the wind blew from that quarter, it was considerably cold, but not comparable to that we felt on the deserts, where every part was covered with ice, hail, or snow. The day we left this place, while our Indians were loading the mules, and we in the tent ready to set out on our journey, an earthquake was felt, which reached four leagues round the country. Our tent rocked from side to side, in conformity to the undulating motion observed in the earth; this shock was only one of the small concussions frequent in those parts.

XIV. Signal of Mulmul.

This signal, and the three following, occasioned several journeys from one to another; as, for the greater accuracy of the observations, auxiliary triangles were to be formed, in order to verify the distances resulting from the principal. The difficulty also of reciprocally distinguishing some signals from others, obliged us to change their position, till they stood in proper places; and consequently laid us under a necessity of going often from one station to another. On the 8th of November, having finished all our observations, the company removed to Riobamba, where I myself had been confined ever since the 20th of October, with a critical disease, which at first attacked

tacked me at Chichichoco, and increasing at Mulmul, I was obliged to remain in a cow-house on that mountain, from whence I was removed to Riobamba; and this accident hindered me from being present at the signals XV. XVI. and XVII. which were those of Guayama, Limal, and Nabuso.

XVIII. Signal of Sisa-pongo.

At the signal of Sisa-pongo we continued from the 9th to the end of November; and here the trigonometrical observations were intermitted till Don George Juan and M. Godin returned from Quito, to which city they repaired in order to take some measures necessary for the continuation of the work. But, that this interval might not be lost, M. Bouguer proposed to make some experiments, in order to demonstrate the system of attraction. The place he made choice of for these experiments was the mountain of Chimbarazo. In this station, and the following of the sandy desert of the same mountain, we suffered more than on any other.

XIX. Signal of Lalanguso.

On the desert of Lalanguso, our observations were continued from the 24th to the 31st of January 1739.

XX. Signal on the desert of Chusay.

The station on the desert of Chusay was one of the most tedious in the whole series of triangles, being unavoidably detained on this disagreeable mountain from the 3d of February to the 24th of March. This delay was occasioned by the difficulty of pitching on proper places for erecting the succeeding signals, that they might stand in full view, be easily distinguished one from another, and form regular triangles. This was indeed a difficult task, the lofty summits of the mountains of the Cordillera of Azuay, where they were to be be placed, intercepting each other from our sight. The tediousness of this station was increased by the rigour of the weather, the strength of the winds, and its great distance from any place where we could procure convenient shelter and refreshments.

XXI. Signal on the desert of Tialoma.

On the desert of Tialoma we continued from the 26th of March to the 25th of April; but had little, except the length of the time, to complain of.

XXII. Signal on the desert of Sinasaguan.

We arrived at the desert of Sinasaguan on the 27th of April, and left it on the 9th of May, the only clear day we had during our stay; but as we have already mentioned our sufferings on this desert, it will be unnecessary to repeat them here.

XXIII. Signal on the desert of Bueran.

We continued on the desert of Bueran from the 10th of May to the 1st of June; but, besides the small height of the mountain, the town of Cannar being only two leagues distant from it, we were in want of nothing. The temperature of the air was also much more mild than on the other deserts; besides, we had the great satisfaction of relieving our solitude by going to hear mass on Sundays, and other days of precept in the town. These comforts had, however, some allay; for while we continued on this desert, the animals, cottages, and Indians, suffered three times in a very melancholy manner by tempests of lightning, which fell on the neighbouring plains; all those countries, especially the desert of Burgay, which borders on that of Bueran, being subject to terrible storms.

XXIV. Sig-

XXIV. Signal on the desert of Yasuay.

Our observations at the signal of Yasuay were not finished till the 16th of July; there being a necessity, before we could conclude them, to pitch on the most convenient place for measuring a second base, in order to prove the accuracy of all the preceding geometrical operations; and, after fixing on a proper spot, to determine where the signals between Yasuay and the base could be most properly placed. In order to this, we went to Cuença, and from thence proceeded to the plains of Talqui and los Bannos. At last it was determined that the base should be measured in the former, by which the result of the triangles was to be verified by my company, and that of the other in the plain of los Bannos. The requisite signals also were erected; and we returned to the desert of Yasuay, where we continued our observations, which employed us from the 7th to the 16th of July. Though this mountain is one of the highest in the whole territory of Cuença, and the ascent so steep that there is no going up but on foot, nor even by that method without great labour, yet the cold is far from being so intolerable as on Sinasaguan and the deserts north of that mountain. So that we cheerfully supported the inconveniencies of this station.

XXV. Signal on the mountain of Borma.

This mountain is but low, as are all the others in the neighbourhood of Cuença, so that here we were not impeded by any cloudy summits. It was also our good fortune that Yasuay, contrary to our apprehensions, was clear and visible the whole 19th of July; so that we finished our observations in two days agreeably.

XXVI.

XXVI. XXVII. XXVIII. XXIX. Signals of Pugin, Pillachiquir, Alparupasca, and Chinan.

The two last being the north and south extremities of the base of Talqui, the four stations of Pugin, Pillachiquir, Alparupasca, and Chinan, did not require our attendance; for being near the base of Talqui, we daily went from the farm-houses where we lodged, and observed the angles, except that of Pillachiquir, to which, on account of its greater distance than that of the other signals, there was a necessity for our visiting; but happily concluding our observations the same day we reached it, there was no reason for our longer stay.

XXX. XXXI. Signals of Guanacauri, and the tower of the great church of Cuença.

The series of triangles, except the two last at the extremities of the second base, being finished, it was necessary to form other triangles, in order to fix the place of the observatory where, when the geometrical observations where finished, the astronomical were to begin. Those which fell to my lot, were a signal on the mountain of Guanacauri, and the tower of the great church of Cuença; and these angles were taken at the same time the astronomical observations were making.

At the north extremity of the arch of the meridian new triangles were afterwards formed, as we have already observed in the foregoing chapter. This rendered it necessary for us to make choice of different places on these mountains for erecting other signals in order to form these triangles. The same order which had been followed during the whole series of mensuration, that each person should take two angles of every triangle, was observed here; and those assigned to me were the following.

XXXII.

XXXII. XXXIII. XXXIV. XXXV. Sɪɢɴᴀʟs on
Guapulo, the mountain of Campanario, and those
of Cosin, and Mira.

Tʜᴇ observations to be made at these four stations
could not be finished till those alarming reasons which
called us to Lima and Chili no longer subsisted, and
we were returned to Quito. The work at the first
and last stations was dispatched without the necessity
of lodging there; for being near Quito and the village of Mira, when the weather promised us a favourable opportunity it was only an easy ride; but we
found it very different with regard to the stations of
Campanario and Cosin. However we left all the four
at the same time, namely, on the 23d of May 1744:
the day when Don George and my elf put the finishing hand to the astronomical observations which we
had re-assumed on the 14th of February of the
same year; and thus the whole process relative to
the mensuration of an arch of the meridian was concluded.

*Signals erected on deserts, &c. where the observations
were conducted by Mr. Godin and Don George
Juan.*

Tʜᴇ stations immediately subsequent to the admea-
surement of the base of Yaruqui, in the year 1736,
and afterwards not made use of, as we have already
observed, were common to both companies; the method which was afterwards followed, for every one to
observe two angles in all the triangles, not having
been thought of; though it both shortened the work,
and, at the same time, rendered it much easier: so
that Don George Juan and Mr. Godin were on the
deserts of Yllahalo and Pambamarca, at the same
time with Mess. Bouguer and Condamine and myself.

I ıı

I. II. SIGNALS on the extremities of the bare of
Yaruqui.

IN order to make the necessary observations relating to these two signals, they left Quito on the
20th of August 1737, and had completely finished
them by the 27th.

III. SIGNAL on the desert of Pambamarca.

AFTER they had concluded all the necessary observations at the extremities of the base, they went without delay to the desert of Pambamarca, and completely finished their operations by the first of September.

IV. SIGNAL on the mountain of Tanlagua.

HAVING finished their observations on the desert,
they came down to the little town of Quenche, in
that neighbourhood, in order to proceed from thence
to Tanlagua; but the Indians, who were to accompany them, being no strangers to the extreme severity of the weather on that desert, discouraged by
their recent sufferings on Pambamarca, and knowing
they should still suffer more on Tanlagua, were not to
be found; and the lowest class of inhabitants in the
town, apprehending that they should be sent on this
painful service, also left their habitations and ab-
sconded; so that the joint endeavours of the alcalde
and priest to discover them proved ineffectual; and
after a delay of two whole days, the curate, with
great difficulty, prevailed on the sacristan, and other
Indians employed in the service of the church, to accompany them, and take care of the loaded mules as
far as the farm-house of Tanlagua, where they arrived the 5th of September. The next day they began
to ascend the mountain, which, being very steep, took
them up a whole day in climbing it. But this being
more

more than the Indians were able to perform, as they carried the field-tent s, baggage, and instruments, they were obliged to stop half way; so that those on the top were under a necessity of passing the night there without any shelter; and a hard frost coming on, they were almost perished with cold; for they were so greatly affected by it, that they had no use of their limbs till they returned to a warmer air. After all these hardships, the gentlemen could not finish their observations, some of the signals being wanting, having either been blown down by the winds, or carried away by the Indian herdsmen: so that, during the interval while persons were employed in erecting others, they returned to Quito, and applied themselves to examine the divisions of the quadrants. These operations, being very tedious, employed them till the month of December, when, all the signals which were wanting being replaced, they again, on the 20th of December, repaired to their post at Tanlagua; and on the 27th finished the observations necessary to be made at that station.

V. SIGNAL on the mountain of Guapulo.

THE signal of Guapulo being erected on a mountain of no great height, and in the neighbourhood of Quito, their residence was not necessary; for, by setting out from the city at day-break, they could reach the field-tent, where the instruments were left, early in the morning. These journeys repeated every day, and though every moment of time was improved to the greatest advantage, it was the 24th of January 1738 before they finished the observations, with that accurate precision so conspicuous in all their operations.

VI. SIGNAL on the Cordiliera and desert of Guamani.

THEY were obliged to make two journeys to the mountain of Guamani, the signal having been first misplaced, so as not to be seen from that erected on Corazon; and consequently there was a necessity for removing it. And though, in order to do this, they ascended the mountain on the 28th of January, they found it necessary to return thithe on the 7th of February, when they were fortunate enough to finish every thing remaining the very next day.

VII. Signal on the desert of Corazon.

This mountain also the gentlemen were obliged to visit twice; the first journey was on the 20th of January, and the second on the 12th of March, 1738.

VIII. Signal of Limpie-pongo, on the desert of Cotopaxi.

They went up to the desert of Cotopaxi on the 16th of March, and remained there till the 31st; when they observed that the signal of Guamani was not visible from thence, and therefore it was necessary to erect another in the intermediate space; which being completed on the 9th of August, they again repaired to the signal of Limpie-pango, on Cotopaxi; where they finished all their operations by the 13th of the same month, and left every thing in exact order. In ascending the mountain in this second journey, the mule on which Don George Juan rode fell down a breach four or five toises deep, but providentially without receiving the least hurt.

As they had been obliged to erect another signal between those of Guamani and Limpie-pongo, in order to continue the series of triangles; so there was also a necessity for returning to some stations, to observe again the angles which had before been determined. These operations, together with the experiments on the velocity of sound, and the observations at the new signal, filled up the interval from

the

the time the operations were suspended on Limpie-pongo, till they returned to finish them.

IX. SIGNAL on the desert of Chinchulagua.

THE operations of the signal of Chinchulagua, erected on the desert of the same name, were completed on the 8th of August; but a doubt arising with regard to one of the angles observed, for the greater certainty, they returned to this station, and again examined the angle in question, after they had finished their observations at Limpie-pongo.

X. SIGNAL on the mountain of Papa-urco.

AFTER verifying the observation on Chinchulagua, they removed to the signal of Papa-urco, and finished their observations in the same month of August. Here they for some time suspended their operations, being called to Quito on affairs of importance, relating to the French academicians.

XI. SIGNAL on the mountain of Milin.

THE affairs which had required M. Godin's presence at Quito being terminated within the month, they returned, on the 1st of September, to make the necessary observations at the signal of Milin, where they continued till the 7th, when they left it, having completely finished their operations.

XII. SIGNAL on the desert of Chulapu.

FROM Milin they proceeded to the desert of Chulapu, where they remained till the 18th of September, when they had finished all their observations. Till this signal, exclusive, each company had observed the three angles of all the triangles; both because they differed from one another, and to prove by this precision the errors in the divisions of the quadrants, be-

fore

fore observed by other methods. But from this signal, inclusive, each company observed two angles only of the other triangles, as had been agreed on.

XIII. SIGNAL in Jivicatsu.

IN Jivicatsu they remained from the 18th to the 26th of September. This station was one of the most agreeable; for, besides the height on which the signal was erected, the temperature of the air, and the cheerful aspect of the country, the town of Pilaro was in the neighbourhood, so that they wanted for nothing.

XIV. XV. SIGNALS on the deserts of Mulmul and Guayama.

THESE two deserts are placed together, because their summits are united by gentle eminences; on one of which is a cow-house, used by the Indians when they go in search of their cattle, which feed on the sides of this mountain. In this cow-house Don George Juan, M. Godin, and their attendants, took up their quarters on the 30th of September, and every morning, when the weather was favourable, repaired to the signal erected on one or other of the eminences. But the distance between the two stations being very small, and the observations made there requiring to be verified by those of other auxiliary triangles, it was absolutely necessary to determine exactly the stations where these triangles were to be formed; and to remain there till the distances were settled, and the observations relating to them concluded; which operations, notwithstanding the greatest diligence was used, employed them till the 20th of October.

EVERY thing at the two preceding stations being finished, they repaired to the village of Riobamba, determining to continue their work without interruption; but meeting with some difficulties concerning the most advantageous position of the subsequent triangles, and
money

money beginning to grow short with our whole society, both Spaniards and French, it was thought necessary to make use of the interval while the proper places for erecting the signals were determining, to procure supplies. Accordingly, M. Godin and Don George Juan again set out from Riobamba for Quito on the 7th of November; but it was the 2d of February following before we had the pleasure of congratulating them on their return, the former having been seized with a fever, which brought him very low, and detained them a considerable time at Quito.

XVI. XVII. Signals on Amula and Sisa-pongo.

The observations necessary at the signal of Amula were finished before the journey to Quito; and from the 2d of February 1739, when they returned to Riobamba, till the 19th, they were employed in those relating to Sisa-pongo.

XVIII. Signal on the mountain of Sesgum.

On the mountain of Sesgum they had occasion to stay only from the 20th to the 23d of February. For this signal stood on the declivity of a mountain, and they vigilantly employed every moment when the other deserts were free from those clouds in which they are usually involved.

XIX. Signal on the desert of Senegualap.

The observations at the signal of Senegualap detained them from the 23d of February to the 13th of March. The length of the time was indeed the most disagreeable part, as otherwise they did not place this among the worst stations they had met with during their course of observations.

XX. Sig-

R 3

XX. Signal on the desert of Chusay.

From Senegualap they proceeded to the desert of Chusay, a station which gave these gentlemen no less trouble than it had done us. Our company had no concern with the station on this desert; for according to the alternative established between the two companies, that of Senegualap was the place to which we were to repair. But after finishing the observations at Lalanguso, being uneasy at the long stay of M. Godin and Don George Juan at Quito, to divert our thoughts by some laborious employment, we divided our company into two, in order to prosecute the mensuration, till those gentlemen returned. Accordingly, M. Bouguer, at the head of one detatchment, went to the signal of Senegualap, and M. de la Condamine and myself repaired to that of Chusay. But M. Godin and Don George Juan joining us there, we returned to our proper company, and the operations were continued in the order agreed on.

XXI. Signal on the desert of Sinasaguan.

This desert was one of those common to both companies; and that of Don George Juan remained on it till the 29th of May, when the observations of both were finished. Thus every member of the two companies equally shared in the fatigues of the operations, and in the hardships unavoidable in such dreary regions.

XXII. Signal on the desert of Quinoaloma.

The desert of Quinoaloma, like the former, may be classed among the most disagreeable stations in the whole series; for though they repaired hither from Sinasaguan, it was the 31st of the same month before they could finish the observations relating to this signal.

In their road from Quinoaloma they passed through the town of Azogues; where leaving their instruments
and

and baggage, they went to Cuença, to survey the plains of Talqui and los Bannos, in order to make choice of one of them for measuring the base; and having pitched on the latter, and consulted with us relating to the signals wanting, they returned to the town of Azogues.

XXIII. Signal on the desert of Yasuay.

On the 15th of June they proceeded to the desert of Yasuay, and continued there till the 11th of July; when, having finished their observations, they returned to Cuença, where they employed themselves in measuring the base on the plain of los Bannos, and in beginning the astronomical observations. This they prosecuted with incessant diligence till the 10th of December following, when, in order to continue them with the greater precision and certainty, a new instrument became necessary, and for this purpose they repaired to Quito.

XXIV. XXV. XXVI. XXVII. Signals of Namarelte, Guanacauri, los Bannos, and the tower of the great church of Cuença.

Whilst they were making the astronomical observations at Cuença, they also, by unwearied diligence, finished those relating to the geometrical mensuration at the four stations of Namarelte, Guanacauri, los Bannos, and the tower of the great church of Cuença. The first three stations were to connect the base (which reached from Guanacauri to los Bannos) with the series of triangles; and the last served for the observatory jointly with the base. The observations at all these were completely finished at this time; for though the next year we found it necessary to go to Cuença to repeat the astronomical observations, yet all the operations relating to the geometrical mensurations were accurately finished at this time.

R 4

XXVIII.

XXVIII. XXIX. XXX. XXXI. XXXII. Signals on the mountains of Guapulo, Pambamarca, Campanario, Cuicocha, and Mira.

In the year 1744, when we returned to the province of Quito, in order to conclude the whole work, having conquered the difficulties which obliged us to intermit the astronomical observations, as we have already observed, Don George Juan added six stations to the series of triangles, there being a necessity for repeating the observations of Guapulo and Pambamarca, in order to extend the series of triangles farther to the northward, and of his repairing again to the mountains of Campanario and Cuicocha. Here, and at Pambamarca, he was obliged to remain amidst all the inconveniencies and hardships of those dreadful regions, till he had completed the necessary observations; all which he bore with great magnanimity; but at those of Guapulo and Mira, which served to connect the observatory, those inconveniencies were avoided; but as the observations at the last station were jointly performed by both companies, the particulars of them have been already mentioned.

CHAP. IV.

Description of the City of Quito.

AS in the preceding descriptions of the several cities and towns, I have not swelled the accounts with chronological and historical remarks, I shall observe the same method with regard to Quito, and only give an accurate account of the present state of this country, the manners and customs of the inhabitants, and the situation of the several places; that such as know them only by name, may avoid those dangerous errors which too often result from forming a judgment of things without a thorough knowledge of them. It

may

may not, however, be amiss to premise, that this province was subjected to the empire of Peru, by Tupac-Inga-Yupanqui, the eleventh Ynca.

GARCILASO, in his history of the Yncas of Peru, the best guide we can follow on this subject, observes, that this conquest was made by the army of that emperor, commanded by his eldest son Hueyna-Capac, who also succeeded him in the empire.   Hueyna-Capac, among other natural children, had one called Ata-Hualpa, by a daughter of the last king of Quito; and being extremely fond of him, on account of his many amiable qualities and accomplishments, in order to procure him an honourable settlement, prevailed on his legitimate and eldest son Huascar to allow him to hold the kingdom of Quito as a fief of the empire; it being an invariable law, that all conquests were to be perpetually annexed to the empire, and not alienated from it on any account whatever.   Thus Hueyna-Capac enjoyed the satisfaction of seeing his favourite a sovereign of large dominions.   But on the death of his father, this prince, of whom such great hopes had been conceived, ungratefully rebelled, seized on the empire, imprisoned his brother, and soon after put him to a violent death.   His prosperity was, however, but of short continuance; for he suffered the same fate by order of Don Francisco Pizarro, who had sent Sebastian de Belalcazar to make a conquest of the kingdom of Quito.   He routed the Indians wherever they ventured to face him; and having soon, by a series of victories, made himself master of the kingdom, and in the year 1534 rebuilt the capital, which had suffered extremely from intestine commotions, called it San Francisco de Quito, a name it still retains, though it was not till seven years after that the title of city was conferred upon it.

WE found from accurate observations, that the city of Quito is situated in the latitude of 0 deg. 13. min. 33. sec. south, and in 298 deg. 15 min. 45. sec. of
longitude

longitude from the meridian of Teneriff. It stands in the inland parts of the continent of South America, and on the eastern skirts of the west Cordillera of the Andes. Its distance from the coast of the South Sea is about 35 leagues west. Contiguous to it, on the north-west, is the mountain and desert of Pichincha, not less famous among strangers for its great height, than among the natives for tne great riches it has been imagined to contain ever since the times of idolatry: and this only from a vague and unsupported tradition. The city is built on the acclivity of that mountain, and surrounded by others of a middling height, among the breaches, or guaycos, as they are called here, which form the eminences of Pichincha. Some of these breaches are of a considerable depth, and run quite through it, so that great part of the buildings stand upon arches. This renders the streets irregular and extremely uneven, some being built on the ascents, descents, and summits of the breaches. This city, with regard to magnitude, may be compared to one of the second order in Europe; but the unevenness of its situation is a great disadvantage to its appearance.

NEAR it are two spacious plains: one on the south called Turu-bamba, three leagues in length; and the other on the north, termed Inna-Quito, about two leagues in extent. Both are interspersed with seats and cultivated lands, which greatly add to the prospect from the city, being continually covered with a lively verdure, and the neighbouring plains and hills always enamelled with flowers, there being here a perpetual spring. This scene is beautifully diversified with large numbers of cattle feeding on the eminences, though the luxuriancy of the soil is such, that they cannot consume all the herbage.

THESE two plains contract as they approach the city, and at their junction form a neck of land, covered with those eminences on which part of Quito
stands.

stands. It may, perhaps, appear strange, that, notwithstanding two such beautiful and extensive plains are
so near the city, a situation so very inconvenientshould
be preferred to either. But the first founders seem to
have had less regard for convenience and beauty, than
for preserving the remembrance of their conquest, by
building on the site of the ancient capital of the Indians, who made choice of such places for erecting
their towns; probably from their being better adapted
to defence. Besides, the Spaniards, during the infancy
of their conquest, little imagined this place would ever
increase to its present magnitude. Quito, however,
was formerly in a much more flourishing condition
than at present; the number of its inhabitants being
considerably decreased, particularly the Indians, whole
streets of whose huts are now forsaken, and in ruins.

South-west from Quito, on the neck of land belonging to the plain of Turu-bamba, is an eminence
called Panecillo, or the Little Loaf, from its figure
resembling a sugar-loaf. Its height is not above a
hundred toises, and between it and the mountains covering the east part of the city is a very narrow road.
From the south and west sides of the Panecillo issue
several streams of excellent water; and from the
eminences of Pichincha several brooks flow down the
breaches, and by means of conduits and pipes plentifully supply the whole city with water; whilst the remainder, joining in one stream, forms a river called
Machangara, which washes the south parts of the
city, and is crossed over by a stone bridge.

Pichincha, in the Pagan times, was a volcano, and
even some fiery eruptions have been known since the
conquest. The mouth, or aperture, was in a pic nearly
of the same height with that on which we took our
station; and the top of it is now covered with sand
and calcined matter. At present no fire is ejected,
nor does there any smoke issue from it. But sometimes the inhabitants are alarmed by dreadful noises,
caused

caused by winds confined in its bowels, which cannot fail of recalling to their minds the terrible destruction formerly caused by its eructations, when the whole city and neighbouring country were often, as it were, buried under a deluge of ashes, and the light of the sun totally intercepted, for three or four days successively, by impenetrable clouds of dust. In the centre of the plain of Inno-Quito is a place called Rumi-bamba, i. e. a stony plain, being full of large fragments of rocks thrown thither by the ejections of the mountain. We have already observed, that the highest part of Pichincha is covered with ice and snow, considerable quantities of which are brought down to the city, and mixed with the liquors drunk by people of fashion.

THE principal square in Quito has four sides, in one of which stands the cathedral, and in the opposite the episcopal palace; the third side is taken up by the town house, and the fourth by the palace of the audience. It is very spacious, and has in the centre an elegant fountain. It is indeed rather disfigured than adorned by the palace of the audience; which instead of being kept in repair conformable to the dignity of government, the greatest part of it has been suffered to fall into ruins, and only a few halls and offices taken any care of; so that even the outward walls continually threaten to demolish the parts now standing. The four streets terminating at the angles of the square are straight, broad, and handsome; but at the distance of three or four quadras (or the distance between every two corners, or stacks of building, and which here consists of about a hundred yards, more or less) begin the troublesome declivities. This inequality deprives the inhabitants of the use of coaches, or any other wheel-carriage. Persons of rank, however, to distinguish themselves, are attended by a servant carrying a large umbrella: and ladies of the first quality are carried in sedans. Except the four streets above men-

mentioned, all the rest are crooked, and destitute both of symmetry and order. Some of them are crossed by breaches, and the houses stand on the sides of their winding course and irregular projections. Thus some parts of the city are situated at the bottom of those breaches, while others stand on their summits. The principal streets are paved; but those which are not, are almost impassable after rain, which is here very common.

Besides the principal square, there are two others very spacious, together with several that are smaller. In these the greatest part of the convents are situated, and make a handsome appearance ; the fronts and portals being adorned with all the embellishments of architecture; particularly the convent of the order of Franciscans, which, being wholly of free-stone, must have cost a prodigious sum; and indeed the justness of the proportions, the disposition of the parts, the elegant taste and execution of the work, render it equal to most of the admired buildings in Europe.

The principal houses are large; some of them have spacious and well-contrived apartments, though none are above one story in height, which is seldom without a balcony toward the street; but their doors and windows, particularly those within, are very low and narrow, following in these particulars the old custom of the Indians, who constantly built their houses among breaches and inequalities, and were also careful to make the doors very narrow. The Spaniards plead in defence of this custom, that the apartments are freer from wind; but be that as it may, I am inclined to think that this peculiarity owed its origin to a blind imitation of the Indians.

The materials made use of in building at Quito are adobes, or unburnt bricks, and clay ; and to the making of the former the earth is so well adapted, that they last a long time, provided they are defended from the rain. They are cemented or joined together by a certain substance called sangagua, a species of mor-

tar

tar of uncommon hardness, used by the ancient Indians for building houses and walls of all kinds, several remains being still to be seen near the city, and
in many other parts of the kingdom, notwithstanding
the remarkable inclemency of the weather; a sufficient proof of its strength and duration.

THE city is divided into seven parishes, the Segra-
rio, (*Plate* II.) St. Sebastian, St Barbaria, St. Roque,
St. Mark, St. Prisca, and St. Blaize. The cathedral,
besides the richness of its furniture, is splendidly
adorned with tapestry hangings and other costly decorations; but in this respect the other parish-churches
are so mean as to have scarce necessaries for performing divine worship. Some of them are without
pavement, and with every other mark of poverty.
The chapel del Sagrario is very large, wholly of
stone, and its architecture executed in an elegant taste;
nor is the disposition of the inside inferior to the
beauty of its external appearance.

THE convents of monks in Quito are those of the
Augustines, Dominicans, and the Fathers of mercy;
which are the heads of provinces; but besides these
there is another of Franciscan recollects, another of
Dominicans, and another of the Fathers of mercy. In
this city is also a college of Jesuits: two colleges for
seculars; one called St. Lewis, of which the Jesuits
have the direction; and the other St. Ferdinand, and
is under the care of the Dominicans. In the first are
twelve royal exhibitions for the sons of auditors and
other officers of the crown. It is also an university
under the patronage of St. Gregory. That of the
second is a royal foundation, and dedicated to St.
Thomas; the salaries of the professors are paid by
the crown. Some of the chairs in this college are
filled by graduates, as those apropriated to the canon
and civil law, and physic; but the latter has been
long vacant for want of a professor, though the degrees
would be dispensed with. The Franciscan convent
has

has a college, called San Buena Ventura, for the religious of its order; and, though under the same roof
with the convent, has a different government and
œconomy.

Quito has also several nunneries; as that of the
Conception, the orders of St. Clare, St. Catharine,
and two of bare-footed Theresians. Of these one was
originally founded in the town of Latacunga; but
having, together with the place itself, been destroyed
by an earthquake, the nuns removed to Quito, where
they have ever since continued.

The college of Jesuits, as well as all the convents
of monks, are very large, well built, and very splendid. The churches also, though the architecture of
some is not modern, are spacious, and magnificently
decorated, especially on solemn festivals, when it
is amazing to behold the vast quantities of wrought
plate, rich hangings, and costly ornaments, which
heighten the solemnity of worship, and increase the
reputation of these churches for magnificence. If those
of the nunneries do not, on those occasions, exhibit
such an amazing quantity of riches, they exceed them
in elegance and delicacy. It is quite otherwise in the
parish-churches, where poverty is conspicuous, even
on the most solemn occasions; though this is partly
imputed to those who have the care of them.

Here is also an hospital, with separate wards for
men and women; and though its revenues are not
large, yet by a proper œconomy they are made to answer all the necessary expences. It was formerly under the direction of particular persons of the city,
who, to the great detriment of the poor, neglected
their duty, and some even embezzled part of the
money received; but it is now under the care of the
order of our Lady of Bethlehem, and by the attention
of these fathers, every thing has put on a different
aspect, the whole convent and infirmary having been
rebuilt,

rebuilt, and a church erected, which, though small, is very beautiful, and finely decorated.

THIS order of our Lady of Bethlehem has been lately founded under the name of a congregation, and had its origin in the province of Guatemala. The name of the founder was Pedro de San Joseph Beta-neur, a native of the town of Chasna (or Villa Fuerte) on the island of Teneriff, in the year 1626. After his death, which happened in the year 1667, his congre-gation was approved of by a bull of Clement X. dated the 16th of May 1672; and still more formally in another of 1674. In 1687, Innocent XI. erected it into a community of regulars; since when it has be-gun to increase in these countries as a religious order. It had indeed before passed from Guatemala to Mex-ico, and from thence, in the year 1671, to Lima, where the fathers had the care of the hospital del Carmen. In the city of St. Miguel de Piura, they took possession of the hospital of St. Ann in the year 1678; and of that of St. Sebastian in Truxillo in 1680. And their probity and diligence in discharg-ing these trusts, induced other places to select them as directors of their hospitals, and among the rest the city of Quito; where, notwithstanding they have been only a few years, they have repaired all former abuses, and put the hospital on a better footing than it had ever known before.

THE fathers of this order go bare-footed, and wear a habit of a dark brown colour, nearly resembling that of the capuchins, which order they also imitate in not shaving their beards. On one side of their cloak is an image of our Lady of Bethlehem. Every sixth year they meet to choose a general, which ceremony is performed alternately at Mexico and Lima.

AMONG the courts, whose sessions are held at Quito, the principal is that of the Royal Audience, which was established there in the year 1563, and consists of a president who is also governor of the province with

regard

regard to matters of law), four auditors, who are at the same time civil and criminal judges, and a royal fiscal, so called, as, besides the causes brought before the audience, he also takes cognizance of every thing relating to the revenue of the crown. Besides this, there is also another fiscal, called Protector de los Indios, "protector of the Indians," who solicits for them, and when injured pleads in their defence. The jurisdiction of this court extends to the utmost limits of the province, with no other appeal than to the council of the Indies, and this only in case of a rejection of a petition, or flagrant injustice.

THE next is the exchequer, or chamber of finances; the chief officers of which are an accomptant, a treasurer, and a royal fiscal. The revenues paid into the receipt of this court are, the tributes of the Indians of this jurisdiction and those of Otabalo, Villa de San Miguel de Ibara, Latacunga, Chimbo, and Riobamba; and also the taxes levied in those parts, and the produce of the customs at Babahoyo, Yaquache, and Caracol: which sums are annually distributed, partly to Carthagena and Santa Martha, for paying the salaries of the presidents, fiscals, corregidors, together with the stipends of the priests, and the governors of Maynas and Quijos; partly for the officers of the Commandries, and partly for the Caciques of the villages.

THE tribunal de Cruzada, or Croisade, has a commissary, who is generally some dignitary of the church; and a treasurer, who is also the accomptant, through whose hands every thing passes relating to the Croisade.

HERE is also a treasury for the effects of persons deceased; an institution long since established all over the Indies, for receiving the goods of those whose lawful heirs were in Spain, that thus they might be secured from those accidents to which, from dishonesty or negligence, they would be liable in private hands, and securely kept for the persons to which they belong: an institution originally very excellent; but now greatly abused, great defalcations being made in the estates before they are restored to their proper owners.

BESIDES these tribunals, here is a commissary of the inquisition, with an alguazil major, and familiars appointed by the holy office at Lima.

THE Corporation consists of a corregidor, two ordinary alcaldes chosen annually, and regidores. These superintend the election of the alcaldes, which is attended with no small disturbance in this city, persons of all ranks being divided into the two parties of Creoles and Europeans or Chapitones, to the great detriment of private repose and sociability. This assembly also nominates the alcalde major of the Indians, who must be a governor of one of the Indian towns within five leagues of the city; and has under him other inferior alcaldes, for the civil government of it. And this alcalde major, together with the others, are little more than the alguazils, or officers of the corregidor or ordinary alcaldes of the city; though, at first, they were invested with much greater power. Besides these, here are others, called alcaldes de harrieros, whose business it is to provide mules, &c. for travellers. And though all these are subordinate to the alcalde major, yet he has very little authority over them.

THE cathedral chapter consists of the bishop, dean, archdeacon, chanter, treasurer, a doctoral, a penitentiary, a magistral, three canons by presentation, four prebends, and two demi-prebends, with the following revenues. That of the bishop 24,000 dollars; the dean 2500; the four succeeding dignities 2000 each; the canons 1500 each; the prebends 600, and the demi-prebends 420. This church was erected into a cathedral in the year 1545; and among other festivals are celebrated in it, with amazing magnificence, those of Corpus Chri ti, and the Conception of our Lady, when all the courts, offices, and persons

sons of eminence, never fail to assist. But the singular pomp of the procession of the host in the former, and the dances of the Indians, must not be omitted. Every house of the streets through which it passes are adorned with rich hangings; and superb triumphal arches are erected, with altars at stated distances, and higher than the houses, on which, as on the triumphal arches, the spectator sees, with admiration, immense quantities of wrought plate, and jewels, disposed in such an elegant manner as to render the whole even more pleasing than the astonishing quantity of riches. This splendor, together with the magnificent dresses of the persons who assist at the procession, render the whole extremely solemn, and the pomp ʼand decorum are both continued to the end of the ceremony.

With regard to the dances, it is a custom, both among the parishes of Quito and all those of the mountains, for the priest, a month before the celebration of the feasts, to select a number of Indians who are to be the dancers. These immediately begin to practise the dances they used before their conversion to Christianity. The music is a pipe and tabor, and the most extraordinary of their motions some awkward capers ; in short, the whole is little to the taste of an European. Within a few days of the solemnity, they dress themselves in a doublet, a shirt, and a woman's petticoat, adorned in the finest manner possible., Over their stockings they wear a kind of pinked buskins, on which are fastened a great number of bells. Their head and face they cover with a kind of mask, formed of ribbands of several colours. Dressed in this fantastical garb, they proudly call themselves angels, unite in companies of eight or ten, and spend the whole day in roving about the streets, highly delighted with the jingling of their bells; and frequently stop and dance, to gain the applauses of the ignorant multitude, who are strangers

to

to elegant dancing. But what is really surprising, is, that without any pay, or view of interest, unless they think it a religious duty, they continue this exercise a whole fortnight before the grand festival, and a month after it, without minding either their labour or families; rambling about, and dancing the whole day, without being either tired or disgusted, though the number of their admirers daily decrease, and the applause is turned into ridicule.

THE same dress is worn by them in other processions, and at the bull-feasts, when they are excused from labour, and therefore highly pleased with them.

THE corporation and cathedral chapter keep, by vow, two annual festivals in honour of two images of the Virgin, which are placed in the villages of Guapulo and Quinche, belonging to this jurisdiction. They are brought with great solemnity to Quito, where a festival is celebrated, with great magnificence and rejoicing, and is succeeded by nine days devotion, the Audience and other courts assisting at the festival. The statues are afterwards returned with the same solemnity to their respective churches, the first of which is one league from Quito, and the other six. These festivals are held in commemoration of the mercy and assistance vouchsafed by the holy Virgin at the time of an earthquake and terrible ejections from Pichincha, by which Latacunga, Hambato, and a great part of Riobamba, were utterly destroyed; while the prayers offered up at Quito to the holy Virgin induced her to interpose in so singular a manner, that not the least misfortune attended this city, though apparently in equal danger with those which suffered.

CHAP.

CHAP. V.

Of the Inhabitants of Quito.

THIS city is very populous, and has, among its inhabitants, some families of high rank and distinction; though their number is but small considering its extent, the poorer class bearing here too great a proportion. The former are the descendants either of the original conquerors, or of presidents, auditors, or other persons of character, who at different times came over from Spain invested with some lucrative post, and have still preserved their lustre, both of wealth and descent, by intermarriages, without intermixing with meaner families though famous for their riches.

THE commonalty may be divided into four classes; Spaniards or Whites, Mestizos, Indians or Natives, and Negroes, with their progeny. These last are not proportionally so numerous as in the other parts of the Indies; occasioned by it being something inconvenient to bring Negroes to Quito, and the different kinds of agriculture being generally performed by Indians.

THE name of Spaniard here has a different meaning from that of Chapitone or European, as properly signifying a person descended from a Spaniard without a mixture of blood. Many Mestizos, from the advantage of a fresh complexion, appear to be Spaniards more than those who are so in reality; and from only this fortuitous advantage are accounted as such. The Whites, according to this construction of the word, may be considered as one sixth part of the inhabitants.

THE Mestizos are the descendants of Spaniards and Indians, and are to be considered here in the same different egrees between the Negroes and Whites ,

S 3   as

as before at Carthagena; but with this difference, that at Quito the degrees of Mestizos are not carried so far back; for, even in the second or third generations, when they acquire the European colour, they are considered as Spaniards. The complexion of the Mestizos is swarthy and reddish, but not of that red common in the fair Mulattos. This is the first degree, or the immediate issue of a Spaniard and Indian. Some are, however, equally tawny with the Indians themselves, though they are distinguished from them by their beards: while others, on the contrary, have so fine a complexion that they might pass for Whites, were it not for some signs which betray them, when viewed attentively. Among these, the most remarkable is the lowness of the forehead, which often leaves but a small space between their hair and eye-brows; at the same time the hair grows remarkably forward on the temples, extending to the lower part of the ear. Besides, the hair itself is harsh, lank, coarse, and very black; their nose very small, thin, and has a little rising on the middle, from whence it forms a small curve, terminating in a point, bending towards the upper lip. These marks, besides some dark spots on the body, are so constant and invariable, as to make it very difficult to conceal the fallacy of their complexion. The Mestizos may be reckoned a third part of the inhabitants.

The next class is the Indians, who form about another third; and the others, who are about one sixth, are the Casts. These four classes, according to the most authentic accounts taken from the parish register, amount to between 50 and 60,000 persons, of all ages, sexes, and ranks. If among these classes the Spaniards, as is natural to think, are the most eminent for riches, rank, and power, it must at the same time be owned, however melancholy the truth may appear, they are in proportion the most poor, miserable

and

and distressed; for they refuse to apply themselves to any mechanic business, considering it as a disgrace to that quality they so highly value themselves upon, which consists in not being black, brown, or of a copper-colour. The Mestizos, whose pride is regulated by prudence, readily apply themselves to arts and trades, but chuse those of the greatest repute, as painting, sculpture, and the like, leaving the meaner sort to the Indians. They are observed to excel in all, particularly painting and sculpture; in the former a Mestizo, called Miguel de Santiago, acquired great reputation, some of his works being still preserved and highly valued, while others were carried even to Rome, where they were honoured with the unanimous applauses of the virtuosi. They are remarkably ready and excellent at imitation, copying being indeed best adapted to their phlegmatic genius. And what renders their exquisite performances still more admirable is, that they are destitute of many of the instruments and tools requisite to perform them with any tolerable degree of accuracy. But, with these talents, they are so excessively indolent and slothful, that, instead of working, they often loiter about the streets during the whole day. The Indians, who are generally shoemakers, bricklayers, weavers, and the like, are not more industrious. Of these the most active and tractable are the barbers and phlebotomists, who, in their respective callings, are equal to the most expert hands in Europe. The shoemakers, on the other hand, distinguish themselves by such supineness and sloth, that very often you have no other way left to obtain the shoes you have bespoke, than to procure materials, seize on the Indian, and lock him up till they are finished. This is indeed partly owing to a wrong custom of paying for the work before it is done; and when the Indian has once got the money, he spends it all in chicha*, so that while it lasts he is never sober; and it is natural

* A kind of beer or ale made of maize, and very intoxicating.

to think that it will not be easy afterwards to prevail on him to work for what he has spent.

THE dress here differs from that used in Spain, but less so with the men than of the women. The former, who wear a black cloak, have under it a long coat, reaching down to their knees, with a close sleeve, open at the sides, without folds; and along the seams, of the body, as well as those of the sleeves, are button-holes, and two rows of buttons, for ornament. In every other particular, people of fortune affect great magnificence in their dress, wearing very commonly the finest gold and silver tissues.

THE Mestizos in general wear blue cloth, manufactured in this country. And though the lowest class of Spaniards are very ambitious of distinguishing themselves from them, either by the colour or fashion of the clothes, little difference is to be observed.

THE most singular dress, with regard to its meanness, is that of the Indians, which consists only of white cotton drawers, made either from the stuffs of the country, or from others brought from Europe. They come down to the calf of the leg, where they hang loose, and are edged with a lace suitable to the stuff. The use of a shirt is supplied by a black cotton frock, wove by the natives. It is made in the form of a sack, with three openings at the bottom, one in the middle for the head, and the others at the corners for the arms, and thus cover their naked bodies down to the knees. Over this is a capisayo, a kind of serge cloak, having a hole in the middle for putting the head through, and a hat, made by the natives. This is their general dress, and which they never lay aside, not even while they sleep. And use has so inured them to the weather, that, without any additional clothing or covering for their legs or feet, they travel in the coldest parts with the same readiness as in the warmest.

THE

The Indians who have acquired some fortune, particularly the barbers and phlebotomists, are very careful to distinguish themselves from their countrymen, both by the fineness of their drawers, and also by wearing a shirt, though without sleeves. Round the neck of the shirt they wear a lace four or five fingers in breadth, hanging entirely round like a kind of ruff or band. One favourite piece of finery is silver or gold buckles for their shoes; but they wear no stockings or other coverings on their legs. Instead of the mean capisayo, they wear a cloak of fine cloth, and often adorned with gold or silver lace.

The dress of the ladies of the first rank consists of a petticoat already described in our account of Guayaquil. On the upper parts of their body they wear a shift, on that a loose jacket laced, and over all a kind of bays, but made into no form, being worn just as cut from the piece. Every part of their dress is, as it were, covered with lace; and those which they wear on days of ceremony are always of the richest stuffs, with a profusion of ornaments. Their hair is generally made up in tresses, which they form into a kind of cross, on the nape of the neck; tying a rich ribband, called balaca, twice round their heads, and with the ends form a kind of rose at their temples. These roses are elegantly intermixed with diamonds and flowers. When they go to church, they sometimes wear a full petticoat; but the most usual dress on these occasions is the veil.

The Mestizo women affect to dress in the same manner as the Spanish, though they cannot equal them in the richness of their stuffs. The meaner sort go barefooted. Two kinds of dresses are worn by the Indian women; but both of them made in the same plain manner with those worn by the men: the whole consisting of a short petticoat, and a veil of American bays. The dress of the lowest class of Indian women is in effect only a bag of the same make and stuff as the

frocks

frocks of the men, and called anaco. This they fasten on the shoulders with two large pins called tupu, or topo. The only particular in which it differs from the frock is, that it is something longer, reaching down to the calf of the leg, and fastened round the waist with a kind of girdle. Instead of a veil, they wear about their neck a piece of the same coarse stuff dyed black, and called Lliella; but their arms and legs are wholly naked. Such is the habit with which the lower class of Indian women are contented.

THE caciquesses, or Indian women, who are married to the alcaldes majors, governors, and others, are careful to distinguish themselves from the common people by their habits, which is a mixture of the two former, being a petticoat of bays adorned with ribbands; over this, instead of the anaco, they wear a kind of black manteau, called acso. It is wholly open on one side, plaited from top to bottom, and generally fastened round the waist with a girdle. Instead of the scanty Lliella which the common Indian women wear hanging from their shoulders, these appear in one much fuller, and all over plaited, hanging down from the back part of their head almost to the bottom of the petticoat. This they fasten before with a large silver bodkin, called also tupu, like those used in the anaco. Their head-dress is a piece of fine linen curiously plaited, and the end hanging down behind: this they call colla, and is worn both for distinction and ornament, and to preserve them from the heat of the sun; and these ladies, that their superiority may not be called in question, never appear abroad without shoes. This dress, together with that universally worn by Indians, men and women, is the same with that used in the time of the Yncas, for the propriety of distinguishing the several classes. The Caciques at present use no other than that of the more wealthy Mestizos, namely, the cloak and hat; but

but the shoes are what chiefly distinguish them from the common Indians.

The men, both Creoles and Spaniards, are well made, of a proper stature, and of a lively and agreeable countenance. The Mestizos in general are also well made, often taller than the ordinary size, very robust, and have an agreeable air. The Indians, both men and women, are generally low, but well proportioned, and very strong; though more natural defects are to be observed among them than in the other classes of the human species: some are remarkably short, some ideots, dumb and blind, and others deficient in some of their limbs. Their hair is generally thick and long, which they wear loose on their shoulders, never tying or tucking it up, even when they go to sleep. But the Indian women plait theirs behind with ribband, and the part before they cut a little above the eye-brows from one ear to another; which form of hair they call urcu, and are so fond of this natural ornament, that the greatest affront possible to be offered to an Indian of either sex, is to cut off their hair; for whatever corporal punishment their masters think proper to inflict on them, they bear with a dutiful tranquillity; but this is a disgrace they never forgive; and accordingly it was found necessary for the government to interpose, and limit this punishment to the most enormous crimes. The colour of their hair is generally a deep black ; it is lank, harsh, and coarse as that of horses. The Mestizos, on the other hand, by way of distinguishing themselves from the Indians, cut off their hair; but the women do not in this respect follow the example of their husbands. The Indians have no beard ; and the greatest alteration occasioned by their arriving at the years of maturity, is only a few straggling hairs on the chin, but so short and thin as never to require the assistance of the razor; nor have

either

either males or females any indications of the age of puberty.

The youths of family are here instructed in philosophy and divinity, and some proceed to the study of the civil law, but follow that profession with reluctance. In these sciences they demonstrate a great deal of judgment and vivacity, but are very deficient in historical and political knowledge, as well as other sciences, which improve the human understanding, and carry it to a certain degree of perfection not otherwise attainable. This is, however, their misfortune, not their fault; being owing to the want of proper persons to instruct them ; for with regard to those who visit this country on commercial affairs, their minds have generally another turn, and their whole time is devoted to acquire riches. Thus after seven or eight years of scholastic instruction, their knowledge is very limited; though endowed with geniuses capable of making the greatest progress in the sciences.

In the women of rank here, their beauty is blended with a graceful carriage and an amiable temper; qualities indeed common to the whole sex in this part of America. Their children are always educated under their own eyes, though little to their advantage, their extreme fondness preventing them from seeing those vices which so often bring youth to ruin and infamy ; nor is it uncommon for them to endeavour to hide the vices of the son from the knowledge of the father; and in case of detection, to interpose passionately in defence of their favourite, in order to prevent his being properly corrected.

This country is observed to abound more in women than men; a circumstance the more remarkable, as those causes which in Europe induce men to leave their country, namely, travelling, commerce, and war, can hardly be said to subsist here. Numbers of families may be found in this country, that have a great variety

riety of daughters, but not one son among them. Nature also in the male sex, especially those who have been tenderly brought up, begin to decay at the age of thirty; whereas the females rather enjoy a more confirmed state of health and vigour. The cause of this may, in a great measure, be owing to the climate; food may also contribute to it; but the principal cause, I make no doubt, is their early intemperance and voluptuousness; this debilitates the stomach, so that the organs of digestion cannot perform their proper office; and accordingly many constantly eject their victuals an hour or two after their meals. Whether this be owing to a custom now become natural, or forced, the day they fail of such ejection, they are sure to find themselves indisposed. But amidst all their weaknesses and indispositions they live the general time, and many even arrive at a very advanced age.

The only employment of persons of rank, who are not ecclesiastics, is from time to time to visit their estates or chacaras, where they reside during the time of harvest; but very few of them ever apply themselves to commerce, indolently permitting that lucrative branch to be possessed entirely by the Chapitones or Europeans, who travel about the country, and pursue their interest with great assiduity. Within the city, however, some few Creoles and Mestizos so far overcome their indolent dispositions as to keep shops.

The want of proper employments, together with the sloth so natural to the inhabitants of this country, and the great neglect of education in the common people, are the natural parents of that fondness so remarkable in these parts for balls and entertainments; and these at Quito are both very frequent, and carried to such a degree of licentiousness and audacity, as cannot be thought of without detestation; not to mention the many tumults and quarrels which thence derive
their

their origin. But such brutality may be considered as the natural consequence of the rum and chicha, which on these occasions are drunk in enormous quantities. It must, however, be remembered, that no person of any rank or character is ever seen at these meetings, their festivity being conducted with the strictest decency and decorum.

Rum is commonly drunk here by persons of all ranks, though very moderately by those of fashion; particularly at entertainments, when it is made into a kind of cordial. They prefer it to wine, which they say disagrees with them. The Chapitones also accustom themselves to this liquor; wine, which is brought from Lima, being very scarce and dear. Their favourite liquor is brandy, brought also from Lima, and is less inflammative than rum. The disorders arising from the excessive use of spirituous liquors are chiefly seen among the Mestizos, who are continually drinking while they are masters of any money. The lower class of women, among the Creoles and Mestizos, are also addicted to the same species of debauchery, and drink excessively.

Another common liquor in this country is the mate, which answers to tea in the East Indies, though the method of preparing and drinking it is something different. It is made from an herb, which, in all these parts of America is known by the name of Paraguay, as being the produce of that country. Some of it is put into a calebash tipped with silver, called here mate or totumo, with a sufficient quantity of sugar, and some cold water, to macerate it. After it has continued in this manner some time, the calebash is filled with boiling water, and the herb being reduced to a powder, they drink the liquor through a pipe fixed in the calebash, and having a strainer before the end of it. In this manner the calebash is filed several times with water and fresh supplies of sugar, till the herb subsides to the bottom, a sufficient indica-

tion

tion that a fresh quantity is wanting. It is also usual to squeeze into the liquor a few drops of the juice of lemons or Seville oranges, mixed with some perfumes from odoriferous flowers. This is their usual drink in the morning fasting, and many use it also as their evening regale. I have nothing to object against the salubrity and use of this liquor; but the manner of drinking it is certainly very indelicate, the whole company drinking successively through the same pipe. Thus the mate is carried several times round the company, till all are satisfied. The Chapitones make very little use of it; but among the Creoles it is the highest enjoyment; so that even when they travel, they never fail to carry with them a sufficient quantity of it. This may indeed be owing in some measure to the dispatch and facility with which it is prepared; but till they have taken their dose of mate, they never eat.

There is no vice to which idleness is not a preliminary; nor is sloth ever unaccompanied with some vice or other. What must then be the state of morality in a country where the greatest part of the people have no work, employment, or calling, to occupy their thoughts: nor any idea of intellectual entertainment? The prevalence of drunkenness has been already mentioned, and the destructive vice of gaming is equally common. But in the latter, persons of rank and opulence, whose example is always followed, have led the way; and their inferiors have universally followed in their destructive paths, to the ruin of families, and the breach of conjugal affection; some losing their stocks in trade, others the very clothes from their backs, and afterwards those belonging to their wives, risking the latter to recover their own. This propensity in the Indians for gaming has by some been imputed to causes, in which I can perceive no manner of relation. To me it plainly appears owing to the leisure of some, who know not how to spend
their

their time, and to the natural sloth and idleness of others.

The common people and Indians are greatly addicted to theft, in which it must be owned they are very artful and dextrous. The domestics also cannot be said to be entirely free from this fault, which is attended with the inconvenience of reserve and suspicion on the part of their master. The Mestizos do not want for audacity in any kind of theft or robbery, though in themselves arrant cowards. Thus, even at an unseasonable hour, they will not venture to attack any one in the street; but their common practice is, to snatch off the person's hat, and immediately seek their safety in their flight; so that before the person robbed can recover himself the thief is out of sight. However trifling this may seem, yet sometimes the capture is very considerable; the hats generally worn by persons of any rank, and even by the wealthy citizens when dressed in their cloaks, are of white beaver, and of themselves worth 15 or 20 dollars, or more, of the Quito currency, besides a hatband of gold or silver lace, fastened with a gold buckle set with diamonds or emeralds. It is very rare that any such things as a robbery on the highway is heard of; and even these may be rather accounted housebreaking, as they are either committed by the carriers themselves or their servants. In order to execute their most remarkable pieces of villany within the city, they set fire, during the darkness of the night, to the doors of such shops or warehouses, where they flatter themselves with the hopes of finding some specie; and having made a hole sufficiently large for a man to creep through, one of them enters the house, while the others stand before the hole to conceal their accomplice, and to receive what he hands out to them. In order to prevent such practices, the principal traders are at the expence of keeping a guard, which patroles all night through the streets where attempts

of

of this kind are most to be apprehended; and thus the shops are secured; for in case any house or shop is broke open, the commander of the guard is obliged to make good the damage received.

Neither the Indians, Mestizos, nor any of the lowest class of people, think the taking any eatables a robbery; and the Indians have a particular rule of conduct in their operations, namely, if one of them happens to be in a room where there are several vessels of silver, or other valuable effects, he advances slowly, and with the utmost circumspection, and usually takes only one piece, and that the least valuable, imagining that it will not be so soon missed as if he had taken one of greater price. If detected in the fact, he resolutely denies it, with a yanga, a very expressive word in his language, and now often used by the Spaniards of this country, signifying that it was done without any necessity, without any profit, without any bad intention. It is indeed a word of such extent in disculpating, that there is no crime to which it is not applicable with regard to the acquittal of the delinquent. If he has not been seen in the very fact, be the circumstances ever so plain against him, the theft can never be ascertained, no Indian having ever been known to confess.

In Quito, and in all the towns and villages of its province, different dialects are spoken, Spanish being no less common than the Inga. The Creoles, in particular, use the latter equally with the former; but both are considerably adulterated with borrowed words and expressions. The first language generally spoken by children is the Inga ; the nurses being Indians, many of whom do not understand a word of Spanish. Thus, the children being first used to the Indian pronunciation, the impression is so strong on their minds, that few can be taught to speak the Spanish language before they are five or six years old; and the corruption adheres so strongly to them, that they speak a jargon composed of both; an impropriety which also gains ground among the Europeans, and even persons of rank, when once they begin to understand the language of the country. But what is still more inconvenient, they use improper words; so that a Spaniard himself, not accustomed to their dialect, has often need of an interpreter

THE sumptuous manner of performing the last offices to the dead, mentioned in the description of Carthagena, is frugal and simple, if compared to that used at Quito and all its jurisdiction. Their ostentation is so enormous in this particular, that many families of credit are ruined by a preposterous emulation of excelling others. The inhabitants may therefore be properly said to toil, scheme, and endure the greatest labour and fatigue, merely to enable their successors to bury them in a pompous manner. The deceased must have died in very mean circumstances indeed, if all the religious communities, together with the chapter of the cathedral, are not invited to his funeral, and during the procession the bells tolled in all the churches. After the body is committed to the earth, the obsequies are performed in the same expensive manner, besides the anniversary which is solemnized at the end of the year. Another remarkable instance of their vanity is, never to bury in their own parish church; so that any one seen to be buried in that manner may be concluded to have been of the lowest class, and to have died wretchedly poor. The custom of making an offering either at the obsequies or anniversary, is still observed, and generally consists of wine, bread, beasts, or fowls, according to the ability or inclination of the survivor.

THOUGH Quito cannot be compared to the other cities in these parts for riches, yet it is far removed from poverty. It appears from several particulars to have been in a much more flourishing state; but at present, though it has many substantial inhabitants,

yet

yet few of them are of distinguished wealth, which in general consists in landed estates, applied to several uses, as I shall show in the sequel.   Here are also no very splendid fortunes raised by trade. Consequently it may be inferred, that the city is neither famous for riches, nor remarkable for poverty   Here are indeed considerable estates, though their produce is not at all equal to their extent: but the commerce, though small, is continual.   It must also be observed, to the credit of this city, that the more wealthy families have large quantities of plate, which is daily made use of; and indeed, through the several classes, their tables are never destitute of one piece of plate at least.

CHAP. VI.

Of the Temperature of the Air at Quito; Distinction between Winter and Summer; Inconveniences, Advantages, and Distempers.

TO form a right judgment of the happy temperature of the air of Quito, experience must be made use of, to correct the errors which would arise from mere speculation; as without that unerring guide, or the information of history, who would imagine, that in the centre of the torrid zone, or rather under the equinoctial, not only the heat is very tolerable, but even, in some parts, the cold painful; and that others enjoy all the delights and advantages of a perpetual spring, their fields being always covered with verdure, and enamelled with flowers of the most lively colours! The mildness of the climate, free from the extremes of cold and heat, and the constant equality of the nights and days, render a country pleasant and fertile, which uninformed reason

T 2   would,

would, from its situation, conclude to be uninhabitable: nature has here scattered her blessings with so liberal a hand, that this country surpasses those of the temperate zones, where the vicissitudes of winter and summer, and the change from heat to cold, cause the extremes of both to be more sensibly felt.

THE method taken by nature to render this country a delightful habitation, consists in an assemblage of circumstances, of which if any were wanting, it would either be utterly uninhabitable, or subject to the greatest inconveniences. But by this extraordinary assemblage, the effect of the rays of the sun is averted, and the heat of that glorious planet moderated. The principal circumstance in this assemblage is its elevated situation above the surface of the sea; or, rather, of the whole earth; and thus not only the reflexion of the heat is diminished, but by the elevation of this country, the winds are more subtile, congelation more natural, and the heat abated. These are such natural effects as must doubtless be attributed to its situation; and is the only circumstance from whence such prodigies of nature, as are observed here, can proceed. In one part are mountains of a stupendous height and magnitude, having their summits covered with snow; on the other, volcanoes flaming within, while their summits, chasms, and apertures, are involved in ice. The plains are temperate; the breaches and valleys hot; and, lastly, according to the disposition of the country, its high or low situation, we find all the variety of gradations of temperature, possible to be conceived between the two extremes of heat and cold.

Quito is so happily situated, that neither the heat nor cold is troublesome, though the extremes of both may be felt in its neighbourhood; a singularity sufficiently demonstrated by the following thremometrical experiments. On the 31st of May, 1736, the liquor in the thermometer stood at 1011: at half an hour

after

after twelve at noon at 1014: on the first of June at six in the morning at 1011: and at noon at 1012¼. But what renders this equality still more delightful is, that it is constant throughout the whole year, the difference between the seasons being scarce perceptible. Thus the mornings are cool, the remainder of the day warm, and the nights of an agreeable temperature. Hence the reason is plain, why the inhabitants of Quito make no difference in their dress during the whole year; some wearing silks or light stuffs, at the same time others are dressed in garments of substantial cloth; and the former as little incommoded by the cold, as the latter are by heat.

The winds are healthy, and blow continually, but never with any violence. Their usual situations are north and south, though they sometimes shift to other quarters, without any regard to the season of the year. Their incessant permanence, notwithstanding their constant variations, preserves the country from any violent or even disagreeable impressions of the rays of the sun. So that, were it not for some inconveniences to which this country is subject, it might be considered as the most happy spot on the whole earth. But when these disagreeable incidents are considered, all its beauties are buried in obscurity; for here are dreadful and amazing tempests of thunder and lightning, and the still more destructive subterraneous earthquakes, which often surprise the inhabitants in the midst of security. The whole morning, till one or two in the afternoon, the weather is generally extremely delightful; a bright sun, serene and clear sky, are commonly seen; but afterwards the vapours begin to rise, the whole atmosphere is covered with black clouds, which bring on such dreadful tempests of thunder and lightning, that all the neighbouring mountains tremble, and the city too often feels their dreadful effects. Lastly, the clouds discharge themselves in such impetuous torrents of rain, that in a very

short

short time the streets appear like rivers, and the squares, though situated on a slope, like lakes. This dreadful scene generally continues till near sun-set, when the weather clears up, and nature again puts on the beautiful appearance of the morning. Sometimes, indeed, the rains continue all the night, and they have been known to last three or four days successively.

On the other hand, this general course of the weather has its exceptions, three, four, or six, or even eight fine days succeeding each other; though, after raining six or eight days in the manner above mentioned, it is rare that any falls during the two or three succeeding. But, from the most judicious observations, it may be concluded, that these intervals of fine or foul weather make up only one fifth of the days of the year.

The distinction of winter and summer consists in a very minute difference observable between the one and the other. The interval between the month of September, and April, May, or June, is here called the winter season; and the other months compose the summer. In the former season the rain chiefly prevails, and in the second the inhabitants frequently enjoy intervals of fine weather; but whenever the rains are discontinued for above a fortnight, the inhabitants are in the utmost consternation, and public prayers are offered up for their return. On the other hand, when they continue any time without intermission, the like fears return, and the churches are again crowded with supplicants for obtaining fine weather. For a long drought here is productive of dangerous distempers; and a continual rain, without any intervals of sunshine, destroys the fruits of the earth: thus the inhabitants are under a continual anxiety. Besides the advantages of the rains for moderating the intense rays of the sun, they are also of the greatest benefit in cleansing the streets and squares

of

of the city, which, by the filthiness of the common people at all hours, are every where full of ordure.

EARTHQUAKES cannot be accounted a less terrible circumstance than any of the former; and if not so frequent as in other cities of these parts, they are far from being uncommon, and often very violent. While we continued in this city and its jurisdiction, I particularly remember two, when several country-seats and farm-houses were thrown down, and the greater part of the numerous inhabitants buried in ruins.

It is doubtless to some unknown quality of the temperature of the air, that the city owes one remarkable convenience, which cannot fail of greatly recommending it; namely, being totally free from moschitos or other insects of that kind, which almost render life a burthen in hot countries. They are not known to the inhabitants; even a flea is seldom seen here; nor are the people molested with venomous reptiles. In short, the only troublesome insect is the pique or nigua, whose noxious effects have been already treated of.

THOUGH the plague or pestilence, in its proper sense, be not known here, no instance of its ravages having appeared in any part of America, yet there are some distempers which have many symptoms of it, but concealed under the names of malignant spotted fevers and pleurisies; and these generally sweep away such prodigious numbers, that, when they prevail, the city may with propriety be said to be visited with a pestilential contagion. Another disease common here is that called mal del valle, or vicho; a distemper so general, that, at the first attack of any malady, they make use of medicines adapted to the cure of it, from its usually seizing a person two or three days after a fever. But M. de Jussieu often observed, that the remedies were generally administered to persons not at all affected by the distemper, which, in his opinion, is a gangrene in the rectum; a disease

T 4   very

very common in that climate, and consequently at the first attack all means should be used to prevent its progress. Persons who labour under a flux are most liable to that.malady; but the inhabitants of this country being firmly persuaded that there can be no distemper that is not accompanied with the vicho, the cure is never delayed. The operation must be attended with no small pain, as a pessary, composed of gun-powder, guinea-pepper, and a lemon peeled, is insinuated into the anus, and changed two or three times a day, till the patient is judged to be out of danger.

THE venereal disease is here so common, that few persons are free from it, though its effects are much more violent in some than in others; and many are afflicted with it, without any of its external symptoms. Even little children, incapable by their age of having contracted it actively, have been known to be attacked in the same manner by it as persons who have acquired it by their debauchery. Accordingly there is no reason for caution in concealing this distemper, its commonness effacing the disgrace that in other countries attends it. The principal cause of its prevalence is, negligence in the cure. For the climate favours the operations of the medicines, and the natural temperature of the air checks the malignity of the virus more than in other countries. And hence few are salivated for it, or will undergo the trouble of a radical cure. This disease must naturally be thought in some measure to shorten their lives; though it is not uncommon to see persons live seventy years or more, without ever having been entirely free from that distemper, either hereditary, or contracted in their early youth.

DURING the continuance of the north and north-east winds, which are the coldest from passing over the frosty deserts, the inhabitants are afflicted with very painful catarrhs, called pechugueras. The air is then
some-

something disagreeable, the mornings being so cold as to require warmer clothing; but the sun soon disperses this inconvenience.

As the pestilence, whose ravages among the human species in Europe, and other parts, are so dreadful, is unknown both at Quito and throughout all America, so is also the madness in dogs. And though they have some idea of the pestilence, and call those diseases similar in their effects by that name, they are entirely ignorant of the canine madness; and express their astonishment when an European relates the melancholy effects of it.   Those inhabitants, on the other hand, are here subject to a distemper unknown in Europe, and may be compared to the small-pox, which few or none escape; but having once got through it, they have nothing more to apprehend from that quarter.   This distemper is one of those called peste; and its symptoms are convulsions in every part of the body, a continual endeavour to bite, delirium, vomiting blood; and those whose constitutions are not capable of supporting the conflicts of the distemper, perish. But this is not peculiar to Quito, being equally common throughout all South America.

CHAP. VII.

Fertility of the Territories of Quito, and the common Food of its Inhabitants.

THOUGH an account of the fruits should naturally succeed that of the climate, I determined, on account of their variety, and their being different in different parts, to defer a circumstantial description, till I come to treat more particularly of each of the jurisdictions.   So that I shall here only take a transient view of the perennial beauty and pleasantness of the country; which has hardly its equal in any

part

part of the known world: the equability of its air exempts it from any sensible changes, whereby the plants, corn, and trees, are stripped of their verdure and ornaments, their vegetative powers checked, and themselves reduced to a torpid inactivity. The fertility of this country, if fully described, would appear to many incredible, did not the consideration of the equality and benignity of the climate inforce its probability. For both the degrees of cold and heat are here so happily determined, that the moisture continues, and the earth seldom fails of being cherished by the fertilizing beams of the sun, some part of every day; and therefore it is no wonder that this country should enjoy a greater degree of fertility than those where the same causes do not concur; especially if we consider, that there is no sensible difference throughout the year; so that the fruits and beauties of the several seasons are here seen at the same time. The curious European observes, with a pleasing admiration, that whilst some herbs of the field are fading, others of the same kind are springing up; and whilst some flowers are losing their beauty, others are blowing, to continue the enameled prospect. When the fruits have obtained their maturity, and the leaves begin to change their colour, fresh leaves, blossoms, and fruits, are seen in their proper gradations on the same tree.

The same incessant fertility is conspicuous in the corn, both reaping and sowing being carried on at the same time. That corn which has been recently sown is coming up; that which has been longer sown is in its blade, and the more advanced begins to blossom. So that the declivities of the neighbouring hills exhibit all the beauties of the four seasons at one single view.

Though all this is generally seen, yet there is a settled time for the grand harvest. But sometimes the most favourable season for sowing in one place, is a month

or

or two after that of another, though their distance is not more than three or four leagues; and the time for another at the same distance not then arrived. Thus, in different spots, sometimes in one and the same, sowing and reaping are performed throughout the whole year, the forwardness or retardment naturally arising from the different situations, as mountains, rising grounds, plains, valleys, and breaches; and the temperature being different in each of these, the times for performing the several operations of husbandry must also differ. Nor is this any contradiction to what I have before advanced, as will be seen in the following account of the jurisdiction.

This remarkable fecundity of the soil is naturally productive of excellent fruits and corn of every kind, as is evident from the delicacy of the beef, veal, mutton, pork, and poultry of Quito. Here is also wheat bread in sufficient plenty; but the fault is, that the Indian women, whose business it is to make it, are ignorant of the best methods both of kneading and baking it; for the wheat of itself is excellent, and the bread baked in private houses equal to any in the known world. The beef, which is not inferior to that of Europe, is sold in the markets by the quarter of the hundred for four rials of that country money, and the buyer has the liberty of choosing what part he pleases. Mutton is sold either by the half or quarter of a sheep; and when fat, and in its prime, the whole carcase is worth about five or six rials. Other species of provisions are sold by the lump, without weight or measure, and the price regulated by custom.

The only commodity of which there is here any scarcity is pulse; but this deficiency is supplied by roots, the principal of which are the camates, arucachas, yucas, ocas, and papas; the three former are the natives of hot countries, and cultivated in the plantations of sugar canes, and such spots are called valles,

les, or yungas, though these names have different senses, the former signifying plains in a bottom, and the latter those on the sides of the Cordillera; but both in a hotter exposure. In these are produced the plantains, guincos, guinea-pepper, chirimogas, aguacates, granadillas, pinas, guayabas, and others natural to such climates, as I have already observed in other countries. The colder parts produce pears, peaches, nectarines, quaitambos, aurimelos, apricots, melons, and water-melons; the last have a particular season, but the others abound equally throughout the whole year. The parts which cannot be denominated either hot or cold, produce frutillas, or Peru strawberries, and apples. The succulent fruits, which require a warm climate, are in great plenty throughout the whole year, as China and Seville oranges, citrons, lemons, limes, cidras, and toronjas. These trees are full of blossoms and fruit all the year round, equally with those which are natives of this climate. These fruits abundantly supply the tables of the inhabitants, where they are always the first served up, and the last taken away. Besides the beautiful contrast they form with the other dishes, they are also used for increasing the pleasure of the palate, it being a custom among the people of rank here, to eat them alternately with their other food, of which there is always a great variety.

THE chirimoyas, aguacates, guabas, granadillas, and Peruvian strawberries, being fruits of which, as well as of the ocos and papas, I have not yet given any description, I shall here give the reader a brief account of them. The chirimoya is universally allowed to be the most delicious of any known fruit either of India or Europe. Its dimensions are various, being from one to five inches in diameter. Its figure is imperfectly round, being flatted towards the stalk; where it forms a kind of navel; but all the other parts nearly circular. It is covered with a thin soft shell, but adhering so closely to the pulp, as not to be
separated

separated without a knife. The outward coat, during its growth, is of a dark green, but on attaining its full maturity becomes somewhat lighter. This coat is variegated with prominent veins, forming a kind of net-work all over it. The pulp is white, intermixed with several almost imperceptible fibres, concentring in the core, which extends from the hollow of the excrescence to the opposite side. As they have their origin near the former, so in that part they are larger and more distinct. The flesh contains a large quantity of juice resembling honey, and its taste sweet mixed with a gentle acid, but of a most exquisite flavour. The seeds are formed in several parts of the flesh, and are about seven lines in length, and three or four in breadth. They are also somewhat flat, and situated longitudinally.

The tree is high and tufted, the stem large and round, but with some inequalities; full of elliptic leaves, terminating in a point. The length is about three inches and a half, and the breadth two or two and a half. But what is very remarkable in this tree is, that it every year sheds and renews its leaves. The blossom, in which is the embryo of the fruit, differs very little from the leaves in colour, which is a darkish green; but when arrived to its full maturity is of a yellowish green. It resembles a caper in figure, but something larger, and composed of four petals. It is far from being beautiful; but this deficiency is abundantly supplied by its incomparable fragrancy. This tree is observed to be very parsimonious in its blossoms, producing only such as would ripen into fruits, did not the extravagant passion of the ladies, for the excellence of the odour, induce them to purchase the blossoms at any rate.

The aguacate, which in Lima and other parts of Peru is known by the antient Indian name plata, may also be classed among the choicest fruits of this country. Its figure in some measure resembles the calabashes

labashes of which snuff-boxes are made; that is, the lower part is round, and tapers away gradually towards the stalk; from whence to its base the length is usually between three and five inches. It is covered with a very thin, glossy, smooth shell, which, when the fruit is thoroughly ripe, is detached from the pulp. The colour, both during its growth and when arrived at perfection, is green, but turns something paler as it ripens; the pulp is solid, but yields to the pressure of the finger; the colour white, tinged with green, and the taste so insipid as to require salt to give it an agreeable relish. It is fibrous, but some more so than others. The stone of this fruit is two inches long, one and a half in thickness, and terminates in a point. The taste is sour. It may be opened with a knife, and consists of two lobes, between which may be distinctly perceived the germ of the tree. Within the shell is a very thin tegument, which separates it from the pulp, though sometimes the tegument adheres to the pulp, and at other times to the shell. The tree is lofty and full of branches; the leaf, both in dimension and figure, something different from that of the chirimoyo.

In the province of Quito they give the name of guabas to a fruit, which, in all the other parts of Peru, is called by its Indian name pacaes. It consists of a pod like that of the algarobo, a little flat on both sides. Its usual length is about a foot, though there are different sizes, some larger and some smaller, according to the country where they grow. Its outward colour is a dark green, and covered with a down, which feels smooth when stroked downwards, and rough when the hand is moved in the contrary direction, as in velvet. The pod, opened longitudinally, is found divided into several cells, each containing a certain spongy medulla, very light, and equal to cotton in whiteness. In this are inclosed some black seeds of a very disproportionate size, the medulla,
whose

whose juice is sweet and cooling, not being above a line and a half in thickness round each seed.

The granadilla resembles a hen's egg in shape, but larger.   The outside of the shell is smooth and glossy, and of a faint carnation colour, and the inside white and soft.   It is about a line and a half in thickness, and pretty hard.   This shell contains a viscous and liquid substance, full of very small and delicate grains less hard than those of the pomegranate.   This me-dullary substance is separated from the shell, by an extreme fine and transparent membrane.   This fruit is of a delightful sweetness, blended with acidity, very cordial and refreshing, and so wholesome that there is no danger in indulging the appetite.   The two former are also of the same innocent quality.   The grana-dilla is not the produce of a tree, but of a plant, the blossom of which resembles the passion-flower *, and of a most delicate fragrance.   But we must observe a remarkable singularity in the fruits of this country, namely, that they do not ripen on the trees, like those of Europe, but must be gathered and kept some time; for if suffered to hang on the trees they would decay.

The last of the fruits I shall mention is the frutilla, or Peru strawberry, very different from that of Eu-rope in size; for though generally not above an inch in length, and two thirds of an inch in thickness, they are much larger in other parts of Peru.   Their taste, though juicy and not unpalatable, is not equal to those of Europe.   The whole difference between the plant and that known in Spain consists in its leaves being somewhat larger.

The papas are natives of a cold climate; and be-ing common in several parts of Europe, where they are known by the name of potatoes, all I shall say of

* This is the identical passion flower, which in England never bears any fruit, the climate being too cold.   A.

them

them is, that they are a favourite food with the in-habitants of these countries, who eat them instead of bread, nor is there a made dish or ragout in which they are not an ingredient. The Creoles prefer them to any kind of meat, or even fowl. A particular dish is made of them, and served up at the best tables, called locro; and is always the last, that water may be drunk after it, which they look upon as otherwise unwholesome. This root is the chief food of the lower class; and they find it so nutritive and strengthening, that they are not desirous of more solid food.

THE oca is a root about two or three inches in length, and about half an inch, or something more, in thickness, though not every where equal, having a kind of knots where they twist and wreathe themselves. This root is covered with a very thin and transparent skin, whose colour is in some yellow, in some red, and others orange. It is eaten either boiled or roasted, and has nearly the same taste as a chesnut; with this difference, however, common to all the fruits of America, that the sweetness predominates. It is both pickled and preserved, the latter being what the Americans are very fond of. This root is also an ingredient in many made dishes. The plant is small, like the camote, yucas, and others already described.

WITH regard to the corn of this country, there is no necessity for enumerating the species, they being the same with those known in Spain. The maize and barley are used by the poor people, and particularly by the Indians, in making bread. They have several methods of preparing the maize; one is by parching, which they call camea. They also make from this grain a drink called chica, used by the Indians in the times of the Yncas, and still very common. The method of making it is this: they steep the maize in water till it begins to sprout, when they spread it in the sun, where it is thoroughly dried; after which
they

they roast and grind it, and of the flour they make a decoction of what strength they please. It is then put into jars or casks, with a proportional quantity of water. On the second or third day it begins to ferment, and when that is completed, which is in two or three days more, they esteem it fit for drinking. It is reckoned very cooling; and that it is inebriating, is sufficiently evident from the Indians: those people have indeed so little government of themselves, that they never give over till they have emptied the cask. Its taste is not unlike cider; but seems in some measure to require the dispatch of the Indians, turning sour in seven or eight days after the fermentation is completed. Besides its supposed quality of being cooling, it is, among other medical properties, confessedly diuretic; and to the use of this liquor the Indians are supposed to be indebted for their being strangers to the strangury or gravel. It is also not surprising that those people who drink it, without any other food than cancha, mote, and muchea, are, with the help of this liquor, healthy, strong, and robust.

MAIZE boiled till the grains begin to split, when it is called mote, serves for food to the Indians, the poor people, and servants in families, who being habituated to it, prefer it to bread.

MAIZE, before it is ripe called chogllos, is sold in the ear, and among the poorer sort of inhabitants esteemed a great dainty.

BESIDES the grains of the same species with those in Spain, this country has one peculiar to itself, and very well deserving to be ranked among the most palatable foods; but still more valuable for its being one of the preservatives against all kinds of abscesses and imposthumes. This useful species of grain, here called quinoa, resembles a lentil in shape, but much less, and very white. When boiled it opens, and out of it comes a spiral fibre, which appears like a small

VOL. I.   U   worm,

worm, but whiter than the husk of the grain. It is an annual plant, being sowed and reaped every year. The stem is about three or four feet in height, and has a large pointed leaf, something like that of the malloro; the flower is of a deep red, and five or six inches in length, and in it are contained the grains or seed. The quinoa is eaten boiled like rice, and has a very pleasant taste; and the water in which it has been boiled, is often used as an apozem. The quinoa is used in external applications, in order to which it is ground and boiled to a proper consistence; and applied to the part affected, from which it soon extracts all corrupt humours occasioned by a contusion.

BESIDES domestic animals, here are great numbers of rabbits caught on the deserts. The partridges are not very plenty, and rather resemble a quail than those of Europe. Turtle-doves abound here, greatly owing to the indolence of the inhabitants in not endeavouring to take them.

BUT one of the principal foods used by the inhabitants is cheese, of which it is computed that the quantity annually confumed amounts to between 70 and 80 thousand dollars of that country money. It is used in various manners, and is the chief ingredient in many dishes. The neighbourhood of Quito also affords excellent butter, and of which there is a great consumption, but falls far short of that of cheese.

THE fondness of these people for sweetmeats exceeds every thing I have ever mentioned of other countries; and this necessarily occasions a great consumption of sugar and honey. One method of indulging this appetite is, to squeeze the juice out of the sugar canes, let it settle, and curdle it, out of which they make small cakes, which they call raspaduras. This is so highly valued by the lower class, that with a slice of it, and another of bread and cheese, they make as hearty a meal as the rich with all their
variety

variety of dishes.　Thus it appears, that if there be some difference between the foods used here and those of Spain, the difference in their preparing them is still greater.

CHAP. VIII.

Of the Commerce of the Province of Quito.

FROM the two preceding chapters, a sufficient judgment may be formed of the products and manufactures in the province of Quito, which are the sources of its commerce.　The persons who are the chief conductors of this commerce, are the Europeans or Chapitones; some settled here, and others coming occasionally.　The latter purchase the country goods, and sell those of Europe.　The manufactures of this province, as we have already noticed, are only cottons, some white called tucuyos, and others striped bays and cloths, which meet with a good market at Lima for supplying all the inward provinces of Peru.　The returns are made partly in silver, partly in gold and silver thread fringes made in that city; wine, brandy, oil, copper, tin, lead, and quicksilver.　The masters of the manufactures either sell their goods to the traders, or employ them as their factors.

On the arrival of the galleons at Carthagena, the traders resort thither either by the way of Popayan or that of Santa Fé, to purchase European goods, which, at their return, they consign to their correspondents all over the province.

The products of the earth are chiefly consumed within the province, except the wheat produced in the jurisdiction of Riobamba and Chimbo, part of which are sent to Guayaquil.　But this is a trade carried on only by Mestizos and poor people.　It would indeed admit of great improvements, were not the freights so

excessively

excessively high, that the trouble and expense of carrying them from Guayaquil to other countries, where there is a scarcity of them, renders it impossible to get a living profit.

GOODS, manufactured by the public, or wove by private Indians, are, together with some kinds of provisions, sent to the jurisdiction of Barbacoas; and this is the commerce in which the chapitones make the first essay of their abilities for trade. These provisions are exchanged for gold, found in that country, and which is afterwards sent to Lima, where it bears a greater price. Their stuffs also find a vent in the governments of Popayan and Santa Fé; and this commerce is perpetually carried on; but the only return in the tiempo muerto, or absence of the galleons, is gold, which, like that from Barbacoas, is sent to Lima.

THE coast of New Spain supplies this province with indigo, of which there is a very large consumption at the manufactories, blue being universally the colour which this people affect in their apparel. They also import, by way of Guayaquil, iron and steel both from Europe and the coast of Guatemala; and though it fetches so high a price, that a quintal of iron sells for above a hundred dollars, and the same quantity of steel for a hundred and fifty, there is a continual demand in order to supply the peasants with the necessary instruments of agriculture.

THE inland, or reciprocal commerce, consists in the consumption of the products of one jurisdiction in another; and is a constant incentive to industry among the inhabitants of the villages, and the lower class. Those of the province of Chimbo purchase home-made tucuyos and bags in those of Riobamba and Quito, in order to vend them at Guayaquil, bringing thence, in return, salt, fish, and cotton; the latter of which, being wove in the looms of Quito, is again sent to Guayaquil in stuffs. The jurisdictions of Riobamba,

Alausi jurisdiction

Alausi and Cuença, by means of the warehouses at Yaguache and Noranjal, carry on a considerable trade with Guayaquil.

THIS trade in the manufactures of that country, which consist only of three sorts, cloth, bags, and linen, is attended with considerable profit to the traders, and advantage to the country, as all the poor people, who are remarkably numerous, and persons of substance, except those of the capital, wear the goods manufactured in the country; those of Europe being so prodigiously dear, that only Spaniards of large fortune, and persons of the highest distinction, can afford to purchase them. The quantity of cloth and stuffs wove in this country, and all by Indians, either in the public manufactures or their own houses, appears from hence to be prodigiously great: and to this, in a great measure, is owing the happy state of this province ; the masters and traders soon raising fortunes, and the servants and dependants contented with the fruits of their industry.

BOOK VI.

Description of the Province of Quito.

CHAP. I.

Extent of the Province of Quito, and the Jurisdiction of its Audience.

IN the five preceding books we have endeavoured, as far as the nature of the subject would permit, to follow the order which the series of our voyage required; and we flatter ourselves it will appear, that, though our principal attention was directed to the astronomical observations, we have not omitted any interesting particular, relating to the towns and provinces through which we passed. We were always persuaded, that if the former tended to the improvement of science, and was agreeable to those who profess it; the latter might prove useful to historians, and be acceptable to those who apply themselves to the study of the constitution, state, customs, and genius of nations. We closed the fifth book with an account of the city of Quito; this we shall employ in treating of the province, which is equally an object of curiosity; and we are enabled to gratify the reader in the most satisfactory manner, having, in the course of our observations, not only surveyed its whole extent, but, by our long stay, obtained the acquaintance of many persons of undoubted judgment and veracity, on whom we could rely for particulars not to be

known

known from ocular inspection. So that we have sufficient reason for warranting the truth of the contents of this history.

THE large province of Quito, at the time when the Spaniards first settled in it, was annexed to the kingdom of Peru, and continued so till the year 1718, when a new viceroyalty being erected at Santa Fé de Bogota, the capital of the new kingdom of Granada, it was dismembered from Peru, and annexed to Granada. At the same time the audience of Quito was suppressed, together with that of Panama, in the kingdom of Terra Firma; though the latter continued dependent on the viceroys of Lima. The intention in this frugal scheme was, that the salaries of the great number of officers in both, which ceased on this abolition, should be applied to the support of the new viceroyalty, in order to prevent any additional burden on the royal revenue; a consequence otherwise unavoidable. But experience has shown the impropriety and insufficiency of this measure; and that the tribunals abolished were of indispensable necessity in their respective cities; an insupportable detriment resulting to the inhabitants from the vast distance of the audiences assigned them; which were, Lima for the kingdom of Terra Firma, and those of the province of Quito were to apply for justice to the audience of Santa Fé. And as the amount of all the salaries suppressed, besides the prejudicing many families, was not sufficient to support the dignity of a viceroy, new ideas succeeded; and rather than keep it up at the expense of the royal revenue, the viceroyalty was suppressed, and things placed again on their antient footing in the year 1722: the officers were restored to their former posts which they had so worthily filled, and the audiences have continued the same as before. But the motives for erecting a new viceroyalty at Santa Fé, being confessedly of the greatest importance, its restitution was again brought on the carpet; and

U 4   the

the great difficulty of supporting it, without detriment either to the public or the audiences, the suppression of which had been so detrimental to the inhabitants, being overcome, the dignity of viceroyalty was again erected in the year 1739, Don Sebastian de Eslaca, lieutenant-general, being appointed the first viceroy, and arrived in the beginning of the year 1740 to take possession of his government; which included the whole kingdom of Terra Firma and the province of Quito.

THIS province is bounded on the north by that of Santa Fé de Bogota, and includes part of the government of Papayan; on the south it is limited by the governments of Peru and Chachapoyas; eastward it extends over the whole government of Maynas, and the river of the Amazons, to the meridian of demarcation, or that which divides the dominions of Spain and Portugal. Its western boundary is the sea, from the coast of Machala, in the gulf of Puna, to the coast of the government of Atacames and the jurisdiction of Barbacoas, in the bay of Gorgona. Its greatest breadth from north to south is about 200 leagues; and its length, from east to west, the whole extent from Cape de Santa Elena, in the south-sea, to the meridian above mentioned; which, by the most accurate computation, is 600 leagues. But a very great part of these vast dominions are, it must be owned, either inhabited by nations of savage Indians, or have not hitherto been thoroughly peopled by the Spaniards, if indeed they have been sufficiently known. All the parts that can properly be said to be peopled, and actually subject to the Spanish government, are those intercepted by the two Cordilleras of the Andes, which, in comparison to the extent of the country, may be termed a street or lane, extending from the jurisdiction of the town of St. Miguel de Ibarra to that of Loga; the country from hence to the government of Popayan, and also that comprehended between the western Cordillera

dillera and the sea. With this limitation the extent of the jurisdictions from east to west will be fifteen leagues or something more, being the distance intercepted between the two Cordilleras. But to this must be added the countries comprehended in the governments of Jaen de Bracamoros, which borders on the jurisdiction of Loja, and the extremity of the whole province, and situated on the east side of the eastern Cordillera; and, to the northward, the government of Quixos, and that of Maynas to the eastward of it: but separated by large tracts of land inhabited by wild Indians; and on the north side of the province from that of Papayan; though the latter is properly a distinct province from that of Quito. Thus on the west side of that interval between the two Cordilleras, lies the lately erected government of Atacames, and the jurisdiction of Guayaquil: on the east side, the three governments above mentioned; and on the north, that of Papayan.

Tʜɪs province, exclusive of these five governments, consists of nine jurisdictions, which in that country are called provinces, that of Quito being subdivided into as many others as there are governments and jurisdictions; which it is necessary for the reader to observe, in order to avoid any perplexity or mistake, when a jurisdiction happens to be called a province; though I shall be careful to avoid it as much as possible. The jurisdictions in the province of Quito, beginning with the most northern, are the following:

     I. The town of San Miguel de Ibarra.
     II. The village of Otabala.
     III. The city of Quito.
     IV. The assiento of Latacunga.
     V. The town of Riobamba.
     VI. The assiento of Chimbo, or Guaranda.
    VII. The city of Guayaquil.

XXII. The

VIII. The city of Cuenca.

IX. The city of Loja.

Of these nine jurisdictions I shall give a succinct account in this and the following chapter, and then proceed to the governments.

I. The town of San Miguel de Ibarra, is the capital of the jurisdiction of that name, which also contains eight principal villages or parishes, the names of
which are,

I. Mira.	V. Salinas.
II. Pimanpiro.	VI. Tumbabiro.
III. Carangue.	VII. Quilca.
IV. San Antonio de Carangue.	VIII. Caguasqui.

This jurisdiction formerly included that of Otabalo; but, on account of its too enormous extent, it
was prudently divided into two.

The town of San Miguel de Ibarra stands on the
extremity of a very large plain or meadow, at a small
distance from a chain of mountains to the eastward of
it, and betwixt two rivers, which keep this whole plain
in a perpetual verdure. The soil is soft and moist,
which not only renders the houses damp, but also
causes the foundations of their buildings often to
sink. It is moderately large, with straight broad
streets, and the greatest part of the houses of stone,
or unburnt bricks, and all tiled. The town is surrounded by suburbs inhabited by the Indians, whose
cottages make the same appearance as in all other mean
places; but the houses are neat and uniform, though
they are but low, having only a ground floor, except
those in the square, which have one story. The parish
church is a large and elegant structure, and of the
same materials as the houses. It is also well ornamented. This town has convents of Franciscans, Dominicans, the Fathers of Mercy, a college of Jesuits
and

and a nunnery of the order of the Conception. Its inhabitants of all ages, sects, and classes, are computed at ten or twelve thousand souls.

WITHIN the limits of this jurisdiction, is the lake of Yagarchoca, famous for being the sepulchre of the inhabitants of Otabalo, on its being taken by Huayna-Capac, the twelfth Ynca, who, instead of showing clemency to their magnanimity, being iritated at the noble resistance they made, ordered them all to be beheaded, both those who had quietly surrendered, and those taken in arms, and their bodies thrown into the lake ; and from the water of the lake being tinged of a bloody hue, it acquired its present name, which signifies a lake of blood.

THE air is very mild, less cold than that of Quito, and at the same time the heat is not at all inconvenient. The temperature of the air is different in all the villages of this jurisdiction, but in most warm, on account of their low situation. These parts are all over this country called Valles, as I have already observed; and the names of those in the jurisdiction of San Miguel de Ibarra are Chotar Carpuela, and several others. Most of the farms in them have plantations of sugar canes, and mills for extracting the juice, from whence they make great quantities of sugar, and very white: some are planted with the fruits common in a hot climate ; and in others cotton only is cultivated, and to the greatest perfection.

THE sugar canes do not ripen here so late as in Quito ; but they may be committed at any time to the mill, there being no necessity for cutting them at any precise time, retaining all their goodness even when suffered to stand two or three months after they are ripe; so that they are cut every quarter, and the mills, by that means, kept at work the whole year.

THE farms situated in a less hot part are employed for cultivating maize, wheat, and barley, in the same manner as in the jurisdiction of Otabalo, and which

we

we shall explain in its proper place.   Here are also
large numbers of goats, but not many sheep; and
though the manufactures here are not so numerous
as in Otabalo, yet the Indians weave a considerable
quantity of cloth and cotton.

In the neighbourhood of the village of Salinas are
salt-mines, which, besides the home consumption,
supply the countries to the northward of it.   This salt
has some mixture of nitre; and though it may thence
be concluded to be less wholesome, yet it is attended
with no ill consequence to those who are accustomed
to it ; but not answering the intention in salting, that
from Guayaquil is used instead of it.

Within the district of the village of Mira, are
great numbers of wild asses, which increase very fast,
and are not easily caught. The owners of the grounds
where they are bred, suffer all persons to take as many as they can, on paying a small acknowledgment in
proportion to the number of days their sport has lasted.   The manner of catching them is as follows: a
number of persons go on horseback, and are attended by Indians on foot.   When arrived at the proper
places, they form a circle, in order to drive them into
some valley ; where, at full speed, they throw the
noose, and halter them : for these creatures, on finding themselves inclosed, make very furious efforts to
escape, and if only one forces his way through, they
all follow with an irresistible impetuosity.   But when
the hunters have noosed them, they throw them down,
and secure them with fetters, and thus leave them till
the hunting is over; when, in order to bring them
away with the greater facility, they pair them with
tame beasts; but this is not easily performed, for
these asses are so remarkably fierce, that they often
hurt the persons who undertake to manage them.
They have all the swiftness of horses, and neither acclivities nor precipices retard them in their career;
when attacked, they defend themselves with their heels
and

and mouth, with such activity, that without slacken-
ing their pace, they often maim their pursuers : but
the most remarkable property in these creatures is,
that after carrying the first load, their celerity leaves
them, their dangerous ferocity is lost, and they soon
contract the stupid look and dullness peculiar to the
asinine species. It is also observable, that these creatures will not permit a horse to live among them; and
if one of them happens to stray into the places where
they feed, they all fall upon him, and, without giving
him the liberty of flying from them, they bite and
kick him till they leave him dead on the spot. They
are very troublesome neighbours, making a most horrid noise; for whenever one or two of them begin to
bray, they are answered in the same vociferous manner by all within the reach of the sound, which is
greatly increased and prolonged by the repercussions
of the valleys and breaches of the mountains.

II. The jurisdiction joining on the south to that
of St. Miguel de Ibarra, is called Otabalo ; in the
jurisdiction of which are the following eight principal villages or parishes :

I. Cayambe.	V. Cotacache.
II. Tabacundo.	VI. San Pablo,
III. Otabalo.	VII. Tocache.
IV. Atontaqui.	VIII. Urquuqui.

The parish of Otabalo is well situated, and so large
and populous, that it is said to contain eighteen or
twenty thousand souls, and among them a considerable number of Spaniards. But the inhabitants of
all the other villages are universally Indians.

The lands of this jurisdiction are laid out in plantations like those of the former, except that here are
not such great numbers of sugar mills; but this is
compensated by its great superiority in manufactures,
a consequence resulting from the multitude of Indians
residing

residing in its villages, who seem to have an innate inclination to weaving; for besides the stuffs made at the common manufactories, such Indians as are not Mitayos, or who are independent, make, on their own account, a variety of goods, as cottons, carpets, pavilions for beds, quilts in damask work, wholly of cotton, either white, blue, or variegated with different colours ; but all in great repute, both in the province of Quito and other parts, where they are sold to great advantage.

THE method of sowing wheat and barley in this jurisdiction, is very different from that used in any of the former; for, instead of scattering the seeds, as is commonly practised, they divide the ground, after it is plowed, into several parts by furrows, and along the sides of them they make little holes a foot distant from one another, putting five or six corns into each. However tedious this may be, it is abundantly made up to the farmer by the uncommon increase, which is usually above a hundred fold.

THIS jurisdiction has a great number of studs of horses, and multitudes of black cattle, from whose milk large quantities of cheese are made. This country is happily situated for pasture, being every where watered with an infinite number of rivulets. It has also large flocks of sheep, though these seem to be neglected, in comparison of the others.

THE village of Cayambe stands in the middle of a spacious plain, at the end of which is the foot of the mountain Cayamburo, one of the largest mountains of the Cordilleras in this part of the country, being equal in height to that of Chimborazo, and its summits covered with snow and ice. Its altitude is so much greater than the rest between it and Quito, that it may be plainly seen from that city. The vicinity of this mountain renders the whole plain of Cayambe cold, which is increased by the violence and continu-

ance

ance of the winds. In the territories of this jurisdiction are two lakes, one called San Pablo, from a village of that name on its banks; it is a league in length, and about half a league in breadth. This lake is every where surrounded with a species of rushes called there totoral, among which are vast flocks of wild geese and gallaretes. This lake receives its water from the mountain of Mojanda; and from it issues one of the branches of the Rio Blanco. The other lake, which has nearly the same dimensions as the former, is called Cuichocha, and is situated in a plain on the side of a mountain of the same name. Near the middle of this are two islands, both which abound with wild cuyes, a species of rabbits, and deer, which often swim to main land; but, when pursued by the hunters, disappoint them by gaining the lake, and swimming back to their retreat. Several small fish are found in this lake, resembling the cray-fish, but without a shell. They are called, by the inhabitants of the adjacent country, prennadillas, and are sent in the pickle to Quito, where they are the more esteemed, as being the only fresh water fish that can be bought in that city. Nor are these caught in any great quantities, though they are also found in the lake of San Pablo.

III. The jurisdiction of Quito consists of the following twenty-five parishes, besides those in the city:

I. St. Juan Evangelista.
II. Santa Maria Magdalena.
III. Chilogalle.
IV. Cono-coto.
V. Zambiza.
VI. Pintac.
VII. Sangolqui.
VIII. Amaguana.
IX. Guapulo.
X. Cumbaya.
XI. Co-collao.
XII. Puembo, and Piso.
XIII. Yaruqui.
XIV. El Quinche.
XV. Guayllabamba.
XVI. Machacha.
XVII. Aloasio.
XVIII. Aloa.
XIX. Yumbicho.
XX. Alangasi.
XXI. Pomasque.
XXII. San

XXII. San Antonio de Lu-   XXIV. Cola-call.
    lum-bamba.   XXV. Tumbaco.
XXIII. Perucho.

THIS jurisdiction, though called Cinco Leguas, five leagues, extends, in some parts, a great deal further, and the lands are as it were covered with plantations, some situated in the plains, some in the capacious breaches, and others on the summit of the mountains; and all producing according to the quality, situation, and exposure of the ground. Those on the temperate plains yield plentiful harvests of maize; those at the bottoms of deep breaches, being in a hot temperature, are planted with sugar canes, from whence they extract great quantities of sugar and rum. From the fruits peculiar to such a temperature, are made a variety of sweetmeats, here called rayados; and of which there is a great consumption among the inhabitants.

THE sugar cane ripens very slowly in this jurisdiction; for though the plantations enjoy a hot air, yet it is not of that degree of heat requisite to its speedy maturity; so that it is three years after they are planted, before they are fit to be cut. Nor are they ever cut but once, the second crop only producing the soca or germ, which serves for replanting.

THE guarapo, which we have had occasion to mention, is nothing more than the juice of the cane, as it flows from the mill, and afterwards suffered to ferment. It is very pleasant, its taste being a sweetish acidity, and, at the same time, very wholesome; but inebriating if drunk to excess. This liquor is a favourite regale among the vulgar.

THE plantations near the summits of the mountains, from their having a variety of temperatures, produce wheat, barley, pot-herbs of all kinds, and potatoes.

ABOVE these plantations are fed numerous flocks of sheep, producing that wool, which, from the several

operations

operations it undergoes, affords employment for such multitudes of people. Some farmers make it their sole business to breed cows, principally for the advantages they derive from their milk in making cheese and butter. In other farm-houses you see various occupations carried on at the same time, namely, the breeding of cattle, agriculture, and manufactures, particularly of cloth, bays, and serges.

From what has been said, it is evident that neither this, nor the preceding jurisdiction, has any general temperature, the degree of cold and heat depending on the situation ; and that to this difference is owing the delightful, and even profitable variety of all kinds of fruits and grains, each finding here a temperature agreeable to its nature. Accordingly, in travelling only half a day, you pass from a climate where the heat sufficiently indicates that you are in the torrid zone, to another where you feel all the horrors of winter. And what is still more singular, and may be esteemed an advantage, no change occurs during the whole year ; the temperate parts never feeling the vicissitudes of cold and heat. This, however, must be allowed not to hold precisely with regard to the mountainous parts, the coldness of which is increased by the violence of the winds, or a change of weather, called tiempo de paramos, when the clouds involve the greatest part of these mountains, and precipitate themselves in a sleet ; at which time the cold becomes intolerable: and, on the other hand, when those frigorific clouds are dispersed, and the wind allayed, so that the rays of the sun reach the earth, they feel the comfortable heat of his cheering beams.

Most of these villages are built with very little regularity. The principal part of them is the church and parsonage, which they call the convent, from the priests being all formerly religious. These structures have some appearance of decency : but the other parts of the village consist of a number of huts with mud-walls, scattered all over the country, where every one has his spot of ground, which he tills for his subsistence. A great part, and in some villages the whole, of the inhabitants are Indians, who live there when out of place : though in some parts the inhabitants are Mestizos, and here and there a Spanish family; but these are extremely poor.

IV. THE first jurisdiction to the southward of that of Quito, is the Assiento Latacunga. The word Assiento implies a place less than a town, but larger than a village. This place stands in a wide plain, having on the east side the eastern Cordillera of the Andes, from whence projects a very high mountain, and at a small distance from its foot is situated Latacunga, in 55 min. 14 sec. 30 thirds, south latitude. On the west side of it is a river, which, though sometimes fordable, on any increase of the waters must be passed over the bridge. This assiento is large and regular; the streets broad and straight; the houses of stone, arched, and well contrived : but, on account of the dangerous consequences so often resulting from earthquakes, without any story. This precaution the inhabitants were taught by a dreadful destruction of all the buildings, on the 20th of June 1698. This terrible concussion was general all over the province of Quito; and its effects, as we shall show in the sequel, in many other places, equally melancholy. Out of six hundred stone houses, the number of which this assiento then consisted, only a part of one, and the church of the Jesuits, were left standing; and even these were so greatly damaged, that there was a necessity for pulling them down. But the greatest misfortune was, that most of the inhabitants were buried under their ruins, the earthquake beginning at one in the morning, a time of general silence and security, and continuing its concussions, at short intervals, the greatest part of the day.

5　　　　　　　　　　　　　　　　THE

The stone of which the houses and churches are built, is a kind of Pumice, or spongy stone, ejected from volcanoes, inexhaustible quarries of it being found in the neighbourhood. It is so light that it will swim in water, and from its great porosity the lime cements the different pieces very strongly together; whence, and from their lowness, the houses are now enabled to support themselves during a concussion much better than before the earthquake, when few were without a story; and if they should be unfortunately thrown down, the crush in all probability would be much less fatal.

The jurisdiction contains these principal villages:

I. Zichios Mayor.
II. Zichios Menor.
III. Yungas, or Colorados.
IV. Ysilimbi.
V. Chisa-Halo, or Toacaso.
VI. Pillaro.
VII. San Phelipe.
VIII. Mula Halo.
IX. Alaquez.
X. San Miguel de Mollcambato.
XI. Saquisili.
XII. Pugili.
XIII. Tanicuchi.
XIV. Cuzubamba.
XV. Tisaleo.
XVI. Angamarca.
XVII. Pila-Halo.

The air of this assiento is the colder, from the place being only six leagues from the mountain of Cotopaxi, which, as it is not less in height and extent than those of Chimborazo and Cayamburo, so it is, like them, covered with ice and snow. The combustible substances within the bowels of this mountain first declared themselves in the year 1533, when Sebastian and Belalcazar, who undertook the conquest of this province, had entered it, and proved very favourable to the enterprise. For the Indians, possessed with the truth of a prediction of their priests, that, on the bursting of this volcano, they would be deprived of their country, and reduced under the government of an unknown prince, were so

X 2   struck

struck with the concurrence of the bursting of this volcano, and the invasion of a foreign army, that the spirit, which universally began to show itself in the preparatives every where made for a vigorous resistance, entirely left them, and the whole province easily conquered, all its caciques submitting to the king of Spain. The large plain in which this assiento stands, is full of fragments of rocks, ejected at that supposed ominous eruption, and some of them to the distance of five leagues from its roots. In the year 1743, while we were on the coast of Chili, a second eruption happened, the particulars of which we shall relate in another place.

The temperature of the air is very different in the several villages of this jurisdiction; being hot in those lying in the valleys; temperate in those situated on the plains; whilst the air in those bordering on the mountains, like that of the assiento, is cold, and sometimes to an excessive degree. The villages are in general larger, and more populous, than those of the other jurisdictions in the same province. Their inhabitants are Indians, Mestizos, and a few Spaniards.

Besides the parish church, which is served by two priests, one for the Spaniards and the other for the Indians, this assiento has convents of Franciscans, Augustines, Dominicans, the Fathers of Mercy, and a college of Jesuits. The churches of these religious are well built, decently ornamented, and kept very neat. The inhabitants, by the nearest computation, amount to between ten and twelve thousand, chiefly Spaniards and Mestizos. Among the former are several families of eminent rank and easy circumstances, and of such virtues and accomplishments as add a lustre to their happy situation. The Indians, as at Quito, live in a separate quarter adjoining to the country.

In this assiento all kinds of trades and mechanic arts are carried on; and, as in all the other parts of this jurisdiction, it has a considerable number of manufactories

ries of cloth, bays, and tucuyos. Great quantities of pork are salted here for exportation to Quito, Guayaquil, and Riobamba, being highly esteemed for the peculiar flavour given to it in the pickling, and which it ever after retains.

ALL the neighbouring country is sowed with clover, and interspersed with plantations of willows, whose perpetual verdure gives a cheerful aspect to the country, and heightens the pleasantness of the assiento.

THE Indians of Pugili and Saquisili are noted for making earthen ware, as jars, pans, pitchers, &c. which are greatly valued all over the province of Quito. The clay of which they are made is of a lively red, very fine, and emits a kind of fragrancy, and the workmanship very neat and ingenious.

V. THE next jurisdiction southwards is Riobamba, the capital of which is the town of the same name. Its jurisdiction is divided into two departments; the corregidor, who resides at Riobamba, appointing a deputy, who lives at the assiento of Hambato, situated between the capital and Latacunga. In the first department are the following principal villages :

I. Calpi.	X. Pungala.
II. Lican.	XI. Lito.
III. Yaruquiz.	XII. Guano.
IV. San Luis.	XIII. Hilapo.
V. Cajabamba.	XIV. Guanando.
VI. San Andres.	XV. Penipe.
VII. Puni.	XVI. Cubijies.
VIII. Chambo.	XVII. Cevadas.
IX. Quimia.	XVIII. Palactanga.

THE department of the assiento of Hamberto has, in its jurisdiction, six principal villages :

I. Isambo.	V. Patate.
II. Quisupincha.	VI. Santo Rosa de Pila-
III. Quero.	guin.
IV. Pelileo.	

THIS assiento stands in the latitude of 1° 41′ 40″ south, and 22′ west, of the city of Quito. In 1533 it was an Indian town, of which Sebastian de Belalcazar having made himself master, the following year marshal Diego de Almagro laid the foundation of the present assiento. It stands in a very large plain surrounded by mountains; particularly on the north side, which is bounded by Chimborazo, from the foot of which it is at no great distance. On the south side is a lake, called Colta, about a league in length and three quarters of a league in breadth, where there are great numbers of wild geese and gallaretas; and its banks covered with plantations.

THE principal square and streets are very regular, straight, and airy; the houses of a light stone, but something heavier than the pumice made use of at Latacunga. Some, especially those in and near the square, have a story; but the others are universally without any, being built low, on account of the earthquakes, which this place has often felt, particularly that already mentioned of 1698, when many of its houses and public buildings were thrown down. The Indians who inhabited this place, and all those to the southward in this jurisdiction, before their conversion to Christianity, were known by the name of Puruayes; and are to this day distinguished from all the other Indians in the whole province.

BESIDES the great church, here is another called St. Sebastian, with convents of the same orders as at Latacunga, and a nunnery of the Conception; contributions are still raised for the use of the hospital, though it is in so ruinous a condition as not to admit of patients.

ON the west side of the assiento is a river cut into small channels or trenches, for watering the adjacent fields: by which means they are rendered so remarkably fertile, that they produce clover the whole year.

THE

THE inhabitants, according to an accurate calcula-
tion, amount to between sixteen and twenty thousand
souls. The manners and customs here are nearly the
same with those at Quito; the greatest part of the
families of distinction in that city owing their origin
to this place. For at the beginning of the conquests,
many of the eminent families which came from Spain
settled here at the conclusion of the war, and have
been very careful not to diminish either the lustre of
their families, or their wealth, by promiscuous alli-
ances, marrying only into one another.

THE magistracy consists of regidores, who are always persons of the first distinction, and from among
those are annually chosen the ordinary alcaldes; with
this singularity, that the validity of the election depends on its being unanimous, a single vote rendering
it void. Besides, the person thus elected is either confirmed or rejected by the townsmen; a privilege
known in no other part of the whole province.

THE air is colder here than at Quito, owing in a
great measure to the neighbourhood of the mountain
of Chimborazo; and, when the wind blows from that
quarter, the weather is so sharp, that the rich families
leave the town, and retire to their estates, situated in
a warmer air, though at no great distance.   This uncomfortable season generally lasts from December to
June, the north and north-west winds then principally prevailing.   It is, however, in a great measure,
free from those violent showers and tempests so common at Quito, that sometimes for many days successively it enjoys serene and delightful weather;
and the same may be said of the greatest part of its
jurisdiction.

HERE are many plantations, or farms, and most of
them considerable; and for the number and largeness
of its manufactories, it surpasses every other part of
this province; though the Indians seem born with an
inclination for weaving, particularly those of the vil-
X 4   lage

lage of Guano, who are famed for their manufactures
of worsted stockings, and it is indeed the only place in
the whole province where they are made.   This industrious disposition probably gave rise to the large
flocks of sheep in this jurisdiction, whence these manufactories are never in want of wool.   The soil is
very fertile, producing all kinds of grain and pulse in
abundance.   And here is most frequently seen what
I have elsewhere observed: that in one part the
husbandmen are sowing, in another reaping; the
landscape also elegantly adorned with such an enchanting variety of colours as painting cannot express. .

In this jurisdiction is a large plain lying south of
the town of Tiocaxas, and famous for a battle between the Spaniards commanded by Belalcazar and
the puruayes Indians, before their courage had been
depressed by the ominous explosion of the mountain.
Both armies fought with great obstinacy, though neither gained the victory.

The assiento of Hambato stands in a wide plain at
the bottom of a mountain.   On the N. side of it
runs a large river, over which a bridge has been built,
it being never fordable on account of its depth and
extreme rapidity.   It is finely situated, and in extent
and populousness nearly equal to Latacunga, the number of its inhabitants amounting to eight or nine
thousand.   The houses are of unburnt bricks, well
contrived, and make a good appearance.   With regard to their lowness, it is owing to a discreet precaution against the melancholy shocks of earthquakes.
It has a parish-church, two chapels of ease, and a
convent of Franciscans.   The earthquake which made
such terrible havock in the assiento of Latacunga,
proved also fatal to this.   The earth near it opened
in several places, of which there still remains an as-
tonishishing monument on the S. side of the assiento,
being a chasm four or five feet broad, and about a
league

league in length, north and south.   And on the N.
side of the river are several openings of the same kind.
The horror of the shock was greatly increased by
terrible eruptions from Mount Carguairaso, from
whence a muddy torrent, formed of ashes, cinders,
and snow melted by the flames from the aperture,
precipitated down the sides of the mountain, over-
flowing the fields, sweeping away the cattle, and every
other object, by its violence.   A track of this impetuous current is still to be seen on the S. side of the
assiento.

THE inhabitants in their manners and customs resemble those of Quito; but with regard to families of
distinction, it is much inferior to Riobamba. Courage
is an innate quality of the natives, but blended with
such vices, that both their neighbours, and the inhabitants of the other parts of the province, will have
no concerns with them, except those absolutely necessary; and, in all dealings with them, take care to
guard equally against their deceit and violence.

THIS jurisdiction in several of its products and manufactures excels all the rest : one of which is bread,
particularly that made at the assiento, which is famous
all over the province; and accordingly it is sent to
Quito, and other parts, without losing any thing of
its goodness by length of time.   The Indian inhabitants of the village of Quero make all sorts of cabinet work, for which there is a great demand all over
the province, as, besides the goodness of the workmanship, this is the only place where goods of this kind
are made.   The jurisdiction of Patate is equally famous for the plenty of sugar canes, and the goodness of the sugar made from them, being of the finest
sort.   That of Santa Rosa de Pilaguin, which, with
its fields, lies on the side of Carguairaso, is famous for
the particular goodness of its barley, as the district
bordering on the assiento is for the exquisiteness of its
fruits; and to this district Quito owes most of the European

ropean kinds sold in that city, the temperature of the air being peculiarly adapted to the perfection of those friuts.

VI. On the W. side of the jurisdiction of Riobamba, between it and Guayaquil, lies that of Chimbo, whose jurisdiction consists of an assiento and seven villages: the former, being the capital, is called Chimbo, and was the residence of the corregidor, till it was thought proper, for the conveniency of commerce, to remove it to Guaranda. This assiento does not contain above eighty families; some of which are Spaniards, but all poor. The names of the villages are,

I. San Lorenzo.	V. Guaranda.
II. Asaneoto.	VI. Guanujo.
III. Chapacoto.	VII. Tomabelas.
IV. San Miguel.	

The most considerable of their villages is that of Guaranda, though the inhabitants are generally Mestizos; there are some Indians, but very few Spaniards.

The jurisdiction of Chimbo, being the first of the Serrania. or ridge of mountains, bordering on that of Guayaquil, carries on, by means of innumerable droves of mules, the whole trade of Quito and the other provinces, by the way of Guayaquil, carrying the bales of cloth, and stuffs, together with the meal, corn, and other products of the country, from the former to the latter; and returning with wine, brandy, salt, cotton, fish, oil, and other goods wanted in the provinces of the mountains. This traffic is of inconceivable benefit to the inhabitants; but it can only be carried on during the summer, the roads in the winter being absolutely impracticable to beasts of any kind. This intermission of trade they call 'Cerrarse la montana,' The shutting up of the mountains.

THE

The temperature of the air at Guaranda, and that of the greatest part of the jurisdiction of Chimbo, from the proximity of Chimborazo, so often mentioned for its frigorific effects, is very cold. The country is large and fertile, like those already mentioned; but the haciendas, or farms, are in general appropriated to the breeding of mules; a few only being sown with different species of grain.

VII. The jurisdiction of Guayaquil is the last; but this has been already treated of at large.

CHAP. II.

Sequel of the Account of the Jurisdictions in the Province of Quito.

VIII. THE jurisdiction bordering on the southern parts of Riobamba, is that of Cuença, whose capital is the city of the same name, founded in the year 1557 by Gil Ramirez Davalos. Its jurisdiction is divided into two departments, of which the capital is one, and that of Alausi the other; the last reaches to Riobamba, and is governed by a deputy of the corregidor. Besides the assiento, it contains only the four following villages:

I. Chumche.	III. Cibambe.
II. Guasuntos.	IV. Ticsan.

But that of the city of Cuença incudes ten:

I. Azogues.	VI. Paccha.
II. Atuncanar.	VII. Gualaseo.
III. Giron.	VIII. Paute.
IV. Canary-bamba.	IX. Delec.
V. Espiritu,	X. Molleturo.

The

THE city of Cuença lies in 2° 53′ 49″ south latitude, and 29′ 25″ west of the meridian of Quito. It stands in a very spacious plain, along which, at about half a league to the northward of the city, runs a little river called Machangara; and close to the south side of the city runs another known by the name of Matadero. Besides these, at the distance of a quarter of a league, runs another called Yanuncay; and at about the same distance is another termed Los Banos, from a village of that name, through which it flows. All these rivers are at some seasons fordable; but at others can only be crossed with safety over the bridges.

THE plain in which this city stands, reaches about six leagues from north to south; and the four rivers, whose courses are nearly in the same direction, form, at a small distance, by the conflux of their streams, a very large river. To the south of the city is another plain of about two leagues in extent, and, with its great variety of regular plantations of trees, and other rural improvements, makes a very delightful appearance all the year round.

THIS city may be classed among those of the fourth order. Its streets are straight, and of a convenient breadth; the houses of unburnt bricks, tiled, and many of them have one story, the owners, from a ridiculous affectation of grandeur, preferring elegance to security. The suburbs, inhabited by the Indians, are, as usual, mean and regular. Several streams of water, by great labour, are brought from the above rivers, and flow through the streets; so that the city is plentifully supplied; and for its admirable situation, and the fertility of the soil, it might be rendered the paradise, not only of the province of Quito, but of all Peru; few cities being capable to boast of so many advantages as concentre here; but, either from supineness or ignorance, they are far from being duly improved. One circumstance, which adds a singular
beauty

beauty to its situation, is, that the mountains are not so high as to intercept the view of a beautiful country; but at a proper distance they rise again to their stupendous height, as is seen in the mountain Azuay, which divides this jurisdiction from that of Alausi.

Cuença contains three parishes; that of the great church consists of Spaniards and Mestizos; the two others, which are called San Blas and San Sebastian, are for the Indians. Here are convents of Franciscans, Dominicans, Augustines, and the Fathers of Mercy; a college of Jesuits, and two nunneries, one of the Conception, and the other of Santa Teresa. Here is also an hospital, but through ill management now almost in ruins.

The magistracy is composed of regidores and ordinary alcaldes, which, according to the general custom, are chosen annually, and at their head is the corregidor. Here is a chamber of finances, under the direction of an accomptant and treasurer. It was formerly kept in the city of Sevilla del Oro, a jurisdiction, and the capital of the department of Macas: but on the loss of the city of Logrono, the village of Cuambaya and other places, it was removed to Loja, and since to Cuença. The revenues paid into it consist of the tribute of the Indians of this department, together with that of Alausi, the jurisdiction of Loja, and the government of Jean de Bracamoros; the duties on provisions, and the customs collected at Naranjal.

The inhabitants here, though of the same classes with those of Quito, differ something in their genius and manners; particularly in a most shameful indolence, which seems so natural to them, that they have a strange aversion to all kinds of work; the vulgar are also rude, vindictive, and, in short, wicked in every sense. From this general reproach, the women must, however, be excepted, being remarkable for an uncommon spirit of industry; as if they were determined

termined to atone for the indolence of the other sex. They spin and weave bays, which, for their goodness, and especially the brilliancy of the colours, are famous in every part of Peru. They also weave some tucuyos; and make bargains with the merchants or traders. They buy and sell; and, in short, manage entirely that little commerce by which their families are supported: whilst their husbands, brothers, and fathers, give themselves up to sloth and idleness, with all its infamous concomitants. The whole number of inhabitants of this city is computed at twenty or thirty thousand souls; and both those of the city and of the jurisdiction are commonly known by the general name of Morlacos.

The pleasures arising from the fertility of the soil are increased by the mildness of the climate, the liquor of the thermometer fluctuating the whole year between 1013 and 1015; so that the cold is very little felt, and the heat very supportable. With regard to rains, and tempests of thunder and lightning, they are as common here as at Quito. In calm weather, the sky is serene, and the inhabitants healthy; nor are malignant fevers and pleurisies, though common to the whole province, so often known as at Quito. The country is finely interspersed with farmhouses and plantations of sugar canes; some parts are cultivated for corn, and others applied to the feeding of sheep and horned cattle, from the last of which they make great quantities of cheese; not inferior to that of Europe; and accordingly there is a very considerable demand for it all over these parts.

The village of Atun-canar, or Great-canar, is famous for its extensive corn fields, and the rich harvest they afford. It is also remarkable for the riches concealed in its mountains, the bravery of its ancient inhabitants, and their unshaken loyalty to Ynca Tupac-Yupanqui, to whom, when his army intended for this country was arrived near the frontiers, sensible of
their

their inability of making any effectual resistance, they submitted, and paid him all the honours which denoted a voluntary subjection; and these marks of loyalty so possessed the emperor in their favour, that, to encourage them to cherish such good dispositions, he ordered several magnificent temples, splendid palaces, and forts, all of stone, to be built here, in the manner of those of Cusco, and the inside of the walls to be plated over with gold. And of these works some monuments still remain in a fort and palace, and of which neither time nor accidents have obliterated their astonishing magnificence; a description will be given of both in another place. These works had such happy effects on the grateful inhabitants, that they fell at last victims to their loyalty; for, having sided with the Ynca Huescar, their lawful sovereign, against his brother Ata Huallpa, and the former losing a decisive battle, the conqueror inhumanly abused his victory, by destroying those unhappy persons who had done no more than their duty, no less than 60,000 of them being massacred in cold blood.

These Indians were united with the Guasantos, and those of Pamallacta, in which district are still to be seen the ruins of another fort, built by the Yncas. The intimacy between the inhabitants of these countries was so remarkable, that they were all called Canarejos, that under one name they might form one body.

The assiento of Alausi, the chief place of the second department, is not very populous, though among its inhabitants are some Spanish families of the first rank. The other inhabitants are Mestizos and Indians, but both classes in mean circumstances. The parochial church is the only ecclesiastical structure; nor has this the ornaments which decency requires.

The village of Ticsan, which stood in this department, was totally destroyed by an earthquake, and the inhabitants removed to a safer situation. The marks

of

of these dreadful convulsions of nature are still visible in several chasms among the mountains, many being two or three feet broad, a convincing proof of the violent concussions in the bowels of the earth at the time of that catastrophe. The temperature of the air is here something colder than at Cuença; but not in a degree sufficient to lessen the exuberant fertility of the soil.

Among the great variety of mines in the jurisdiction of Cuença, and which I shall consider more at large in the sequel, those of gold and silver, according to the common opinion, are not the least numerous. Report has indeed magnified them to such a degree, that, to prove the astonishing quantity of those metals, the inhabitants relate the following story, the truth of which I do not pretend to warrant. It exhibits indeed an instance so contrary to the common order of things, as to be scarce reconcileable to reason. I shall, however, venture to relate it, because, if the reader should think it incredible, it will at least serve to convey an idea of the riches supposed to be concealed in the bowels of the mountains.

Between the valleys of Chugui-pata, which extend from the village and jurisdiction of Los Azogues southward, and that of Poute running eastward along the banks of the same name, are several eminences which divide the two plains, and among these one higher than the rest called Supay-urco, a name said to have been given it on the following account. An inhabitant of the province of Estramadura in Spain, from the extremity of his distress, abandoned himself to despair; and, in the frenzy of his wild imagination, sometimes implored the assistance of Satan, and sometimes cursed the moment that gave birth to his wretched being, and was for laying violent hands upon himself. The devil taking advantage of his condition appeared to him, but in a dress which sufficiently concealed his nature, and courteously asked the cause of his excessive

cessive melancholy; and being informed that it was owing to an unhappy change of circumstances, from a plentiful fortune to the most extreme poverty, the devil, with a cheerful air, told him, that he would show him a spot from whence he might have what quantity of gold he pleased, the mine being absolutely inexhaustible. The Spaniard embraced the offer with the greatest transport of joy; and, concluding that it would at least prove a journey of some days, purchased, with the penurious remains of his substance, a few loaves, which he packed up in his wallet; and, his mind being something easier from these flattering promises, laid himself down to rest till the time appointed, when he was to call upon his guide. But when he awaked, he found himself in a country absolutely unknown, the plain of Chequipata lying before him, and himself reclined on the eminence of Supay-urco. His astonishment, at viewing such multitudes of strange objects, can be much better conceived than expressed. For some time indeed he doubted whether they were real or illusive, till tired with uncertainties, and determined to know in what country he was, he directed his way to a house of some figure, which he saw at a distance. This happened fortunately to belong to a Spaniard, who was a native of the same province of Estramadura; and being informed by his servants that a stranger of the same country was at the gate, the master, pleasing himself with the hopes of hearing some news from his native land, ordered him to be brought in, received him with great marks of friendship, and, being at breakfast, made him sit down with him, and began to enter on the pleasing inquiry after his friends and relations; but his guest taking out one of his loaves, which the gentleman knew was baked in Spain, and finding it quite new, was so lost in astonishment, that he forgot both his breakfast and relations, insisting (though afraid to hear) that his apparent countryman should inform him how it was possible to make so long a voyage in so short a time. The other readily satisfying his desire, they both agreed that this must have been an action of that enemy to mankind, who had brought the poor Spaniard thither to enrich himself from the treasures concealed in the bowels of the hill on which he had laid him; and ever since it has been called Supay-urco, or the Devil's Hill. This story is well known throughout all the jurisdiction of Cuença, even the children are acquainted with it; and father Manuel Rodriguez, in his ' Historia del Maranon, y Amazonas, lib. ii. cap. 4.' mentions it. From all which it may be inferred, that it is, in reality, of as ancient a date as the inhabitants of Cuenca pretend; that it has descended through a long series of time without alteration; and from this story, though destitute of proof, the notion that this hill contains an inexhaustible treasure had its rise.

IX. The last jurisdiction of the province of Quito, on this side, is that of Loja, the capital of which is called by the same name, and was founded in the year 1546, by captain Alonso de Mercadillo. It resembles, in extent, form, and buildings, the city of Cuença; but the temperature of the air is considerably hotter. In its district are the following fourteen villages:

I. Saraguro and Ona.	VIII. Zororonga.
II. San Juan del Valle.	IX. Dominguillo.
III. Zaruma.	X. Catacocha.
IV. Yuluc.	XI. San Lucas deAmboca.
V. Guachanana.	XII. El Sisne.
VI. Gonzanama.	XIII. Malacatos.
VII. Cariumanga.	XIV. San Pedro del Valle.

Loja, besides two churches, has several convents, a nunnery, a college of Jesuits, and an hospital.

In theterritory of this jurisdiction is produced that famous specific for intermitting fevers, known by the name

name of Cascarilla de Loja, or Quinquina. Of this specific there are different kinds, one of which is more efficacious than the others. M. de Jussieu, whom I have already had occasion to mention more than once, being sent to make botanical observations, and take care of the health of the academicians, took the trouble of making a journey to Loja, purely to examine the tree which produces it; and in a full description, which he drew up for the satisfaction of botanists and other curious persons, enters, with his known skill and accuracy, into a very minute distinction of the several species, and enumerates the smallest circumstances. At the same time he was pleased to inform the corregidor of the differences, and to instruct the Indians employed in cutting it to distinguish each species, that the best sort only might be sent unmixed to Europe. Nor was this all; he farther instructed them how to make an extract of it, and prevailed on the inhabitants of that territory to use it, where its virtues had till that time been neglected, though intermitting fevers are there as common as in any other parts. Before he undeceived them, the natives imagined that it was exported to Europe only as an ingredient in dyeing; and though they were not entirely ignorant of its virtues, they made no use of it, little imagining that a simple of so hot a nature could be good for them. But this ingenious physician convinced them of their mistake by many happy effects; so that now it is generally used in all kinds of fevers: and persons of undoubted veracity, who have since visited Loja, have given me very pleasing accounts of its salutary effects.

THE tree which produces the cascarilla is not of the largest size, its usual height being about two toises and a half, and the body and branches of a proportionate thickness. In this, however, there is some difference, and in that consists the goodness of the cascarilla, the largest branches not yielding the best.

Ther

There is also a difference both in the blossom and seed. The Indians, in order to take off the cassarilla or cortex, cut down the tree; after which they bark it, and dry the quinquina. There are here large and thick forests of this tree; but notwithstanding this, there is a very sensible diminution of them, occasioned by the Indians not sowing the seed; those which grow spontaneously not being by any means equal to those which have been cut down.

In the jurisdiction of Cuença have also been discovered many woody parts, in which this valuable tree is found: and when I was in that country, a priest at Cuença procured a large quantity of cascarilla, and sent it to Panama, the only place from whence it is exported. This instance, together with his assurances that it was of the same kind with that of Loja, induced several of the inhabitants of Cuença to attempt the discovery, and were soon convinced that the jurisdiction contained large forests of this tree, which had been neglected by them, whilst their neighbours reaped no small advantages from it.

The jurisdiction of Loja has also a very great advantage from breeding the cochineal, and which intelligent persons reckon of equal goodness with that of Oaxaca in New Spain; but the inhabitants are so far from applying themselves to the breeding of that insect, sufficient to supply the demands of a particular trade, that they breed no more than what they imagine will be sufficient for the dyers in that and the neighbouring jurisdiction of Cuença. To this elegant and lasting colour it is probably owing that the bays of Cuença, and the carpets of Loja, are preferred to all others: though the beauty of the colours may in some measure proceed from the superior skill of the workmen of Loja and Cuença, over those of Quito and other parts of the province where the same goods are manufactured. The cochineal is also bred in the department of Hambato, though without any constant gatherings

therings of that insect. It is not however to be
doubted, but that a more careful attention would
ensure them the same success in great as in smalll
quantities.

Having mentioned this insect, so highly valued in
every part of the world for the incomparable beauty
of its red, which it equally communicates to wool,
silk, linen, and cotton, it may be expected that I
should give some farther account of it; and as I
should be sorry to disappoint any rational curiosity of
my readers, and at the same time to insert any thing
that is not strictly true, I was unwilling to rely wholly
on my own experience; together with the accounts I
procured at Loja and Hambato, especially as Oaxaca
is the principal place where this insect is produced, I
made it my business to consult persons well acquainted
with the subject, and received the following account,
in which they all unanimously agreed.

The cochineal is bred on a plant known in Oaxaca,
and all those parts where it abounds, by the name of
ncspal*, or nopalleca, the Indian fig-tree, which, except in the difference of the foliage, resembles the
tunos, so common in the kingdom of Andalusia. The
leaf of the tuna being broad, flat, and prickly; and
that of the nopal, oblong, with several eminences;
and instead of spines, has a fine smooth membrane,
of a fine permanent and lively green.

The method of planting the nopal is by making
rows of holes, about half a yard deep, and about two
yards distant from one another. In each of these holes
is placed one or two leaves of the nopal, in a flat position, and then covered with earth. This leaf soon
after shoots up into a single stem, which during its
growth divides into several branches, and these succes-

* This plant is called by botanists, *Opuntia maxima, folio oblongo
rotundo majore, spinulis obtusis mollibus et innocentibus obsito, flore striis
rubris variegato.* Sloane's Catalogue.

Y 3   sively

sively produce fresh leaves, the largest being nearest to
the stem, which is full of knots, as are also the branches,
and from these the leaves have their origin. The
usual height of this plant is about three yards, which
it seldom exceeds. The season when the nopal displays all its beauty and vigour is, like that of other
plants, from the spring to the autumn, which at
Oaxaca, and other parts of North America, is at the
same time as in Spain. Its blossom is small, of a
bright red, and in the shape of a bud, from the centre
of which proceeds the tuna, a name given to its
fruit; and as this increases, the blossom fades, till at
length it falls. When the tuna, or fig, is ripe, the
outward skin becomes white; but the pulp is so fully
impregnated with a deep red, that it tinges of a blood
colour the urine of those who eat it; a circumstance
of no small uneasiness to those who are unacquainted
with this particular. Few fruits, however, are either
more wholesome or pleasant.

THE ground where the nopal is intended to be
planted must be carefully cleansed from all kinds of
weeds, as they drain the soil of those juices which
the nopal requires. Also after the cochineal is taken
from the plant, which is never done till the insects
are arrived at perfection, all the superfluous leaves
are plucked off, that they may be succeeded by others
the following year. For it must be observed, that
the chineal which are bred on young plants thrive
much better, and are of a finer quality, than those
produced on such as have stood some years.

THE cochineal was formerly imagined to be a fruit
or seed of some particular plant; an error which probably arose from an ignorance of the manner in
which it is propagated; but, at present, every one is
convinced of its being an insect, agreeably to its
name, signifying a wood-louse, which generally breeds
in damp places, especially in gardens. These insects,
by rolling themselves up, form a little ball, something

thing less than a pea, and in some places are known by the name of Baquilas de San Anton, i. e. St. Anthony's little cows: and such is the figure of the cochineal, except that it has not the faculty of rolling itself up; and its magnitude, when at its full growth, does not exceed that of a tick, common in dogs and other animals.

These insects breed and are nourished on the nopals, where their eggs are placed among the leaves ; the juice of the plant, which is their sole nourishment, becomes converted into their substance; when, instead of being thin and waterish, and, to all outward appearance, of little or no use, it is rendered a most beautiful crimson colour.   The plant is in May or June in its most vigorous state, and at this favourable season the eggs are deposited ; and in the short space of two months, from an animalcule, the insect grows up to the size abovementioned: but its infant state is exposed to a variety of dangers; the violent blasts of the north wind sweep away the eggs from the foliage of the plant ; and, what is equally fatal to their tender constitutions, showers, fogs, and frosts, often attack them, and destroy the leaves, leaving the careful cultivator this only resource, namely, that of making fires at certain distances, and filling the air with smoke, which frequently preserve them from the fatal effects of the inclemency of the weather

The breeding of cochineal is also greatly obstructed by birds of different kinds, which are very fond of these insects ; and the same danger is to be apprehended from the worms, &c. which are found among the plantations of nopals : so that, unless constant care be taken to fright the birds away from the plantation, and to clear the ground of those various kinds of vermin which multiply so fast in it, the owner will be greatly disappointed in his expectations.

When the insects are at their full growth, they are gathered and put into pots of earthen ware; but great

Y 4   attention

attention is requisite to prevent them from getting out, as in that case great numbers of them would be lost: though there is no danger of it, where they are at liberty on the nopal leaves, those being their natural habitation, and where they enjoy a plenty of delicious food; for though they often remove from one leaf to another, they never quit the plant; nor is it uncommon to see the leaves entirely covered with them, especially when they are arrived at maturity. When they have been confined some time in these pots, they are killed and put into bags. The Indians have three different methods of killing these insects; one by hot water, another by fire, and a third by the rays of the sun; and to these are owing the several gradations of the colour, which in some is dark, and in others bright; but all require a certain degree of heat. Those, therefore, who use hot water are very careful to give it the requisite heat, and that the quantity of water be proportioned to the number of insects. The method of killing them by fire, is to put them on shovels into an oven, moderately heated for that intention; the fine quality of the cochineal depending on its not being over-dried at the time of killing the insects: and it must be owned, that among the several ways made use of to destroy this valuable creature, that of the rays of the sun seems to bid fairest for performing it in the most perfect manner.

BESIDES the precaution requisite in killing the cochineal, in order to preserve its quality, it is equally necessary to know when it is in a proper state for being removed from the leaves of the nopal; but, as experience only can teach the cultivator this necessary criterion, no fixed rule can be laid down. Accordingly, in these provinces where the cultivation of these insects is chiefly carried on, those gathered by Indians of one village differ from those gathered in another; and even those gathered by one person in the same village
are

are often different from those gathered by another ; every individual adhering to his own method.

The cochineal insect may in some circumstances be compared to the silkworm, particularly in the manner of depositing its eggs.   The insects destined for this particular are taken at a proper time of their growth, and put into a box well closed, and lined with a coarse cloth, that none of them be lost.   In this confinement they lay their eggs and die.   The box is kept close shut till the time of placing the eggs on the nopal, when, if any motion is perceived, it is a sufficient indication that the animalcule has life, though the egg is so minute as hardly to be perceived; and this is the seed placed on the foliage of the nopal, and the quantity contained in the shell of a hen's egg is sufficient for covering a whole plant. It is remarkable that this insect does not, or at least in any visible manner, injure the plant ; but extracts its nourishment from the most succulent juice, which it sucks by means of its proboscis through the fine teguments of the leaves.

The principal countries where the cochineal insects are bred, are Oaxaca, Flascala, Ceulula, Nueva Gallicia, and Chiapa in the kingdom of New Spain; and Hambato, Loja, and Tucuman in Peru.   And though the nopal thrives equally in all, yet it is only in Oaxaca that they are gathered in large quantities, and form a branch of commerce, the cultivation of these little creatures being there the chief employment of the Indians ; whereas in others, where the inhabitants take but little trouble in their cultivation, they breed wild, and those gathered in them are accordingly called grana sylvestria*.   Not that either the insects or nopals are of different species ; for with regard to the disadvantageous difference between the

---

* This wild cochineal is generally known in England by the name cochineal mestique.

colour

colour of the wild cochineal and that of Oaxaca, it does not proceed from a difference of species, but from a want of proper care in its improvement; and were the culture every where alike, this difference would no longer subsist. But the Indiansneglectit, either because no commerce of that kind has been opened among them ; or from an aversion to the trouble and attention requisite to bring those insects to perfection ; or, lastly, from the apprehension that the fruits of all their time and care may be destroyed by one of the above-mentioned accidents.

The temperature best adapted to the production of this insect cannot be precisely determined, there being in Oaxaca, as well as in the province of Quito, parts of very different temperatures, some hot, some temperate, and others cold ; yet all ·breed the cochineal. It is, however, very probable, that the most proper climate is the temperate and dry; because in these the nopal thrives the best. And agreeably to this observation it is remarked, that Hambato and Loja are the countries in the province of Quito where they most abound ; though they are.also seen in other parts, where both the heat and cold are greater.

Here I cannot help observing, that Andalusia in Spain appears to me extremely well situated for breeding cochineal, both from the nature of the climate, and the plantation of fig-trees, which there attain so great perfection. Here also neither frosts, fogs, or snows, are to be apprehended, particularly in spring: and the happy medium between cold and heat is, as I have before observed, that which this creature is particularly fond of.

The inhabitants of Loja, who are known all over this province by the name of Lojanos, do not exceed ten thousand souls, though formerly, when the city was in its greatest prosperity, they were much more numerous. Their character is much better than that
of

of the inhabitants of Cuença; and besides their affinity in customs and tempers to the other villages,
they cannot be branded with the character of being
slothful.   In this jurisdiction, such numerous droves
of horned cattle and mulesare bred, that it supplies
the others of this province, and that of Piura in
Valles.   The carpets also manufactured here are of
such remarkable fineness, that they find a ready sale
wherever they are sent.

Tʜᴇ corregidor of Loja is also governor of Yaguarsongo, and principal alcalde of the mines of Zaruma;
and, as such, a chair of state is placed for him at all
public solemnities of the church, where he is present;
a distinguishing honour allowed only to the presidents
or governors of those provinces.   The post of governor of Yaguarsongo is at present a mere title without any jurisdiction; part of the villages which formed
it being lost by the revolt of the Indians, and the
others added to the government of Jaen; so that the
corregidor of Loja enjoys only those honours intended to continue the remembrance of that government.

Tʜᴇ town of Zeruma, in the jurisdiction of which
are those mines of gold I shall mention in another
part, has presented the corregidor of Loja with the
title of its alcalde major.   It was one of the first
towns founded in this province, and at the same time
one of the most opulent; but is at present in a mean
condition, owing chiefly to the decay of its mines, on
which account most of the Spanish families have retired, some to Cuença, and others to Loja; so that
at present its inhabitants are said not to exceed six
thousand.   The declension of these mines, which is
not so much to be imputed to a scarcity of metal, as
to the negligence of those concerned in working
them, has been disadvantageous to the whole department of Loja; and consequently diminished the number of its inhabitants.

Hᴀᴠɪɴɢ

HAVING thus described those nine jurisdictions
which form the most wealthy part of the province of
Quito, I shall, in the following chapters, treat of the
governments.

CHAP. III.

Account of the Governments of Popayan and Atacames,
belonging to the Province of Quito.

WE have already given a just account of every
thing worthy notice in the jurisdictions within
the audience of Quito. To render the narrative complete, it is necessary that we now proceed to the governments within the limits of that audience; as they
jointly form the vast country of the province of Quito. And though they generally give the name of
province to every government, and even to the departments into which both are subdivided, we shall not
here follow this vulgar acceptation, it being in reality
founded only on the difference of the notions of Indians who formerly inhabited this country, every one
being governed by its curaca, or despotic sovereign.
These nations the Yncas subdued, and obliged them
to receive the laws of their empire: but the curacas
were confirmed in all those hereditary rights of sove
reignty compatible with the supreme prerogative.
Were we indeed to use the name of province in this
sense, every village must be called so; for it may be
easily shown, that, in the time of heathenism, every
village had its particular curaca: and sometimes, as
in Valles, in this jurisdiction of Popayan, in Maynas,
and the Moragnon, there was not only a curaca in each
village, with all the appendages of government, but
the inhabitants spoke a different language, had different laws and customs, and lived totally independent
of each other. But these villages and ancient pro-

2   vinces

vinces being now comprehended under the jurisdiction of one single tribunal; and those which before were under a multitude of curacas acknowledging one sovereign, and composing one province, where justice is administered to them in the name of the prince; and the governments being in juridical affairs dependent on the audience of Quito; they can only be considered as parts of its province. It is therefore requisite, in order to form a proper idea of this country, that I should treat of them in the same circumstantial manner I have already observed in describing the jurisdictions.

I. The first government in the province of Quito, and which terminates it on the north, is that of Popayan. It is not indeed wholly dependent on it, being divided into two jurisdictions, of which that on the north and east belong to the audience of Santa Fé, or the new kingdom of Grenado; Quito having only those parts lying towards the south and west; so that, without omitting any thing remarkable in the whole government, I shall be a little more explicit in my account of the department belonging to Quito.

The conquest of the whole country now containing the government of Popayan, or at least the greater part of it, was performed by that famous commander Sebastian de Belalcazar, who, being governor of the province of Quito, where he had settled a perfect tranquillity, and finished the building of that city, being informed that on the north side of his government lay a country of great extent, and richer than the parts he already possessed, prompted by that spirit which had animated the Spaniards to extend their reputation by a series of amazing conquests in this part of the globe, he set out on his enterprise in 1536, at the head of 300 Spaniards; and after several sharp encounters with the Indians of Pasto, who first opposed his march, he proceeded in his conquests, and reduced the two principal curacas of that country, Calambas

bas and Popayan (after whom both the country and
chief town were called), two brothers equally respected for their power and military talents.   This defeat
opened him a passage to future conquests; and the
neighbouring nations, terrified at the success of those
illustrious warriors, submitted to the king of Spain.
Belalcazar, after these exploits, in the prosecution of
his conquests, had several other encounters with Indians, fired with the disdain of submitting to a foreign
yoke. His conquests were, however, at last so rapid, that at the close of the same year he pitched his
camp in the centre of that country, where the mildness of the climate, the fertility of the soil, and salubrity of the air, conspired to induce him to render it
the seat of the Spanish government. Accordingly, in
1537, he laid the foundation of the first city, which
still retains the name of Popayan ; and whilst the
place was building, he, to keep his people in exercise, and prevent the Indians he had conquered from
forming themselves into a new army, or carrying on
any clandestine correspondence with those whom his
arms had not reached, sent out detachments different
ways, with orders to march into the neighbouring
countries, that they might prevent the rising of some,
and reduce others to obedience.

Belalcazar had scarce finished his new town,
when the officers of these corps, on their return, made
such a report of the riches and fertility of the country,
that he determined to view it in person, increase the
number of towns, and by that means secure the possession of it. Accordingly he continued his march to
Cali, where he built a town, which still retains the
same name, though in a different country ; for after it was finished in the country of the Gorrones Indians, captain Miguel Munoz soon after removed
it, on account of the unhealthiness of the air. Belalcazar founded also another town, called Santa Fé
de Antioquia ; and, charmed with the fertility and
richness

richness of the country, he proceeded to people it every where.

Nor was this all; for Belalcazar, in order to enhance the glory and importance of this country, was very desirous of making a road from Quito to the North Sea, as he had before done to the Pacifick Ocean. Among the discoveries made by his captains whilst he was employed in superintending the building of Popayan, one was, that at no great distance from that place were two of the principal sources of the great river of Magdalena; whence he conceived they might easily find a passage to the North Sea. This opinion the general had the pleasure of finding unanimously agreed to, which induced him to make every disposition for the security and welfare of his conquests, being determined to return by way of that river to Spain, in order to solicit the title of governor of the country which he had discovered, conquered, and peopled. Accordingly the title was conferred on him, and in his government were comprehended all the territories then considered as within his conquests; but in the year 1730 the country of Choco was separated from it, and made a particular government, though the order was not carried into execution till the year 1735. This part, belonging to the province of the new kingdom of Granada, does not come within our description.

The city of Popayan, one of the most ancient in these parts, that title having been granted it on the fifth of July 1538, stands in a large plain, having on the north side an interrupted prospect of the country. Its latitude is 2° 28' north; lies about two degrees east of the meridian of Quito, on the east side of a mountain of a middling height called M, from the resemblance it bears to that letter; and, being covered with a variety of trees, affords an entertaining prospect: the west side is also diversified with small eminences.

The

THE city is moderately large, with broad, straight, level streets; and, though not every where paved, are equally convenient, the foot-path near the houses being paved in all parts; and the middle of the streets, being composed of a hard small gravel, is never dirty in rainy weather, nor dusty in the great droughts of this climate; hence the middle of the streets are more convenient for walking than even the pavement itself.

THE houses are built of unburnt bricks, as at Quito, and entirely of the same construction: all the houses of note have a story; but the others only a ground floor. An idea of the largeness and convenience of the offices and apartments may be formed by their outward appearance, as well as the magnificence of the furniture, which is all brought from Europe; the expence of which must be enormously great, as, beside the long voyage, there is a necessity for bringing it a prodigious distance by land carriage, and subject to unknown dangers in these countries.

THE church was erected into a cathedral in the year 1547, and is the only parochial church in the city. Not that its extent is too small for maintaining others; but, having originally been the only church, the prebends could never be brought to allow of its being subdivided, and part of its revenues applied to the support of other parishes. Here are also convens of Franciscans, Dominicans, and Augustines, with a college of Jesuits; all of them having churches. In the latter is also a grammar school. The plan of an university, under the direction of the same fathers, is in such forwardness, that the charter is already granted. The number of religious belonging to each of these convents is but small, some of them amounting to no more than six or eight. It is, however, very different with regard to one of the nunneries, that of the Incarnation, the professed nuns being between forty and fifty; but the whole number, nuns, seculars, and servants included,

included, exceeds four hundred. The other nunnery
is of the order of Santa Teresa. All these convents
and their churches are pretty large; and if the latter
do not dazzle the sight with the splendor of their ornaments, they do not want any which decency requires.
Here was formerly a convent of bare-footed Carme-
lites, built on a spacious plain, near the top of the
mountain of M, from whence, on account of the
sharpness of the winds, the fathers some time after
removed to the foot of the mountain. But they were
also soon disgusted with their new situation; the dry
and salted fish, salads, and such like, being the only
particulars which this country affords, suitable to the
perpetual abstinence of their order; and accordingly
they again retired to their original situation, chusing
rather to suffer the inclemency of the weather, than
be confined to disagreeable food. This was the case
of another convent of the same order founded at Latacunga, where there is also no fresh fish of any sort
to be had. It must, however, he observed, that the
Teresian convents, who are under the same vow of
abstinence, are not discouraged by these inconveniences; nor is there a single instance of any deficiency
in the appointed number of nuns.

F<small>ROM</small> the mountain of M issues a river, which
by running through the city, besides other conveniences, carries away all its soil. Two bridges are
erected over it, one of stone and the other of wood.
The name of this river is del Molino. Its waters
have a particular medicinal virtue, which they are
thought to derive from the many briars through
which they flow. In this mountain is also a spring
of very charming water; but, not being sufficient to
supply the whole city, it is conveyed to the nunneries,
and the houses of men of rank. A little above a league
to the north of Popayan runs the river Cauca. It is
very large and deep, its current rapid, and subject to
dangerous swellings in the months of June, July, and

August; the season when the horrors of the mountains of Cuanacas, where it has its source, are at their height; so that the passage of it is extremely dangerous, as many travellers, rashly exposing themselves to the intenseness of its cold, amidst thick snows and violent winds, have fatally experienced.

THE inhabitants of Popayan and Quito differ very sensibly in their casts; for as at Quito and the other towns and villages of its jurisdictions, the most numerous class of people is that of the casts which sprung from the intermarriages of Spaniards and Indians; so at Popayan, Carthagena, and other parts where Negroes abound, the lower class consists of casts resulting from the marriages of the Whites and Negroes; but very few Indian casts. This is owing to the great multitude of Negro slaves kept as labourers at the plantations in the country, the mines, and to do the servile offices in the city: so that the number of Indians here are very few, compared with the other parts of the province. This government has, however, many large villages of them; and it is only in the capital, and other Spanish towns, that they are so greatly out-numbered by the Negroes.

THE inhabitants of Popayan are computed at between twenty and twenty-five thousand; and among these are many Spanish families, particularly sixty, known to have been originally descended from very noble families in Spain. It is worth observing here, that, whilst other towns see their inhabitants constantly decreasing, Popayan may boast of a daily increase. This has indeed nothing mysterious in it; the many gold mines worked all over its jurisdiction, afford employment to the indigent, and, consequently, occasion a great resort of people to these parts.

POPAYAN is the constant residence of the governor; whose office being purely civil, it is not requisite, as in many others, that he should be acquainted with military affairs. Within the jurisdiction of his government,

ment, all matters, civil, political, and military, are under his direction. He is also the chief magistrate of the city; the others are the two ordinary alcaldes, chosen annually, and a proper number of regidores, the constitution being the same as in other cities.

HERE is a chamber of finances, into which are paid the several branches of the royal revenue; as the tribute of the Indians, the duties on goods, the fifth of the metals, and the like.

THE ecclesiastical chapter is composed of the bishop, whose revenue is settled at six thousand dollars annually; the dean, who has five hundred; the archdeacon, chanter, rector, and treasurer, who have each four hundred. This see is a suffragan of the archbishoprick of Santa Fé de Bogota.

POPAYAN, lying within the jurisdiction of the inquisition of Carthagena, has a commissary from thence. Here is also another of the Croisade; but the authority of these two judges extends not beyond the diocese, which is far less than that of the government, a considerable part of it belonging to the archbishoprick of Quito.

THE jurisdiction of the government of Popayan reaches southward to the river Mayo, and to Ipiales, where it borders on the jurisdiction of the town of San Miguel de Ibarra; north-east it terminates with the province of Antioquia, the last of its provinces, and contiguous to that of Santa Fé; and northward borders on the government of Carthagena. Its ancient western bounds were the South sea, but it has since been so contracted by the new government of Choco, that the territory of Barbacoas is the only part of it which reaches to the sea; eastward it spreads itself to the sources of the river Coqueta, which are also thought to be those of the river Oronoco and Negro: its extent is not precisely determined: but a probable conjecture may be made, that from east to west it is about 80 leagues, and little less from north to south.

This

This jurisdiction being so large, and containing many towns and villages, is divided into several departments, over each of which the principal governor nominates a deputy for the administration of justice, and introduces them to the audience to which they belong, where his nomination is confirmed; a circumstance necessary to procure them all the weight and security in the several departments which are conferred on them. Those which form the government of Popayan are,

I. Santiago di Cali.	VII. Almaguer.
II. Santa Fé de Antioquia.	VIII. Caloto.
III. Las Quatro Ciudades.	IX. San Juan de Pasto.
IV. Timana.	X. El Raposo.
V. Guadalajara de Buga.	XI. Barbacoas.
VI. S. Sebastian de la Plata.	

In each of these departments, besides the chief town, are several others very large and well peopled; and great numbers of seats and farm-houses, where the number of people employed gives them the appearance of villages rather than private dwellings.

Of the above-mentioned departments, those towards the north and east of the city of Popayan, as Santa Fé de Antioquia, Las quatro Ciudades, Timana, and S. Sebastian de la Plata, belong to the audience and province of Santa Fé; the others lying nearer to Quito belong to its province; and those of San Juan de Pasto, and Barbacoas, are within its diocese.

The departments of Cali and Buga, lying betwixt the governments of Popayan and Choco, thrive, as being the channel of the commerce which is carried on continually betwixt those two governments: whereas it is otherwise with that of Almaguer, from the smallness of its jurisdiction, and the little traffic there. That of Caloto, as its extent is considerable,

so

so is it rich, and abounds in the products of the earth, the soil being fertile, and the country every where interspersed with farms. That of El Raposo is on the same happy footing as the two first. That of Pasto is also large, but less wealthy. Barbacoas is very small ; and in such a general want of provisions, that, except a few roots and grains peculiar to hot and moist climates, it is supplied with every thing from other provinces.

The temperature of this government is entirely the same as that already spoken of in the other parts of the province of Quito; that is, it varies according to the situation of places; some being more cold than hot, others the reverse; and some, throughout the whole year, enjoy a continual spring, as particularly Popayan the capital. The like may be said of the soil, which exuberantly produces the grains and fruits proper to its situation : and the farms breed great numbers of horned cattle and sheep, for the consumption of the towns and country people: and in the territory of Pasto grasiery is a very profitable article, large herds and flocks being driven to Quito, where they always find a good market. The jurisdiction of Popayan is more subject to tempests of thunder and lightning, and earthquakes, than even Quito ; though in the latter, as we have observed, they are so very frequent. No longer ago than 1735, at one in the afternoon on the second of February, the greatest part of the town was ruined by one. This remarkable frequency of tempests and earthquakes, in the country of Popayan, may be conjectured to proceed from the great number of mines, in which it exceeds all the others within the province of Quito.

But of all the parts in this jurisdiction Caloto is accounted to be the most subject to tempests of thunder and lightning; this has brought into vogue Caloto bells, which not a few persons use, being firmly persuaded that they have a special virtue against light-

Z 3   ning.

ning. And indeed so many stories are told on this head, that one is at a loss what to believe. Without giving credit to, or absolutely rejecting all that is reported, leaving every one to the free decision of his own judgment, I shall only relate the most received opinion here. The town of Caloto, the territory of which contains a great number of Indians, of a nation called Paezes, was formerly very large, but those Indians suddenly assaulting it, soon forced their way in, set fire to the houses, and massacred the inhabitants: among the slain was the priest of the parish, who was particularly the object of their rage, as preaching the gospel, with which they were sensible their savage manner of living did not agree, exposing the folly and wickedness of their idolatry, and laying before them the turpitude of their vices. Even the bell of the church could not escape their rancour, as by its sound it reminded them of their duty to come and receive divine instruction. After many fruitless endeavours to break it, they thought they could do nothing better than to bury it under ground, that, by the sight of it, they might never be put in mind of the precepts of the gospel, which tended to abridge them of their liberty. On the news of their revolt, the Spaniards in the neighbourhood of Caloto armed; and, having taken a smart revenge of the insurgents in a battle, they rebuilt the town, and having taken up the bell, they placed it in the steeple of the new church; since which the inhabitants, to their great joy and astonishment, observed, that, when a tempest appeared brooding in the air, the tolling of the bell dispersed it; and if the weather did not every where grow clear and fair, at least the tempest discharged itself in some other part. The news of this miracle spreading every where, great solicitations were made for procuring pieces of it to make clappers for little bells, in order to enjoy the benefit of its virtue, which, in a country where tempests are both so dreadful and

frequent,

frequent, must be of the highest advantage.   And to this Caloto, owes its reputation for bells.

Iɴ the valleys of Neyba, and others within the jurisdiction of Popayan, is a very remarkable insect, particularly famous for the power of the small quantity of venom in it.   This insect, which is shaped like a spider, is much less than a bug   Its common name is coya, but others call it coyba; its colour is of a fiery red, and, like spiders, it is generally found in the corners of walls, and among the herbage.   Its venom is of such a malignity, that, on squeezing the insect, if any happen to fall on the skin of either man or beast, it immediately penetrates into the flesh, and causes large tumours, which are soon succeeded by death.   The only remedy hitherto known, is, on the first appearance of a swelling, to singe the party all over the body with the flame of straw, or long grass, growing in those plains.   In order to this, the Indians of that country lay hold of the patient, some by the feet, and others by the hands, and with great dexterity perform the operation, after which the person is reckoned to be out of danger.   But it is to be observed, that though this insect be so very noxious, yet squeezing it between the palms of the hands is attended with no bad consequence: from whence the plain inference is, that the callus, usual on the hands of most people, prevents the venom from reaching the blood.   Accordingly the Indian muleteers, to please the curiosity of the passengers, squeeze them betwixt the palms of their hands, though unquestionably, should a person of a delicate hand make a trial, the effects would be the same as on any other part of the body *.

Nᴀᴛᴜʀᴇ is equally admirable in her works, and in her care of them.   Man is endued with discernment,

* The Brazilians say, oil and salt is a certain cure for the poison of the coyba. A.

know-

knowledge, and observation, that he may avoid whatever is hurtful to his being; and the irrational species
receive the like notices from instinct, and are not less
observant than man. The people who travel along
these valleys, where they are so much in danger of
these coyas, according to the warning before given
them by the Indians who attend them, though they feel
something stinging them or crawling on their neck or
face, are careful not to scratch the part, nor even so
much as lift up their hands to it, the coya being of
such a delicate texture that it would immediately
burst: and as there is no danger whilst they do not
eject the humour in them, the person acquaints some
one of the company with what he feels, and points
to the place; if it be a coya, the other blows it away.
The beasts, who are not capable of such warning, are
yet by instinct taught a precaution against the danger
which may result from these insects in the pastures;
for before they offer to touch the herbage, they blow
on it with all their force in order to disperse any of
these pernicious vermin; and when their smell acquaints them that they are near a nest of coyas, they
immediately leap back and run to some other part.
Thus they secure themselves from the venom of these
insects, though sometimes a mule, after all its blowing, has been known to take in some with its pasture,
on which, after swelling to a frightful degree, they
have expired on the spot.

AMONG the plants of the country of Popayan, in
the jurisdiction of Timana, grows the cuca or coca,
an herb so esteemed by the Indians in some provinces
of Peru, that they would part with any kind of provisions, the most valuable metals, gems, or any thing
else, rather than want it. It grows on a weak stem,
which for support twists itself round another stronger
vegetable, like the vine. Its leaf is about an inch and
a half or two inches in length, and extremely smooth;
the use the Indians make of it is for chewing, mixing
it

it with a kind of chalk or whitish earth called mambi. They put into their mouth a few cuca leaves, and a suitable portion of mambi, and, chewing these together, at first spit out the saliva which that manducation causes, but afterwards swallow it; and thus move it from one side of the mouth to the other, till its substance be quite drained; then it is thrown away, but immediately replaced by fresh leaves. This herb is so nutritive and invigorating, that they labour whole days without any thing else; and on the want of it, they find a decay in their strength: they also add, that it preserves the teeth sound, and fortifies the stomach.

In the southern provinces of Peru great quantities of it are produced, being cultivated by the Indians; but that growing wild in the neighbourhood of Cusco is accounted the best of any. It makes no small article in trade, particularly vast quantities of it are carried to the mine-towns, that the owners of the mines may have wherewithal to furnish the Indians, who otherwise could not be brought to work, or would not have strength to go through it.

This coca is exactly the same with the betel of the East Indies. The plant, the leaf, the manner of using it, its qualities, are all the same: and the eastern nations are no less fond of their betel than the Indians of Peru and Popayan are of their coca; But in the other parts of the province of Quito, as it is not produced, so neither is it used.

In Pasto, one of the most southern districts of Popayan, are certain trees which yield a resin called mopa-mopa; and of this is made a varnish, which besides its exquisite beauty will bear boiling water, and even acids. The method of applying it is, to dissolve some of the resin in one's mouth, and then wet the pencil with it; afterwards it is dipped in the colour which is to be laid on, and when dried has all the lustre of the Chinese laque, but with this superior
quality,

quality, that it never wears off, nor becomes moist, though rubbed with spittle. The cabinets, tables, &c. made by the Indians of this country, and thus varnished, are carried to Quito, where they are highly valued.

Popayan is one of the best trading countries within the province of Quito, as all the vast variety of Spanish goods from Carthagena are consigned thither and forwarded to Quito; and great numbers of traders go their rounds through the several jurisdictions, to the great conveniency of the towns and villages, which thus supply themselves. Besides this transitory commerce, it has another reciprocal with Quito, to which it exports horned cattle and mules, and receives in return cloths and bays. Its active commerce consists in dried beef, salted pork, roll-tobacco, hogs-lard, rum, cotton, pita, ribbons, and other small wares, which are brought to Choco, and there exchanged for gold; sugar and snuff are imported from Santa Fé and sent to Quito; and the returns to Santa Fé are home-made cloths and bays. Here is also another traffic, which consists in bartering silver for gold; for, there being an abundance of the latter, and a scarcity of the former, silver is brought to exchange for gold; of which great profit is made by converting it into doubloons: the like is also practised at Choco and Barbacoas, which are in the same case as to metals.

Popayan being the centre of all these several kinds of commerce, the most wealthy persons of the whole jurisdiction are here, and five or six of its inhabitants are reckoned to be masters of above 100,000 dollars; twenty to be worth betwixt 40 and 80,000, besides many of smaller, yet handsome, fortunes: and this exclusive of their farms and mines, with which this country abounds. The former are the same with those I have had occasion to mention in the other parts of this province, according to the quality of the temperature.

WEST

WEST of the western Cordillera of the Andes, is the government of Atacames, which on this quarter borders on the jurisdictions of the corregmientos of Quito and the town of San Miguel de Ibarra; northward on the department of Barbacoas in the government of Popayan; its western boundary is the South Sea; and southward it joins the territory of Guayaquil. Thus it reaches along the coast from the island of Tumaco, and the house of Husmal, which lie in one degree and a half north latitude, to the bay of Caracas, and the mountains of Balsamo, in 34 min. south latitude.

THE country of this jurisdiction lay a long time uncultivated; and if not wholly, at least the greatest part of it, unknown; for, after its conquests by Sebastian de Belalcazar, the peopling of it was neglected, either because the Spaniards were more intent in regulating their conquests than in improving what they had got, or because the country did not seem to them so proper for a settlement as the sierra or mountainous parts; or perhaps they judged it barren and unhealthy. And though care was taken to furnish Quito with priests, to preserve its Indian inhabitants in an adherence to those precious truths they had embraced; yet it was with the total neglect of that improvement of the country, which was seen in all the other parts where the Spaniards had settled. Thus these people, though Christians by profession, remain in that rusticity and savageness natural to men who are out of the way of rational conversation and commerce to civilize them; an Indian only coming now and then from their woods with aji, achote, and fruits, to sell at Quito, where they seem struck with amazement at the sight of such a concourse of people at one place; it being indeed far beyond what could be imagined by such as seldom or never came to any distance from their poor cottages, dispersed and shut up in the woods, and living among the wild beasts.

THOUGH

THOUGH the country of Atacames lay thus neglected for some years after the introduction of the Christian religion, and its inhabitants had performed homage to the king of Spain; yet the importance of making settlements here, and cultivating the ground, for facilitating the commerce betwixt the province of Quito and the kingdom of Terra Firma, was not unknown, as thereby an end would be put to the inconveniences of carrying it on by the way of Guayaquil; which being a great circuit, the trade suffered in many particulars; and indeed could not long have subsisted, without making a settlement of Spaniards in Atacames; as thus the way would be much shorter for the commerce betwixt Terra Firma and Quito, which now conveniently supplies it with provisions of all kinds, and receives European goods in return.

PURSUANT to these views, Paul Durango Delgadillo was in the year 1621 appointed governor of Atacames and Rio de las Esmaraldas. He had some years before entered into a contract with the Marquis de Montes Claros for opening a way from the town of San Miguel de Ibarra to the river Santiago, one of those which traversed the country belonging to the jurisdiction of this government; and likewise to people and cultivate it. But failing of fulfilling the agreement, though he was not wanting in endeavours, the government in the year 1626 was taken from him and conferred on Francisco Perez Menacho, who however had no better success than he who had been displaced.

AFTER these two, came Juan Vincencio Justiniani in the same character; but he, seeing the insuperable difficulties according to the methods of his predecessors, confidently offered to make the way by the river Mira, but also failed in the execution; and Don Hernando de Soto Calderon, who began it in the year 1713, and rather more sanguine in his assurances of success than the former, also disappointed the general expecta-

expectation ; and thus the so much desired communication of the province of Quito and the kingdom of Terra Firma remained as it was till the year 1733, when Don Pedro Vicente Maldonado, being invested with the same powers as his predecessors, surpassed them in execution ; and in 1741 laid open a direct communication betwixt Quito and the Rio de las Esmaraldas; and having verified his proceeding before the audiences, and obtained their approbation, he returned to Spain, to solicit the confirmation of his employment as governor, and the rewards specified in the contract. On the favourable report of the supreme council of the Indies, his majesty, in 1746, confirmed him as governor of that country, which, in 1747, was formally erected into a government, by the commission then given to the above gentleman, who by his skill and resolution had so well deserved it.

The towns within the government of Atacames are at present but small and poor, having hitherto lain out of the way of traffic, and the country but little cultivated. However, this governor takes such measures for the improvement of it, that already the face of things begins to alter greatly for the better; and the fertility of the soil will naturally invite settlers, and the communication being opened through it betwixt the kingdom of Terra Firma and the province of Quito, will cause a circulation of money. In the mean time this government contains 20 towns, five of which are on the sea-coast, and stand the first in the following list : the others are inland places.

I. Tumáco,	VI. Lachas.
II. Tola.	VII. Cayàpas.
III. San Matheo de Esme-	VIII. Inta.
raldas.	IX. Gualéa.
IV. Atacàmes.	X. Nanegàl.
V. La Canoa.	XI. Tambillo.
	XII. Ni-

The inhabitants of the five towns are Spaniards, Mestizos, Negroes, and Casts, which sprung from these three species. Those of the other fifteen are in general Indians, having few Spaniards, Mulattos, or Negroes, among them. With the spiritual concerns eleven priests are invested, who continually reside in the great towns, and occasionally visit the others, where are chapels of ease.

The temperature of Atacames is like that of Guayaquil, and accordingly produces the same kinds of vegetables, grains, and fruits, though some of them to a much greater perfection; for, by lying higher, it is not subject to the inundations proceeding from the swellings of rivers: and thus the cacaco, in its plantations and forests, having all the moisture that plant delights in, without being drowned, is much superior to the other in size, oiliness, and delicacy of flavour. It likewise produces in great abundance vanillas, achote, sarsaparilla, and indigo; also a great deal of wax is made here: and the forests are so thick set with trees of a surprising bulk and loftiness, as to be impenetrable; and these trees, as in the forests of Guayaquil, are of an infinite variety; some fitter for land works, others for naval uses, and some excellent for both.

CHAP.

CHAP. IV.

*Description of the Governments of Quixos, and Macas;
with an Account of Jean de Bracamoros, the Discovery and Conquest of it.*

NEXT to the government of Popayan, which
has been treated of in the foregoing chapter,
follow those of Quixos and Macas, on the east side of
the Cordillera of the Andes; it is divided into two
districts, Quixos being the north part of the government, and Macas the south, with the country of
Camelos lying betwixt them. As their situation and
other circumstances require that each should be treated distinctly, I shall begin with Quixos, which on
the north side borders on the jurisdiction of Popayan;
eastward it reaches to the river Aguarico, and westward is separated from the jurisdictions of Quito, Latacunga, and the town of San Miguel de Ibarra, by
the Cordilleras of Cotopaxi and Cayamburo. The
first discovery of the country of Quixos is owing to
Gonzalo Diaz de Pineda, in the year 1536, who,
among the officers sent from Popayan by Sebastian de
Belalcazar, to trace the course of the river of Magdalena, and take a survey of the country adjacent to
that which had been conquered, was appointed to make
discoveries in these parts, which he performed with
great care and dispatch; and finding it to abound in
gold, and cassia trees, he returned to his commander;
and on his report, Gonzalo Pizaro, in the year 1539,
at that time governor of Quito, marched to it with
a design of reconnoitring its whole extent, and making settlements. But, his expedition miscarrying,
the conquest of this country, though from Pineda's
report very desirable, was suspended till the year
1549, when the marquis de Canete, viceroy of Peru,
gave a commission to Gil Ramirez Davalos, a man of
undaunted

undaunted courage when interest was in view, for reducing the Indians and making settlements in the country; which he accordingly accomplished, and founded the town of Baeza, the capital of the government, in the year 1559; and it was soon followed by other towns and villages, still existing; but with very little improvement beyond their first state.

THE town of Baeza, though the most ancient of the country, and long the residence of the governors, has always remained very small, which is owing to the building of the two cities of Avila and Archidona, still subsisting, and at that time the chief object of the attention of the settlers, Baeza being left as first built; and these, so far from having increased suitably to the title of cities, which was given them at their foundation, remain on their first footing. The cause of the low state of the places here is the nature of the country, which, in air, fertility, and other enjoyments of life, being inferior to that of Quito, few settle here who can live in the other. Baeza is indeed extremely declined, consisting only of eight or nine thatched houses, with about twenty inhabitants of all ages, so that from the capital it is become annexed to the parish of Papallacta, in which town resides the priest, who has besides under his care another town called Maspu. This decay was no more than a consequence of the removal of the governor, who of late has resided at Archidona.

THE city of Archidona is a small place, lying in one degree and a few minutes S. of the equinoctial, and about one degree 50 minutes E. of the meridian of Quito. The houses are of wood, covered with straw, and the whole number of its inhabitants is reckoned at betwixt 650 and 700, consisting of Spaniards, Indians, Mestizos, and Mulattos; it has only one priest, under whose care are also three other towns, called Misagualli, Tena, and Napo; the last receives its name from the river on the borders of which it

stands;

stands; and this situation proved its ruin on the 30th of Nov. 1744, when, by the explosion of the Volcano, or Cotopaxi, of which a more particular account shall be given in another place, this river became so swelled by the torrents of melted snow and ice, that it entirely bore down the town of Napo, and the houses were carried along by the impetuosity of the current.

THE city of Avila, but very much below that title, stands on oo degr. 44 min. S. lat. and near 2 degr. 20 min. E. of Quito. It is so much less than the former, that its inhabitants scarce amount to 300 of both sexes. Its houses are built of the same materials. It has also a priest, whose ecclesiastical jurisdiction comprehends six towns, some of them in largeness and number of inhabitants not inferior to the city. These are,

I. La Conception.   IV. Motte.
II. Loreto.   V. Cota Pini.
III. San Salvador.   VI. Santa Rosa.

THE foregoing towns constitute the chief part of this government; but it also includes the towns of the mission of Sucumbios, the chief of which is San Miguel. At the beginning of this century they were ten, but are now reduced to these five :

I. San Diego de los Palmares.
II. San Francisco de los Curiquaxes.
III. San Joseph de los Abuccees.
IV. San Christoval de los Yaguages.
V. San Pedro de Alcantara de la Cocao, or Nariguera.

THE inhabitants of the two cities, and the villages in the dependencies, and those of Baeza, are obliged to be constantly upon their guard against the infidel Indians, who frequently commit depredations among their houses and plantations. They compose different

VOL. I.   A a   and

and numerous nations; and are so dispersed all over the country, that every village is under continual apprehensions from those which live in its neighbourhood: and when an action happens between the inhabitants and those Indians to the advantage of the former, all they get by it is to return quietly to their dwellings with a few prisoners, no booty being to be had from a people who live without any settlement; and from mere savageness make no account of those things in which the bulk of mankind place their happiness. Their method in these incursions is, after an interval of apparent quiet and submission, to steal up to the Spanish settlements at a time when they have reason to conclude that the inhabitants are off their guard; and if their intent be answered, they fall to pillaging and plundering; and, having got what is nearest at hand, retire with all speed. This perpetual danger may also be reckoned among the causes which have hitherto kept the government in such low circumstances.

The temperature of all this country is hot and very moist. The rains are almost continual; so that the only difference betwixt it, Guayaquil, and Porto Bello, is, that the summer is not so long: but the distempers and inconveniencies of the climate are the same. The country is covered with thick woods; and in these are some trees of a prodigious magnitude. In the south and west part of the jurisdiction of Quixos is the canela or cinnamon-tree, which, as I have before observed, being discovered by Gonzalo Diaz de Pineda, he from them called the country Canelos, which name it still retains. A great quantity of it is cut for the necessary consumption, both in the province of Quito and in Valles. The quality of this cinnamon does not come up to that of the East Indies; but in every other particular very much resembles it; the smell, its circumference, and thickness, being nearly the same: the colour is something browner,

browner, the great difference lying in the taste, that of Quixos being more pungent, and without the exquisite flavour of that of the East Indies. The leaf is the same, and has all the delicate smell of the bark; but the flower and seed surpass even those of India; the former particularly is of an incomparable fragrancy, from the abundance of aromatic parts it contains; and this favours an opinion, that the trees duly culti- vated might be made in every respect equal to those of the island of Ceylon.

The other products in the island of Quixos are the very same with those in all the other lands in the same climate as this government. The like may be said of fruits, roots, and grains, as wheat, barley, and others, which, requiring a cold air, seldom thrive much in any of an opposite quality.

The other district of Macas is bounded on the east by the government of Maynas; southward by that of Bracamoros and Yaguarsongo; and westward, the east Cordillera of the Andes divides it from the jurisdiction of Rio Bamba and Cuença. Its chief town bears the splendid title of the city of Mucas, being the com- mon name given to the whole country. And this is better known than its proper ancient name of Sevilla del Oro. It lies in two degrees thirty minutes S. latitude, and forty minutes E. of Quito. Its houses, which do not exceed 130, are built of timber, and thatched. Its inhabitants are reckoned at about 1200; but these, and it is the same all over this district, are generally Mestizos with Spaniards. The other towns belonging to this jurisdiction are:

I. San Miguel de Narbaes.	V. Zuna.
II. Barahonas.	VI. Payra.
III. Yuquipa.	VII. Copueno.
IV. Juan Lopez.	VIII. Aguayos.

The spiritual government of them all is lodged in two priests; one of whom residing in the city has the

care

care of the four first: and to the latter, who lives at Zuna, belong that town and the three others. At the conquest, and for some time after, this country was very populous, and, in honour of the great riches drawn from its capital, was distinguished by the name of Sevilla del Oro; but at present only the memory of its former opulence remains. Such an extreme declension proceeded from an insurrection of the natives, who, after swearing allegiance to the king of Spain, took arms, and made themselves masters of the city of Logrono, and a town called Guamboya, both in the same jurisdiction, and very rich. These devastations have so discouraged any further settlement there, that the whole country lies as a waste; no money goes current in it, and the only way the wretched inhabitants have to provide themselves with necessaries is by bartering their home products.

THE nearness of Macas to the Cordillera of the Andes causes a sensible difference betwixt its temperature and that of Quixos: for though it be also a woody country, the diversity betwixt the two most distant seasons of the year is manifest; and as its territory is different from that of the jurisdiction of Quito, so the variety in the periods of the season is also great. Thus winter begins here in April, and lasts till September, which is the time of summer betwixt the Cordilleras: and at Macas the fine season is in September, and is the more delightful on account of the winds which are then mostly northward; and thus charged with the frigorific particles which they have swept away from the snowy mountains over which they have passed. The atmosphere is clear; the sky serene; the earth clothed in its various beauties; and the inhabitants, gladdened by such pleasing objects, rejoice that the horrors of winter are passed, as they are no less dreadful and detrimental here than at Guayaquil.

IN grains and other products which require a hot and

and moist temperature, the country is very fruitful ; but one of the chief occupations of the country people here, is the culture of tobacco, which, being of an excellent kind, is exported in rolls all over Peru. Sugar-canes also thrive well here ; and consequently cotton. But the dread of the wild Indians, who have often ravaged their country, discourages them from planting any more than what just suffices for present use; they being here in the same unhappy situation as in Quixos, the villages having in their neighbourhood bands of those savage Indians ; and when they imagine them to be furthest off, are often suddenly assaulted by them, so that they must be ready at every instant to take arms.

Among the infinite variety of trees which crowd the woods of this country, one of the most remarkable is the storax, the gum of which is of a most exquisite fragrancy ; but is rare, the trees growing in places at some distance from the villages ; and it is dangerous going to them, by reason of the savage Indians, who lie in wait like wild beasts. The like may be said with regard to the mines of Polvos Azules, or Ultra-marine, from which, by reason of that danger, very little is brought ; but a finer colour cannot be imagined.

The territory belonging to Macas also produces cinnamon trees, which, as the reverend Don Juan Joseph de Lozay Acuna, priest of Zuna, a person of eminent learning, and perfectly versed in natural history, told me, is of a superior quality to that of Ceylon, here known by the name of Spanish cinnamon; and this was confirmed to me by many other persons of judgment. This cinnamon visibly differs from that of Quixos, which, as the same person informed me, proceeds from the full exposure of the Maca trees to the sun, its rays not being intercepted by the foliage of any other trees near them ; and these

also

also are at a distance from the roots of other trees, which deprive them of part of the nourishment necessary to bring it to perfection. And this opinion is confirmed by a cinnamon tree planted either accidentally or by design, near the city of Macas, the bark of which, and especially the blossom, in its taste, fragrancy and aromatic power, far exceeds that of the East Indies.

GREAT quantities of copal are brought from Macas, also wild wax; but the latter of little value, for, besides being reddish, it never indurates; and the smell of it, when made into candles, and these lighted, is very strong and disagreeable; and that of Guayaquil and Valles no better. Indeed all the wax in those countries cannot come into competition with those of Europe; though it must be observed, that there is no small difference in the bee, which in this country is much larger, and its colour inclinable to black. However, it might be made something better, if the inhabitants were acquainted with the art of cleansing and working it as in Europe; and if it could not be brought to equal the European, a greater consistence might be given to it, which would be no small advantage.

THE government, which on the south limits the jurisdiction of the audience of Quito, and follows next to Macas, is that of Jaen, which was discovered and subdued by Pedro de Vargara in the year 1538, whom Hernando Pizarro had appointed to command in that expedition. Afterwards Juan de Salinas entered the country, with the title of governor of it; and he having by his courage and courtesy reduced the Indians, and ingratiated himself with them, a more formal settlement was made, and several towns built, which are still existing, though in no better condition than those of Macas and Quixos. Some still retain the appellation of city, not that their largeness, number of inhabitants,

habitants, or wealth, become the title, but on account of the privileges annexed to it.

At the time of the conquest this government was known by the names of Igualsongo and Pacamoros, since corrupted into Yaguarsongo and Bracamoros; the names of the government conferred on Juan de Salinas. And thus they continued to be called for many years, till the Indians of both territories in a sudden revolt destroyed the principal towns. Those which were spared, after passing near an age in wretchedness and barbarism, happily recovered themselves, became united to the city of Jaen, as part of a government, with the title of Jaen de Bracamoros; and the title of governor of Yaguarsongo was, as before related, kept up by being annexed to the corregidor of Loja.

The town of Jaen, with the addition of Pacamoros, or Bracamoros, from the reunion of the towns of that country to it, was founded in the year 1549, by Diego Palomino. It stands in the jurisdiction of Chaca-Inga, belonging to the province of Chuquimayo, and is the residence of the governor. It is situated on the north shore of the river Chinchipe, at its conflux into the Maranon. It lies in about five degrees 25 min. S. lat. and its long. may be conjectured to be very little distant from the meridian of Quito, if not under it. The account given of the mean condition of the cities of Macas and Quixos also suits Jaen. We must however observe, that it is much more populous, its inhabitants being, of all ages and sexes, computed at 3 or 4000; though these for the most part are Mestizos, with some Indians, but very few Spaniards.

Juan de Salinas likewise found in his government of Yaguarsongo three other cities, still subsisting, but small, mean, and defenceless, like Jaen. Their names are Valladolid, Loyola, and Saniago de las Montagnas: the last borders on the government of Minas, and is only separated from its capital, the city of Borja, by the Pongo de Manceriche. In this country of Jaen de Bracamoros are several small villages:

I. San Joseph.	VI. Chinchipe.
II. Chito.	VII. Chyrìnos.
III. Sànder.	VIII. Pomàca.
IV. Charape.	IX. Tomepènda.
V. Pucarà.	X. Chuchùnga.

The inhabitants of which are mostly Indians, with some Mestizos, but no great number of either.

Though Jaen stands on the bank of the river Chinchipe, and so near the Maranon, yet the latter is not navigable up to it; so that those who are to embark on it go by land from Jaen to Chuchunga, a small place on another river of that name, and in 25 deg. 29 min. lat. whence they fall down into the Maranon. This town, which may be accounted the port for Jaen, lies four days journey from the city, which is the method of calculating the distances here; the difficulties of the road increasing them far beyond what they are in reality, that not seldom that which on good ground might be travelled in an hour or two, takes up a half and sometimes a whole day.

The climate of Jaen, and the same may be said of the whole jurisdiction of this government, is like that of Quixos, except that the rains are neither so lasting nor violent; and, like that of Macas, it enjoys some interval of summer; when the heats, tempests, and all the inconveniences of winter, abate. The soil is fruitful in all the grains and products agreeable to its temperature. The country is full of wild trees, particularly the cacao, the fruit of which, besides the exuberance of it on all the trees, is equal to that cultivated in plantations; but is of little use here for want of consumption: and the carriage of it to distant parts would be attended with such charges, as to prejudice

judice

judice its sale.   Thus the fruits rot on the trees, or are eaten by monkeys or other creatures.

At the time of its first discovery, and the succeeding conquest, this country was in great repute for its riches ; and not undeservedly, vast quantities of gold being brought from it.   But these gains were soon brought to a period by the revolt of the Indians, though in the opinion of many, who look upon those people to be a part of the human species no less than themselves, the insurrection was owing to the excessive rigour of the Spaniards, in making them work in the mines under insupportable fatigues.   At present, all the gold collected here is by Indians washing the sands of the rivers during the time of the inundations ; and thus find gold dust, or small grains of gold, with which they pay the tributes, and purchase necessaries ; and they make so little account of this metal, that, though by a proper industry they might get a considerable quantity, it is only the poorest Indians that live near the settlements who practise it : as for the independent Indians, they give themselves no concern about it.

The jurisdiction of this government produces in particular vast quantities of tobacco ; the cultivation of it indeed is the chief occupation of all the inhabitants.   After steeping the plant in hot mead, or decoctions of fragrant herbs, in order to improve its flavour, and the better to preserve its strength, it is dried, and tied up in the form of a saucisson, each of a hundred leaves.   Thus it is exported into Peru, all over the province of Quito, and the kingdom of Chili, where no other is used for smoking, in cornets of paper, according to the custom of all these countries.   This great vogue it owes to the manner of preparing the leaves, which gives it a particular relish, and a strength to its smoke, that is very agreeable to those who are fond of that amusement.   The country also produces a great deal of cotton ; likewise large

breeds

breeds of mules; and these three articles constitute the advantageous traffic which this government carries on with the jurisdiction of its province and the other parts of Peru.

In the countries of Jaen de Bracamoros, Quixos, and Macas, are seen great numbers of those wild animals, a description of which has been given in treating of other countries of a like climate. But these, besides tigers, are infested with bastard lions, bears, dantas or grand bestias (an animal of the bigness of a bullock, and very swift, its colour generally white, and its skin very much valued for making buff leather; in the middle of its head is a horn bending inward). Those three kinds of wild beasts are unknown in the other countries; and that they are known here, is owing to the proximity to the Cordilleras, where they breed, as in a cold climate adapted to their nature: whence they sometimes come down into the neighbouring countries; but without this circumstance of lying so near the mountains, they would never be seen. Among the reptiles in the country is the maca, a snake which the Indians distinguish by the name of curi-mullinvo, having a shining spotted skin like that of the tiger, curi in the Indian language signifying gold; it is wholly covered with scales, and makes a frightful appearance, its head being out of all proportion to the body, and has two rows of teeth, and fangs like those of a large dog. The wild Indians, as an ostentatious mark of their intrepidity, and to give them a more terrible appearance, paint on their targets figures of this snake, the bite of which is incurable; and wherever it has seized, it never lets go its hold; which the Indians would also intimate by their device.

CHAP.

CHAP V.

Government of Maynas, and of the River Maranon, or that of the Amazons ; its Discovery, Course, and that of the Rivers running into it.

HAVING treated of the governments of Popayan and Jaen de Bracamoros, which are the northern limits of the province of Quito ; as also of Atacames, which is its western boundary ; I now proceed to the government of Maynas, the eastern limit of its jurisdiction.   This is particularly entitled to a separate and succinct description, as the great river Maranon flows through it.

THE government of Maynas lies contiguous to those of Quixos and Jaen de Bracamoros, towards the east. In its territories are the sources of those rivers, which, after rapidly traversing a vast extent, form, by their conflux, the famous river of the Amazons, known also by the name of Maranon.   The shores of this and many rivers which pay it the tribute of their waters, environ and pervade the government of Maynas.   Its limits, both towards the north and south, are little known, being extended far among the countries of infidel Indians ; so that all the account which can be expected is from the missionaries employed in the conversion and spiritual government of the wild nations which inhabit it.   Eastward it joins the possessions of the Portuguese, from which it is separated by the famous line of demarcation, the boundary between the Spanish and Portuguese possessions.

WERE I to confine myself in general to the extent of the government of Maynas, my description would be very imperfect, and want the noblest object of the reader's curiosity, a description of the river of the Amazons; a subject no less entertaining than unknown; and the more difficult of obtaining a thorough know-

knowledge of, from its lying so very remote. This description I shall divide into the three following heads, which shall contain its source, and the principal rivers whereof it is composed ; its course through the vast tracts of land it waters ; its first discoveries, and the subsequent voyages made on it ; in order to give an adequate idea of this prince of rivers ; and at the same time a more circumstantial account of the government of Maynas.

I. *Of the Source of the River Maranon, and of the many others which compose it.*

As, among the great number of roots by which nourishment is conveyed to a stately tree, it is difficult from the great length of some, and the magnitude of others, to determine precisely that from which the product is derived ; so the same perplexity occurs in discovering the spring of the river Maranon ; all the provinces of Peru as it were emulating each other in sending it supplies for its increase, together with many torrents which precipitate themselves from the Cordilleras, and, increased by the snow and ice, join to form a kind of sea of that which at first hardly deserves the name of a river

The sources by which this river is increased are so numerous, that very properly every one which issues out of the eastern Cordillera of the Andes, from the government of Popayan, where the river Caqueta or Yupura has its source, to the province of Guanuco, within thirty leagues of Lima, may be reckoned among the number. For all the streams that run eastward from this chain of mountains, widening as they advance from the source by the conflux of others, form those mighty rivers, which afterwards unite in the Maranon ; and though some traverse a larger distance from their source, yet others, which rise nearer, by receiving in their short course a greater number of
brooks,

brooks, and consequently discharging a quantity of water, may have an equal claim to be called the principal source. But without confidently determining this intricate point, I shall first consider the sources of those which run into it from the more remote distances, and next those which precipitate themselves down several cascades formed by the crags of the Andes, and, after being augmented by others it receives, join the Maranon in a more copious stream; leaving it to the reader to determine which is the original source.

The most received opinion, concerning the remotest source of the river Maranon, is that which places it in the jurisdiction of Tarma, issuing from the lake of Lauricocha, near the city of Guanuco, in 11 deg. S. lat. whence it directs its course S. almost to 12 deg. through the country belonging to this jurisdiction; and, forming insensibly a circuit, flows eastward through the country of Juaxa; where, after being precipitated from the east side of the Cordillera of the Andes, proceeds northward; and, leaving the jurisdictions of Mayabamba and Chacha-poyas, it continues its course to the city of Jaen, the lat. of which in the foregoing chapter has been placed in 5 deg. 21 min. There, by a second circuit, it runs toward the E. in a continual direction; till at length it falls into the ocean, where its mouth is of such an enormous breadth, that it reaches from the equinoctial to beyond the first deg. of north lat. Its distance from Lauricocha lake to Jaen, its windings included, is about 200 leagues; and this city being 30 deg. to the W. of its mouth, is 600 leagues from it, which, with the several circuits and windings, may without excess be computed at 900 such leagues : so that its whole course, from Lauricocha to its influx into the ocean, is at least 1100 leagues.

Yet the branch which issues from Lauricocha is not the only one flowing from these parts into the Maranon;

ranon; nor is it the most southern river which discharges its waters into that of the Amazons; for S. of that lake, not far from Asangara, is the source of the river which passes through Guamanga: "Also in the jurisdictions of Vilcas and Andaguaylas are two others, which, after running for some time separately, unite their streams, and discharge themselves into the river issuing from the lake Lauricocha. Another rises in the province of Chimbi-Vilcas. And lastly, one still further to the south, is the river Apurimac, which, directing its course to the northward, passes through the country of Cusco, not far from Lima-Tambo; and after being joined by others, falls into the Maranon about 120 leagues east of the junction of the latter with the river Santiago. But here it is of such a width and depth, as to leave a doubt whether it insinuates itself into the Maranon, or the Maranon pays tribute to the Ucayale, as it is called in that part; since at the conflux its impetuosity forces the former to alter the straight direction of its course, and form a curve. Some will have the Ucayale to be the true Maranon, and found their opinion on the remoteness of its source, and the quantity of its waters, which equals at least, if it does not exceed, that of Lauricocha.

In the space intercepted between the junction of the Maranon and the river Santiago, are the Pongo de Manzeriche, and the mouth of the river Ucayale; and about mid-way betwixt them the river Guallaga, which has also its source in the Cordilleras, east of the province of Guamanga, and falls into the Maranon. One of the rivers contributing to its increase has its rise in the mountains of Moyo-Bamba; and on its banks, in the middle of its course towards the Guallaga, stands a small village called Llamas; which, according to the most credible accounts, was the place where Pedro de Orsica embarked with his people on

his

his expedition for the discovery of the Maranon, and the conquest of the adjacent countries.

EASTWARD of Ucayale, the Maranon receives the river Yabari, and afterwards four others, namely, the Yutay, Yurua, Tefe, and Coari; all running from the south, where they have their source nearly in the same Cordilleras as that of the Ucayale; but the countries through which the latter passes being inhabited by wild Indians, and consequently but little known to the Spaniards, its course, till its junction with the Maranon, cannot be ascertained: and it is only from vague accounts of some Indians, that in certain months of the year it is navigable. There is indeed a tradition of voyages made up it, and by which it was perceived to run very near the provinces of Peru.

BEYOND the Rio Coari eastward, the Cuchibara, also called the Purus, joins the Maranon; and after that likewise the Madera, one of the largest rivers that unite their waters with it. In 1741, the Portuguese sailed up it, till they found themselves not far from Santa Cruz de la Sierra, betwixt 17 and 18 deg. of south lat. From this river downwards, the Maranon is known among the Portuguese by the name of the river of the Amazons: upwards they give it the name of the river of Solimoes. Within a small distance follows the river of Topayos, likewise very considerable; and which has its source among the mines of Brazil. After these it is further joined by the rivers Zingu, dos Bocas, Tocantines, and Muju, all issuing from the mines and mountains of Brazil; and on the eastern shore of the latter stands the city of Gran Para.

HAVING thus given an account of the most distant branches of the stately river of Maranon, and of the principal ones which join it from the south, I proceed to those, the sources of which are nearer, issuing from the Cordilleras, and which immediately run in-

to

to the eastern direction; and also those which join it from the north.

In the mountains and Cordilleras of Loja and Zamora rise several little rivers, the conflux of which forms that of Santiago; and from these of Cuença, others which unite in the Paute: but this, on its union with the former, loses its name, being absorbed by the Santiago, (so called from a city of that name) near which it joins the two others from Lauricocha and Apurimac. The river Morona issues from the lofty deserts of Sangay; and passing very near the city of Macas, runs in a S. E. course, till it loses itself in the principal channel of the Maranon; which happens at the distance of about 20 leagues E. of Borja, the capital of the government of Maynas.

In the mountains of the jurisdiction of Riobamba, those of Latacunga, and the town of San Miguel de Ibarra, are the sources of the rivers Pastaza and Tigre; and from Cotopaxi and its Cordillera issue the first branches of the rivers Coca and Napo. These, though their sources are at no remarkable distance, run to a great extent before they join; and retaining the name of Napo, fall into the Maranon, after a course of above 200 leagues in a direct line from E. to W. with some, though insensible, inclinations to the S. This is the river which father Christopher de Acuna, who will be mentioned hereafter, takes for the true Maranon, to which, as exceeding all the rest in largeness, the others may be said to add their waters.

From the mountains of the jurisdiction of San Miguel de Ibarra, and those of Pasto, issues the river Putú-mayo, called also Ica, which, after running S. E. and E. about 300 leagues, joins the Maranon much more eastward than the river Napo: lastly, in the jurisdiction of Popayan, the river Caqueta has its origin, which becomes divided into two branches; the western, called Yupura, disembogues itself into the
Maranon

Maranon like another Nile, through seven or eight mouths, and these are at such a distance, that the intermediate space betwixt the first and the last is not less than 100 leagues; and the other, which runs to the eastward, is not less famous under the name of Negro. M. de la Condamine, in the narrative of his voyage, confirms the opinion of its being one of the communications betwixt the Oronoque and Maranon; and corroborates his assertion, by the authority of a map composed by father John Ferreira, rector of the college of Jesuits in the city of Gran Para; in which he observes, that in the year 1744 a flying camp of Portuguese, posted on the banks of the Negro, having embarked on that river, went up it, till they found themselves near the Spanish missions on the river Oronoque, and meeting with the superior of them, returned with him to the flying camp on the river Negro, without going a step by land; on which the author makes this remark, That the river Caqueta, (already mentioned, and so called from a small place by which it passes, near its source) issuing from Mocoa, a country joining eastward to Almaguar in the jurisdiction of Popayan, after running eastward with a small declension towards the south, divides itself into two branches; one of which declining a little more southward, forms the river Yupura, and afterwards separating into several arms, runs, as we have noted above, into the Maranon, through seven or eight mouths; and the other, after a course eastward, subdivides itself into two branches, one of which, running north-east, joins the Oronoque; and the other, in a south-east direction, is the river Negro. This subdivision in the branches of large rivers, and their opposite courses, though something extraordinary, is not destitute of probability; for a river flowing through a country every way level, may very naturally divide into two or more branches, in those parts where it meets with any inclination, though almost insensible, in the ground. If this declivity be not very great, and the river large and deep, it will easily become navigable every where, with a free passage from one arm into the other. And in this manner the marshes are formed in a level country, as we have particularly remarked in the coast of Tumbez: for the sea-water on the flood running into these various mouths, which sometimes àre 20 leagues distant or more, a vessel enters one arm by the favour of the tide; but coming to a place where the soil rises, the stream runs against her, being the water which the same flood had impelled through another channel. Thus the ebb causes the waters to separate at that point; and each portion of water takes the same course at going out as at its entrance; yet the place where the separation is made is not left dry. But even though the place where the waters of the river Caqueta are separated should not be level, or nearly horizontal, but lie on a considerable declivity; yet if this fall be equal on both sides, one part of the waters may take its course to the Oronoque, and the other to the Negro, without any other consequence than that the great rapidity would render them impracticable to navigation; but this has nothing to do with the division of the waters, it being no more than forming an island either large or small.

FROM the province of Quito there are three ways to the river Maranon; but all extremely troublesome and fatiguing, from the nature of the climate, and being full of rocks, that a great part of the distance must be travelled on foot; for being so little frequented, no care has been taken to mend them, whence they are even more dangerous than the others in South America, of which we have given a description.

THE first of these roads, which is the nearest to the town of Quito, runs through Baza and Archidona: where you embark on the river Napo. The second

is

is by Hambato and Papate, at the foot of the mountain of Tunguragua; and from thence the road lies
through the country of Canelos, watered by the river
Bobonaza, which joining the Pastaza, both discharge
themselves into the Maranon. The third lies through
Cuença, Loia, Valladolid, and Jaen, from whence at
the village of Chuchunga, which is as it were its port,
this river becomes navigable; and here all embark
who are either going to Manas, or a longer voyage
on this river. Of the three, this alone is practicable
to beasts; but the tediousness of the distance from
Quito renders it the least frequented; for the missionaries, who take these journeys oftener than any
other set of men, in order to avoid its circuit, and the
danger of the pass of Manzeriche, prefer the difficulties and dangers to the others.

In the long course of this river from Chuchunga,
are some parts where the banks, contracting themselves, form streights, which, from the rapidity of the
waters, are dangerous to pass. In others, by a sudden turn of its direction, the waters are violently carried against the rocks; and in their repercussion, form
dangerous whirlpools, the apparent smoothness of
which is no less dangerous than the rapidity in the
streights. Among these, one of the most dangerous
is that betwixt Santiago de las Montanas and Borja,
called Pongo de Manzeriche; the first word of which
signifies a door or entrance, and by the Indians is applied to all narrow places; the second is the name of
the adjacent country.

The Spaniards who have passed this streight make
the breadth of it to be no more than twenty-five yards,
and its length three leagues; and that, without any
other help than merely the current of the water, they
were carried through it in a quarter of an hour. If
this be true, they must move at the rate of twelve
leagues an hour; a most astonishing velocity! But
M. de la Condamine, who examined it with par-
B b 2　　　　　　　　ticular

ticular attention, and to whose judgment the greatest
deference is due, is of opinion, that the breadth of
the Pongo, even in its narrowest part, is twenty-five
toises; and the length of the Pongo about two leagues,
reckoning from the place where the shores begin to
approach, as far as the city of Borga. And this
distance he was carried in fifty-seven minutes. He
observes also, that the wind was contrary;- and consequently his balza did not go so far as the current
would otherwise have carried her; so that, making
allowance for this obstruction, the current may be
stated at two leagues and a half or at three leagues
an hour.

THE breadth and depth of this river is answerable
to its vast length; and in the pongos or streights, and
other parts where its breadth is contracted, its depth
is augmented proportionally. And hence many are
deceived by the appearance of other rivers which join
it, their breadth causing them to be taken for the real
Maranon; but the mind is soon convinced of its error,
by observing the little increase which the Maranon receives from the influx of them. This large river, by
continuing its course without any visible change in its
breadth or rapidity, demonstrates that the others,
though before the object of astonishment, are not comparable with it. In other parts it displays its whole
grandeur; dividing itself into several large branches,
including a multitude of islands, particularly in the
intermediate space between the mouth of the Napo and
that of the Coari, which lies something to the westward of the river Negro; where, dividing itself into
many branches, it forms an infinite number of islands.
Betwixt the mission of Peba, which is at present the
last of the Spanish, and that of San Pablo the first
of the Portuguese, M. de la Condamine, and Don
Pedro Maldonado, having measured the breadth of
some of these branches, found them nearly equal to
nine hundred toises, that is, almost a sea league. At

8   the

the influx of the river of Chuchunga, the place where the Maranon becomes navigable, and where M. de la Condamine first embarked on it, he found its breadth to be one hundred and thirty-five toises: and though this was near its beginning, the lead did not reach the bottom at twenty-eight toises, notwithstanding this sounding was made at a great distance from the middle of the river.

The islands formed by the Maranon east of the Napo, terminate at the river Coari, where it again reunites its waters, and flows in one stream: but here its breadth is from one thousand to twelve hundred toises, or near half a league; and here the same ingenious gentleman, after taking all possible precautions against the current, as he had before at the mouth of the river Chuchunga, sounded, but found no bottom with one hundred and three fathom of line. The river Negro, at the distance of two leagues from its mouth, measured twelve hundred toises in breadth, which being nearly equal to that of the principal river, and some of those we have named, Ucayale, the Madera, and others, were found to be nearly of the same width.

About one hundred leagues below the mouth of the river Negro, the shores of the Maranon begin to approach each other near the efflux of the river Trumbetas, which part is called the Estrecho de Pauxis, where, as also at the posts of Peru, Curupa and Macapa, along its banks, and on these east of the rivers Negro and Popayos, the Portuguese have forts. At the Estrecho de Pauxis, where the breadth of the river is near nine hundred toises, the effect of the tides may be perceived; though the distance from the sea-coasts be not less than two hundred leagues. This effect consists in the waters, which, without any change in the direction of their course, decrease in their velocity, and gradually swell over their banks. The flux and reflux are constant every twelve hours, with

the natural differences of time. But Mode la Condamine with his usual accuracy, as may be seen in
the narrative of his own voyage, observed that the
flux and reflux perceived in the ocean, on any certain
day and hour, is different from that which is felt at
the same day and hour, in the intermediate space between the mouth of the river and Pauxis, being rather the effect of the tides of the preceding days;
proportional to the distance of the place from the
river's mouth; for as the water of one tide cannot
flow two hundred leagues within the twelve hours, it
follows, that having produced its effect to a determined distance during the space of one day, and renewing it in the following by the impulse of the succeeding tides, it moves through that long space with
the usual alternation in the hours of flood and ebb;
and in several parts these hours coincide with those
of the flux and reflux of the ocean.

After flowing through such a vast extent of country, receiving the tribute of other rivers precipitated
from the Cordilleras, or gliding in a more gentle
course from remote provinces; after forming many
circuits, cataracts, and streights; dividing itself into
various branches, forming a multitude of islands of
different magnitudes, the Maranon at length, from
the mouth of the river Xingu, directs its course N. E.
and enlarging its channel in a prodigious manner, as
it were to facilitate its discharge into the ocean, forms
in this astonishing space several very large and fertile
islands; of which the chief is that of Joanes or Ma-
rayo, formed by a branch of the great river which separates from it twenty-five leagues below the mouth
of the Xingu; and directing its course to the southward, in a direction opposite to that of the principal
stream, opens a communication between the Maranon and the river of D s Bocas, which has before
received the waters of the Guanapu and Pacayas, and
flows into it through a mouth of above two leagues in
breadth.

breadth. These are afterwards joined by the river
Tocantines; the outlet of which is still broader than
the former, and at a still greater distance: the river
of Muju, on the eastern side of which stands the city
of Gran Para, discharges its waters into the same
stream; and it afterwards receives the river Capi,
which washes the city of the same name.

THE river of Dos Bocas, after joining that of Ta-
gipuru, runs eastward, forming an arch as far as the
river of Tocantines, from which it continues N. E.
like the Maranon, leaving in the middle the island of
Joanes, which is nearly of a triangular figure, except
the south side about one hundred and fifty leagues
in length, and forms the arch of a circle. This island
divides the Maranon into the two mouths, by which
that river disembogues itself into the sea. The principal of these two mouths from Cape Maguari in
this island, and the North Cape, is about forty-five
leagues broad; and that of the channel of Tagi-
puru, as likewise of the rivers which have joined it,
from the same Cape Maguari to Tigioca point, is
twelve leagues.

THIS river, which exceeds any one mentioned either
in sacred or profane history, has three names; and is
equally known by them all, each implying its stupendous majesty, and importing its superiority to any
other in Europe, Africa, or Asia. And this seems
to have been intended by the singularity of its having
three different names; each of them enigmatically
comprehending those of the most famous in the other
three parts of the world; the Danube in Europe, the
Ganges in Asia, and the Nile in Africa.

THE names which express the grandeur of this river,
are the Maranon, the Amazons, and Orellana. But
it is not known with certainty that either of them was
the original, before its discovery by the Spaniards,
given it by the Indians; though very probably it was
not without many; for as various nations inhabited

its banks, it was natural for every one to call it by a particular name, or at least to make use of that which had been previously given it. But either the first Spaniards who sailed on it neglected this inquiry, or the former names became confounded with others given it since that epocha, so that now no vestiges of them remain.

THE general opinion prefers, in point of antiquity, that of Maranon, though some authors will have it posterior to the two others; but we conceive they are mistaken, both in their assertion, and in the cause of that name. They suppose that it was first given to this river by the Spaniards, who sailed down it under the conduct of Pedro de Orsua, in 1560 or 1559; whereas it had been known by that name many years before: for Pedro Martyr in his Decades, speaking of the discovery of the coast of Brazil, in the year 1500, by Vincente Yanez Pinzon, relates, among other things, that they came to a river called Maranon. This book was printed in the year 1516, long before Gonzalo Pizarro undertook the discovery of the river, and conquest of the adjacent nations who inhabited its banks; or Francisco de Orellana had sailed on it. This demonstrates the antiquity of the name of Maranon; but leaves us under the same difficulties with regard to its date and etymology. Some, following Augustine de Zarate, attribute the origin of this name to a Spanish commander called Maranan, from whom, as being the first that displayed the Spanish ensign on this river, it was thence called after his own name. But this opinion is rather specious than solid; being founded only on the similarity of the names, a very exceptionable inference; especially as no mention is made of any such officer in any history published of these discoveries and conquests; whence it seems natural to conclude, that Zarate, on hearing that the river was called Maranon, inferred that the name was taken from some person of eminence who had made

an

an expedition on it. . For had he known any thing further, he doubtless would have enriched his history with some of the adventures of the discovery of it; for if he had not thought them sufficiently interesting, it is something strange that all the Spanish historians should be in the same way of thinking, and concur to suppress the memory of a Spaniard whose name was thought worthy to be given to the most distinguished river in the world. But what carries along with it a much greater air of probability is, that Vicente Yanez Pinzon, upon his arrival in the river, heard it called by the Indians who inhabited its islands and banks, Maranon, or some name of a similar sound; and thence Vicente Yanez concluded that its name was Maranon. Hence it is undeniable, that the preference in antiquity belongs to the name of Maranon; and that this name was not given it by Orsua or his men, in allusion to some feuds and confusions among them, called in Spanish maranas, or from being bewildered among the great number of islands, forming enmaranado, or an intricate labyrinth of channels, according to the opinion of some historians.

THE second name is that of the river of the Amazons, which was given it by Francisco Orellana, from the troops of women who made part of the body of Indians who opposed his passage; and who were not inferior either in courage or the dexterous use of the bow, to the men; so that, instead of landing where he intended, he was obliged to keep at a distance from the shore, and often in the middle of the channel, to be out of their reach. However, on his return to Spain, and laying before the ministry an account of his proceedings, and of the female warriors that opposed him, he was by patent created governor of these parts, in recompense, as it was expressed, for his having subdued the Amazons: and ever since the river has been called by that name.

SOME

SOME have indeed doubted, whether the Maranon and the Amazons were the same river; and many seem to be strongly persuaded, that they were really different. But this opinion proceeds only from the river's not having been completely reconnoitred till the close of the last century.

THIS particular of the Amazons is confirmed by all writers, who have given a succinct account of the river, and Orellana's expedition : and though this proof is abundantly sufficient, if not of its reality, at least of its probability, it is additionally confirmed by the tradition still subsisting among the natives, which we may believe on the authority of one of the most eminent geniuses the province of Quito ever produced; I mean Don Pedro Maldonado, who was a native of the town of Riobamba, but lived at Quito, and whose performances are well known in the republic of letters. In 1743, this gentleman and M. de la Condamine agreed to return to Europe in company, by the way of the river Maranon; and among their other inquiries towards a complete knowledge of it, and the countries through which it flows, they did not forget the famous Amazons; and were informed by some old Indians, that it was an undoubted truth, that there had formerly been several communities of women, who formed a kind of republic, without admitting any men into the government: and that one of these female states still subsisted; but had withdrawn from the banks of the river to a considerable distance up the country; adding, that they had often seen some of these female warriors in their country. M. de la Condamine, in the narrative of his voyage down this river, printed at Paris in the year 1745, and who had all the rational curiosity of his fellow-traveller Don Pedro Maldonado, relates some of the facts told him by the Indians, concerning the Amazons whom they had seen. But I shall only here insert what historians have said

on

on this head, leaving every one to give what degree of credit he pleases to the adventure of Orellana, and the actual existence of the Amazons.

SOME who are firmly persuaded of the truth of the adventure of the Amazons with Orellana, and believe that their valour might be equal to that of the men, in defence of their country and families, will not hear of a female republic separated from the intercourse of men. They say, and not without sufficient reason, that the women who so gallantly opposed Orellana were of the Yurimagua nation, at that time the most powerful tribe inhabiting the banks of the Maranon, and particularly celebrated for their courage. It is therefore, say they, very natural to think, that the women should, in some degree, inherit the general valour of their husbands, and join them in opposing an invader, from whom they imagined they had every thing to fear, which might inflame their ardour; as likewise from an emulation of military glory, of which there are undeniable instances in the other parts of the Indies.

THE third and last name is that of the Orellana, deservedly given to it in honour of Francisco de Orellana, the first who sailed on it, surveyed a great part of it, and had several encounters with the Indians who lived in its islands or along its banks. Some have been at a great deal of pains to assign certain distances through its long course, and to appropriate to each of these one of the three names. Thus they call Orellana all that space from the part where this officer sailed down in his armed ship till it joins the Maranon. The name of Amazons begins at the influx of another river, at the mouth of which Orellana met with a stout resistance from the women or Amazons; and this name reaches to the sea: and lastly, the name of Maranon comprehends the river from its source a considerable way beyond the Pongo downwards

wards all along the part of the descent of this river through Peru; alleging that this was the part through which Pedro de Orsua entered the river; supporting their opinion by a derivation, to which we cannot subscribe, namely, that he gave it this name on account of the disturbances which happened among his men. The truth is, that the Maranon, the Amazons, and the Orellana, are one individual river; and that what is meant by each of these names, is the vast common channel into which those many rivers fall, which contribute to its greatness. And that to the original name of Maranon the two others have been added for the causes already mentioned. The Portuguese have been the most strenuous supporters of this opinion, calling it by no other name than that of the Amazons, and transferring that of Maranon to one of the captainships of Brazil, lying betwixt Grand Para and Siara; and whose capital is the city of San Luis del Maranon,

II. *Account of the first Discoveries and of the most famous Expeditions on the Maranon, in order to obtain a more adequate Idea of this famous River.*

AFTER this account of the course and names of this river, I shall proceed to the discovery of it, and the most remarkable voyages made thereon. Vicente Yanez Pinzon, one of those who had accompanied the admiral Don Christopher Columbus in his first voyage, was the person who discovered the mouth through which this river, as I have before taken notice, discharges itself into the ocean. This adventurer, at his own expense, in 1499, fitted out four ships, discoveries being the reigning taste of that time. With this view he steered for the Canary Islands; and after passing by those of Cape de Verd, continued his course directly west, till on the 26th of January, in

the

the year 1500, he had sight of land; and called it
Cabo de Consolacion, having just weathered a most
violent storm.   This promontory is now called Cabo
de San Augustin.   Here he landed; and, after taking
a view of the country, coasted along it northward;
sometimes he lost sight of it, when on a sudden he
found himself in a fresh-water sea, out of which he
supplied himself with what he wanted; and being
determined to trace it to its source, he sailed upwards,
and came to the mouth of the river Maranon, where
the islands made a most charming appearance.   Here
he staid some time, carrying on a friendly traffic
with the Indians, who were courteous and humane
to these strangers.   He continued advancing up
the river, new countries appearing still as he sailed
further.

To this maritime discovery succeeded that by land
in the year 1540, under the conduct of Gonzalo
Pizarro, who was commissioned for this enterprise
by his brother the Marquis Don Francisco Pizarro,
on the report which Gonzalo Diaz de Pineda had made
of the country of La Canela, in the year 1536; at
the same time making him governor of Quito.
Gonzalo Pizarro arrived at the country of Los Canelos; and following the course of a river, either the
Napo or Coca, it is not certain which, though more
probably the first, met with unsurmountable difficulties
and hardships; and seeing himself destitute of provisions of every kind, and that his people, by feeding
on the buds and rinds of trees, snakes, and other creatures, wasted away one after another, he determined
to build a vessel, in order to seek provisions at the
place where this river joined another; the Indians
having informed him that there he would meet with a
great plenty.   The command of this vessel he gave
to Francisco de Orellana, his lieutenant-general and
confident, recommending to him all the diligence
and punctuality which their extremity required.   After

ter sailing eighty leagues, Orellana arrived at the junction of the two rivers, but met with nothing of what he had been sent for; being disappointed in the provisions he sought, the trees not bearing any fruit, or the Indians having already gathered it. His return to Pizarro seemed very difficult, if not impracticable, on account of the rapidity of the current; besides, he could not think of returning, without bringing with him that relief so earnestly expected; so that, after long debating the matter with himself, he determined, without the privity of his companions, to sail with the current to the sea. But this could not long remain a secret, the hoisting the sails sufficiently demonstrating his intentions; and some vehemently opposing such a desertion, as they called it, were near coming to blows. But at length Orellana, by plausible reasons and magnificent promises, pacified them; and the opposition ceasing, he continued his voyage, after setting ashore Hernando Sanchez de Vargas to perish with hunger, as being the ring-leader of the malecontents; and persisting in his invectives against Orellana's project.

PIZARRO, surprised at having no account of Orellana, marched by land to the place where he had ordered him, and near it met with Hernando Sanchez de Vargas, who acquainted him with the whole affair of the vessel; at which Pizarro seeing himself without resource, a considerable part of his men dead, the other so exhausted with fatigue and hunger that they dropt down as they marched, and those in the best state reduced to mere skeletons; he determined to return to Quito, which, after fatigues and hardships even greater than the former, he at last reached with a handful of men in the year 1542, having only reconnoitred some rivers, and the adjacent country; a service disproportionate to the loss of so many men, and the miseries suffered in this enterprise.

THIS

Tʜɪs was the first expedition of any consequence, to make discovery of the river Maranon : and if the success of Pizarro was not equal to his force and zeal, he was at least the instrument of its being entirely accomplished by another ; and to his resolution in pressing forward through difficulties and dangers, and by his expedient of building the armed vessel, must, in some measure, be attributed the happy event of Orellana's voyage, who, with a constancy which showed him worthy of his general's favour, reconnoitred the famous river of the Amazons through its whole extent, the adjacent country, its innumerable islands, and the multitude and difference of nations inhabiting its banks.   But this remarkable expedition deserves a more particular detail.

Oʀᴇʟʟᴀɴᴀ began to sail down the river in the year 1541 ; and in his progress through the several nations along its banks, entered into a friendly conference with many, having prevailed upon them to acknowledge the sovereignty of the kings of Spain formally, and with the consent of the caciques took possession of it.   Others, not so docile, endeavoured to oppose, with a large fleet of canoes, his further navigation : and with these he had several sharp encounters.   In one Indian nation bravery was so general, that the women fought with no less intrepidity than the men ; and by their dexterity showed that they were trained up to the exercise of arms.   This occasioned Orellana to call them Amazons ; which name also passed to the river.   The scene of this action, according to Orellana's own account, and the description of the place, is thought to have been at some distance below the junction of the Negro and Maranon.   Thus he continued his voyage till the 26th of August, in the same year ; when, having passed a prodigious number of islands, he saw himself in the ocean.   He now proceeded to the isle of Cubagua, or, according to others, to that of La Trinidad, with a design of going

ing to Spain, to solicit for a patent as governor of these countries. The distance he sailed on this river, according to his own computation, was eighteen hundred leagues.

This discovery was followed by another, but not so complete; it was undertaken in the year 1559 or 1560, under Pedro de Orsua, by commission from the marquis de Canete, viceroy at Peru, who at the same time conferred on him the title of governor of all his conquests. But the first news of Orsua was, that he and the greatest part of his men were killed in an ambuscade by the Indians; a catastrophe entirely owing to his own ill conduct, which destroyed the great armament made for this enterprise, and created an aversion to designs liable to such dangers.

In the year 1602, the reverend Raphael Ferrer, a Jesuit, having undertaken the mission of Cofanes, fell down the Maranon, and attentively surveyed the country as far as the conflux of the two rivers where Orellana had left Hernando Sanchez de Vargas; and at his return to Quito gave a very circumstantial account of what he had seen, and the different nations he had discovered.

Another, but fortuitous, view of the river was taken in 1616. Twenty Spanish soldiers, quartered in Santiago de las Montanas, in the province of Yaguarsongo, pursued a company of Indians, who, after murdering some of their countrymen in the city, fled up the country, and embarked on the Maranon in their canoes. The soldiers, in falling down the river, came to the nation of the Maynas, who received them in a friendly manner; and after some discourse showed a disposition of submitting to the king of Spain, and desired missionaries might be sent them. The soldiers, on their return to Santiago, having made a report of the good inclination of the Maynas, and their desire of being instructed in the Christian religion, an account was sent to the prince of Esquiloche, viceroy

of

of Peru: and in 1618, Don Diego Baca de Vega was appointed governor of Maynas and Maranon; and may be said to have been in reality the first, as neither Pizarro, Orellana, nor Orsua, though invested with the title, were ever in possession of it, having made no absolute conquests; a necessary circumstance towards realizing the title.

THIS expedition was performed in 1635 and 1636, and was succeeded by that of two Franciscans, with others of the same order, who set out from Quito with a determined zeal for propagating Christianity among the nations on the Maranon. But many of them, unable to support themselves under the fatigues and hardships natural in such a country, and discouraged with the little fruit their good desires produced, after wandering among mountains, woods, and deserts, returned to Quito, leaving only two, Dominico de Brieda and Andrew de Toledo, both lay-brothers. These, either from a religious zeal, or naturally more brave and hardy, or of greater curiosity, ventured to penetrate further into those dreary wastes. They were indeed attended by six soldiers, remaining of a whole company who had been sent, under captain Juan de Palacio, for the safeguard of the missionaries; but so many of them had returned with the religious to Quito, that these six and the captain were all that remained: and that officer, a few days after, lost his life in an action against the Indians.

THE six soldiers and two lay-brothers, however, continued with undaunted resolution to travel through countries inhabited by savages, unknown, and full of precipices on all sides; at length they committed themselves to the stream, in a kind of launch; and after many fatigues, hardships, and here and there a rencounter, reached the city of Para, at that time dependent on, or united with, the captainship of the Maranon, the governor of which resided at San Louis,

VOL. I.   C c   whither

whither they went, and gave him an account of what they had observed in this navigation.

At that time the crown of Portugal was annexed to Spain; and the governor of the captainship, or Maranon, for the sovereign of both kingdoms, was Jacome Reymundo de Norona, who, zealous for the improvement of this discovery, as of the highest importance to his prince, fitted out a fleet of canoes, under the command of captain Texera, to go up the river, and survey the country with greater form and accuracy. This flotilla departed from the neighbourhood of Para, on the 28th of October, 1637, with the two religious on board; and after an incessant fatigue in making way against the stream, they arrived at Pahamino on the 24th of June, 1638. This place belongs to the Jurisdiction of the government of Quixos; whence Texera, with the soldiers and the two religious, went to Quito, where he gave an account of the expedition to the audiencia, which transmitted the particulars to the count de Chinchon, viceroy of Peru; and he, agreeably to the zeal he had always manifested for enlarging his majesty's dominions, held a council about making more particular discoveries along the shores of that river.

Among other things, the count de Chinchon gave orders, that the Portuguese flotilla should return to Para; and with it sent some intelligent persons, whose zeal might be depended on, with orders to take an accurate survey of the river and its banks; and after discharging this commission, to proceed to Spain, and make a report of their expedition to the council of the Indies, in order to be laid before his majesty, that measures might in consequence be taken for securing the conquest of these nations. The persons chosen were, the Reverend Fathers Christopher de Accuna and Andrez de Artieda, Jesuits, and persons every way equal to the service. They left Quito on the 16th of February, 1639; and having embarked with the armadilla,

madilla, after a voyage of ten months, they arrived at Gran Para on the 12th of December, whence, according to their instructions, they passed over to Spain, and completely acquitted themselves of the trust reposed in them.

At the end of the last century, another expedition was undertaken, for making discoveries on the Maranon ; but at that time it was already so well known, that most of the adjacent lands had been improved by the missions which the Jesuits had settled there: and the government of Manas now includes many nations, who, on the fervent preaching of the Jesuits, having embraced Christianity, vowed obedience to the kings of Spain ; and a happy alteration was seen in their morals and customs.   The banks of this river, where before only wild Indians were seen living in the manner of beasts, were now turned into plantations and regular towns, the inhabitants of which shewed that they were not destitute of reason and humanity. These improvements were in a great measure owing to Father Samuel Fritz, who, in 1686, preached the Gospel among those people, and in a short time was the instrument of the conversion of many nations : but the continual fatigues and hardships, both by land and water, affected his health to such a degree, that he was obliged to set out for Pará in January 1689, and arrived there on the 11th of September of the same year.   Here he remained in a disagreeable inactivity, till his health was restored, and some affairs settled which required instructions from the court of Lisbon.

July the 8th, 1691, Father Fritz left Para, in order to return to his mission, which then reached from the mouth of the river Napo to some distance beyond the Negro, and included the Omaguas, Yurimaguas, Aysuares, and many other adjacent nations, the most numerous of the whole river.   October the 13th, in the same year, he returned to the town of Nuestra Se

C c 2   nora

nora de las Nieves, the capital of the Yurimagua nation; and having visited the rest under his charge, to the number of forty-one, all large and populous, he went, on other public affairs, to the town of Laguna, the capital of all the missions on the Maranon, where the superior resided; and afterwards repaired to the city of Lima, in order to communicate to the count de Moncloa, at that time viceroy, a full account of all those countries. This last journey he undertook by the way of the rivers Guallaga, Patanapura, Moyobamba, Chachapoyas, Caxamarca, Truxillo, and Lima.

The affairs which brought this indefatigable missionary to Lima, where he was received with great honour, being finished; Father Fritz, in August 1693, set out, on his return to his missions, by the way of the city of Jaen de Bracamoros, with a view of reconnoitring the course and situation of the rivers which, from those southern parts, fall into the Maranon. By the help of these additional lights, he drew a map of that river, which was engraved at Quito, in the year 1707: and though it had not all the accuracy which could be desired, the father being without instruments for observing the latitudes and longitudes of the chief places, taking the course of the rivers, and determining the distances; yet it was received with very great applause, as being the only one in which were laid down the source and direction of all the rivers which join the Maranon, and the whole course of the latter till its junction with the ocean.

III. *Account of the Conquest, Missions, and Nations, established on the Maranon.*

The discovery of this famous river, and the survey of the adjacent countries and nations, was followed by the conquest of the nations who inhabited its banks and

and islands. The miscarriage of the expedition under Gonzalo Pizarro has already been mentioned: Orellana was not more fortunate; when, pursuant to the grant of the government, he returned to settle in it; and Orsua's fate was still more deplorable, perishing himself, with the greatest part of his followers. But we are now to speak of the more successful enterprize of Don Diego Baca de Vega, whom we have already mentioned, but in a cursory manner.

THE government of Maynas, and the Maranon, having been conferred on De Vega; confident of the good dispositions of the Maynas Indians, as it had been carefully cultivated, since its first commencement with the Santiago soldiers, he entered the country with a little colony, and founded the city of San Francisco de Borga, in 1634, as the capital of the whole government; a title which it justly deserved, for being the first erected in that vast country; and also on account of the friendship which the Indians had shewn for the Spaniards ever since their first arrival. The new governor being a person of judgement and penetration, was not long in observing that these nations were rather to be governed by moderation and gentleness, with a proper firmness to create respect, than by rigour or austerity: and accordingly informed the audiencia of Quito and the Jesuits of their disposition. Missionaries were accordingly sent them, in the persons of Gaspar de Cuxia and Lucas de Cuebas, who came to Maynas in the year 1637; and their preaching had such remarkable success, that, being not of themselves sufficient for instructing the multitudes of new converts, they sent to Quito for assistance: and thus the number of missions continually increased, and whole nations resorted from their forests in search of the light of the Gospel. By this means the king's dominions were extended, every proselyte with joy acknowledging himself in his new state a subject of the

C c 3   king

king of Spain, as he owed to his bounty the inesti-
mable felicity of his conversion.

Thus the missions and the number of towns increased together, and the propagation of the Christian
religion in those remote countries, and the aggrandise-
ment of the Spanish monarchy, went hand in hand.
But the most distinguished æra of these progressions
was the year 1686, by the zeal and activity of Father
Fritz, whom we have had occasion before to mention
with honour: he went directly among the nation of
the Omaguas, who having by the Cocamas Indians
been informed of the mildness and wisdom with
which the missionaries taught them to live under just
and wholesome laws, and a police hitherto unknown
among them ; together with the many happy effects
it had produced in those nations which had conformed
to their instructions ; animated with these pleasing
relations, they sent, in 1681, a deputation to the town
of Laguna, belonging to Cocamas, where Father Lorenzo Lucero, superior of the missions, resided, en-
treating him to send among them persons for their
instruction: but the father at that time was not in a
capacity of complying with their request, all the missionaries being employed elsewhere. He therefore
dismissed them, with commending their good intentions ; promising them, that he would send to Quito
for a proper person to instruct them in those salutary
doctrines embraced by the other nations.

The Omaguas, full of anxiety, did not give Father Lorenzo Lucero time to neglect his promise ; for,
on hearing that new missionaries, and among them
Father Samuel Fritz, were just arrived at Laguna
from Quito, the same deputation returned to request
the immediate performance of the promise ; and having the greatest reason to expect it would be com-
plied with, great part of the people came in canoes
to the town of Laguna, as a testimony of respect to
Father Fritz, in order to conduct him to their country,

where

where they treated him with such veneration, that in his progress through the towns they would not suffer him to walk, but carried him on their shoulders; an honour which the caciques reserved to themselves alone. The effects of his preaching were answerable to these marks of ardour and esteem, so that in a short time the whole nation was brought to a serious profession of Christianity, deploring their former ignorance and brutality, and forming themselves into a political community, under laws calculated for the happiness of society. And their example so influenced several other adjacent nations, that the Yurimaguas, Asuares, Banomas, and others, unanimously and voluntarily came and addressed themselves to Father Fritz, desiring him to instruct them how to live in the same order and regularity as the Omaguas. Thus whole nations, on embracing Christianity, submitted to the sovereignty of the Spanish monarchs: and all the countries from the Napo to a considerable distance below the Negro, were reduced without the least force throughout the whole extent of the government of Maynas: and such, at the end of the last century, was the number of the nations thus converted, that Father Fritz, though without indulging himself in any respite, was not able to visit every single town and village within the compass of a year, exclusively of the nations under the care of other missionaries, as those of the Maynas, Xebaros, Cocamas, Panos, Chamicuros, Aguanos, Muniches, Otanabes, Roamaynas, Gaes, and many more. The other missions were in the same flourishing condition.

THE city of San Francisco de Boria, which we have already mentioned as the capital of Maynas, stands in 4 deg. 28 min. S. lat. and 1 deg. 54 min. E. of the meridian of Quito: but of its largeness and appearance we can only add, that it resembles the cities of the government of Jaen: and its inhabitants, though consisting of Mestizos and Indians, and the place is

the

the residence of the governor of Maynes and Maranon: yet they are not equal in number to those of
Jaen de Bracamaros. The principal town of the missions, and in which the superior is obliged to reside,
is Santiago de la Laguna, lying on the eastern bank
of the river Guallaga. The places which at present
compose those missions in the government of Maynas,
and diocese of Quito, are:

On the River Napo.

I. San Bartholome de Necoya.
II. San Pedra de Aguarico.
III. San Estanislao de Aguatico.
IV. San Luis Gonzaga.
V. Santa Cruz.
VI. El Nombre de Jesus.
VII. San Pablo de Guajoya.
VIII. El Nombre de Maria.
IX. San Xavier de Icaguates.
X. San Juan Bautista de los Encabellados.
XI. La Reyna de los Angeles.
XII. San Xavier de Urarines.

On the River Maranon, or Amazons.

I. La Ciudad de San Francisco de Borja.
II. La Certaon, or inland country towards St.
Teresa.
III. San Ignacio de Maynas.
IV. San Andres del Alto.
V. Santo Thomas Apostol de Andoas.
VI. Simigaes.
VII. San Joseph de Pinches.
VIII. La Concepcion de Cagua-panes.
IX. La Presentacion de Chayabitas.
X. La Incarnacion de Paranapuras.
XI. La Conception de Xebaros.

XII. San

XII. San Antonio de la Laguna.
XIII. San Xavier de Chamicuro.
XIV. San Antonio Adad de Aguanos.
XV. Nuestra Senora de las Neves de Yurimaguas.
XVI. San Antonio de Padua.
XVII. San Joaquin de la Grande Omagua.
XVIII. San Pablo Apostol de Napeanos.
XIX. San Phelipe de Amaonas.
XX. San Simon de Nahuapo.
XXI. San Francisco Regis de Yameos.
XXII. San Ignacio de Bevas 'y Caumares.
XXIII. Nuestra Senora de las Nieves.
XXIV. San Francisco Regis del Baradero.

BESIDES these towns, which have existed for some time, there are several others yet in their infancy ; and the Indians, by whom they are inhabited, of different nations from those above mentioned : likewise many others, both large and populous ; some on the banks of the rivers which fall into the Maranon, and others up the country. Many of the inhabitants of both nations hold a friendly intercourse with the Spanish missionaries, and with the inhabitants of the Christian villages, with whom they traffic, as well as with the Spaniards and Mestizos, settled at Borja and Laguna. All these nations of Indians have some resemblance in their customs; but in their languages very different, every one seeming to have a particular dialect, though there are some of a nearer affinity than others to the general language of Peru. The most difficult to be pronounced is that of the Yameos Indians: while, on the other hand, none is so easy and agreeable to the ear as that of the Omaguas: and the genius and tempers of these two nations were found to be as different as their language. Thus the Omaguas, even before their submission, gave many surprizing proofs of the clearness of their intellects; but were surpassed by the Yurimaguas, both in wit and penetration. The

former

former lived in villages under some kind of government, peacefully obeying their curacas or chiefs They were less barbarous; their manners less turbulent and corrupt than those of most other Indians The Yurimaguas formed a kind of republic: and had some laws which were strictly observed, and the breach of them punished in an exemplary manner. But in police the preference doubtless belongs to the Omaguas: for, besides living in society, there was an appearance of decency among them, their nudities being covered, which by others were totally neglected. This disposition in those two nations for. making approaches, however small, to civil customs and a rational life, not a little contributed to the speedy progress of their conversion. They were more easily. convinced, from the light of nature, of the truth and propriety of the doctrines preached by the missionaries; and were convinced, that happiness, both public and private, was intimately connected with an uniform observance of such precepts, instead of the innumerable evils resulting from the manner of living hitherto preached by them.

Among the variety of singular customs prevailing in these nations, one cannot help being surprized at the odd taste of the Omaguas, a people otherwise so sensible, who, to render their children what they call beautiful, flat the fore and hind parts of the head, which gives them a monstrous appearance; for the forehead grows upwards in proportion as it is flatted; so that the distance from the rising of the nose, to the beginning of the hair, exceeds that from the lower part of the nose to the bottom of the chin : and the same is observable in the back part of the head. The sides also are very narrow, from a natural consequence of the pressure; as thus the parts pressed, instead of spreading, conformably to the common course of nature, grows upwards. This practice is of great antiquity among them; and kept up so strictly, that

3　　　　　　　　　　　　　　they

they make a jest of other nations, calling them cala-bash heads.

In order to give children this beautiful flatness, the upper part of the head is put, soon after the birth, betwixt two pieces of board; and repeated, from time to time, till they have brought it to the fashionable form.

Another nation of these Indians, affecting a striking appearance, make several holes in both their upper and under lips, both sides of the cartilage of their nose, their chins, and jaws: and in these they stick fine feathers, or little arrows, eight or nine inches long. The reader's own imagination will sufficiently paint the strange appearance they must make with these decorations. Others place a great beauty in long ears; and accordingly extend them by art to such a degree, that in some the inferior lobe touches the shoulder: and they value themselves on the nickname of long ears, which has been given them in ridicule. The method they make use of to extend their ears, is this: they bore a hole in the lobe, and fasten to it a small weight, which they from time to time increase, till the ear is stretched to nearly the length above-mentioned: and as the lobe increases in length, so likewise does it in magnitude. Others paint some parts of their bodies; some the whole. All have something peculiar in their modes and customs, but generally of such a nature, that Europeans wonder how they could ever enter the thoughts of rational creatures *.

After describing this great river, and giving an account of the villages and nations near its banks, I shall proceed to some other particularities relating to it; as the extraordinary species of fish found in its waters, and likewise the birds and other animals seen in the adjacent countries through which it flows. Among

---

* Another remarkable custom is, that of their tying their privities in a bladder before they go into the water. A.

the

the various kinds of fish, are two of an amphibious nature; the caymans or alligators, and the tortoise, which swarm on the shores and islands. Its tortoises, for taste, are preferred to those of the sea. Another remarkable fish here is the pexe-buey, or sea-cow, so called from its resembling the land quadruped of that name. This is one of the largest species known in the river, being generally three or four yards in length, and of a proportional thickness: the flesh is very palateable, and, according to some, has pretty much the taste of beef. It feeds on the herbage growing along the shore, but the structure of its body does not admit of its coming out of the water. The female has dugs for suckling its young; and whatever some may have said of any farther resemblance to the terrestrial species of that name, it has neither horns nor legs. It has indeed two fins, which serve equally for swimming, and supporting itself on the banks whilst feeding. The general method of the Indians for fishing, is with inebriating herbs, like that I have mentioned on the river Guayaquil. On some occasions they make use of arrows dipped in poison, of such an activity, that the slightest wound immediately kills the fish. This is also their method of hunting; and in both they are so very expert and active, that they are very seldom known to miss their aim. This powerful venom is principally the juice of a bejuco, near six fingers broad, and flat on both sides, of a brownish colour, and growing in very damp marshy places. In order to prepare the poison, they cut it into pieces, which they bruise and boil in water. On taking it off the fire, they add to it a particular ingredient which causes a coagulation. With this they rub the point of their arrows; and when dry, for want of fresh unction, they moisten it with their spittle: the quality of it is so frigorific, that it immediately repels all the blood to the heart, where the vessels burst, being unable to contain such a torrent

as

as suddenly rushes into them. But what is most surprizing here, is, that the creature thus killed, and
its coagulated blood, are eaten without any inconveniency. The most powerful antidote to this venom is, immediately to eat sugar: but this specific,
though often salutary, is not infallible, as several
melancholy instances have demonstrated.

THE borders and parts adjacent to this famous river, as well as those contiguous to the others which
discharge their waters into it, abound with large and
lofty trees, the wood of which is of different colours;
some white, others of a dark brown; some red, or
veined with variety of colours. Some of another
species distil balsams of an exquisite fragrancy, or
rare and medicinal gums; others are noted for their
delicious and salubrious fruits. Among these the
wild cacoa, by the mere goodness of the soil, without any culture, grows in the greatest plenty, and yields
fruit of a goodness equal to that in the jurisdiction
of Jean and Quixos. Here also are gathered great
quantities of sarsaparilla, vanillas, and a bark called
declavo or cloves: for though it resembles cinnamon
in appearance, except its colour which is something
darker, its taste and smell are very different, being
nearly the same with that of the East-India clove.

As to quadrupeds, birds, reptiles, and insects, they
are nearly the same, and in as great numbers as those
already mentioned in the description of other hot
countries. One reptile of a very extraordinary nature, and known only here and in the provinces of
New Spain, I shall, as a conclusion of my account of
the Maranon, add a description of.

IN the countries watered by that vast river, is bred
a serpent of a frightful magnitude, and most deleterious nature. Some, in order to give an idea of its
largeness, affirm that it will swallow any beast whole;
and that this has been the miserable end of many a
man. But what seems still a greater wonder, is the
attractive

attractive quality attributed to its breath, which ir-
resistibly draws any creature to it, which happens to
be within the sphere of its attraction: but this, I
must own, seems to surpass all belief. The Indians
call it jacumama, i. e. mother of water: for as it delights in lakes and marshy places, it may in some
sense be considered as amphibious. I have taken a
great deal of pains to enquire into this particular;
and all I can say is, that the reptile's magnitude is
really surprizing. Some persons whose veracity is not
to be questioned,· and who have seen it in the provinces of New Spain, agreed in their account of the
enormous corpulency of this serpent, but with regard to its attractive quality could say nothing decisive*.

SUSPENDING therefore for the present all positive
judgement, without giving entire credit to all the
qualities vulgarly attributed to this animal, especially
the more suspected, as not improbably flowing from
astonishment, which frequently adopts absurdities, it
being impossible, in so great a perturbation, to consult reason; let me be indulged with some variation of

* I have seen three of these serpents killed; out of the body of one
of them was taken a hog about 10 stone in weight. The largest
was about 11 feet long, and 23 inches in circumference; the
smallest about 9 feet long, and 19 in circumference. They generally lie coiled up, and wait till their prey passes near enough to
be seized. As they are not easily distinguished from the large rotten wood (which lies about in plenty in these parts), they have opportunities enough to seize their prey and satiate their hunger. The
Indians watch this opportunity, and when they have half gorged
their prey, kill them without danger. As I was walking in the
woods one day, attended by two Indians and a Negro boy, we
were within 10 yards of one of these serpents, when the Negro
cried out, Cobra, Senhor! Cobra, Senhor! on which it made away
into a neighbouring thicket, which concealed from our sight the most
hideous creature I at that time had ever seen. In its motion, which
was slow and peculiar to that serpent, it appeared like a serpentine
log, with two bright gems for eyes, placed within three or four inches from the end which was farthest from us; from which rays of
azure light seemed to dart. A

the

the accidents, to investigate the cause, in order to come at the knowledge of its properties, which it is difficult to ascertain, unless supported by undoubted experiments. Not that I would offer my opinion as a decisive rule; I desire that the judgement of others may declare for that which appears most conformable to truth. I would also further acquaint the reader, that I only speak from the testimony of those who have seen this famous serpent, having never myself had an opportunity of examining it with my own eyes.

First, it is said, that this serpent, in the length and thickness of its body, very much resembles the trunk of an old tree, whose roots have for some time ceased to convey the usual nourishment; and that on every part of it grows a kind of moss, like that seen on the bark of wild trees. This is accounted for by the dust and mud adhering to it; and alternately moistened and dried by the water and sun. This forms a slight crust over the thick scales; and this crust is increased by the sluggishness and slow motion of the serpent; which, unless when forced by hunger to go in quest of food, continues motionless in one place for several days together; and even then its motion is almost imperceptible, leaving a track like that of a log of timber drawn along the ground.

Its breath is asserted to be of such a nature as to cause a kind of drunkenness or stupidity in man or beast, which has the misfortune of being within the bounds of its activity; and thus causes the animal involuntarily to move till it unhappily comes within the reach of the serpent, which immediately swallows it. This is the vulgar report: and it is added, that the only method of averting the danger, is on first feeling the breath to cut it, that is, to stop it by the interposition of another body, which hastily intervening, cuts the current of the blast and dissipates it. Thus the person, who was moving on to certain destruction, is enabled to take another path, and avoid the fatal catastrophe.

tastrophe. These particulars, if thoroughly considered, seem mere fables: as indeed the learned M.
de la Condamine intimates; and the very circumstances with which they are decorated, increase their
improbability.

BUT, in my opinion, with a little alteration in the
circumstances, what seems to shock credibility, will
appear natural and founded on truth.

THAT its breath is of such a quality as to produce a kind of inebriation in those whom it reaches,
is far from being impossible; the urine of the fox is
well known to have the same effect; and the breath
of the whale is frequently attended with such an insupportable fœtor as to bring on a disorder in the
brain. I therefore see no manner of difficulty in admitting that the breath of this serpent may be of that
intoxicating quality attributed to it; and may be considered as an expedient for catching its prey, as otherwise the creature, from the slow movement of its
body, would be utterly incapable of providing itself
with food; whereas, by this deleterious smell, the
animal may be thrown into such horror and perplexity,
as to be unable to move, but remain fixed like a
statue, or faint away, whilst the snake gradually approaches and seizes it. As to what is related of cutting the breath, and that the danger is limited to the
direction in which the serpent breathes; these are
tales, which to believe, would imply an utter ignorance of the origin and progress of odours. In short,
the vulgar errors, propagated by these rude nations,
have gained credit among the Spaniards, merely because none has had the curiosity or resolution to put
them to the test of experience.

CHAP.

CHAP. VI.

Of the Genius, Customs, and Manners of the Indians who are Natives of the Province of Quito.

THE subject of this chapter, and its circum-stances, are of such a nature, that, if what ancient histories deliver concerning them should recur to the memory, they will appear totally different. Indeed the disproportion between what I read, and what I am going to relate, is so remarkable, that, on a retrospect towards past times, I am utterly at a loss to account for the universal change of things; especially when surrounded by such visible monuments of the industry, polity, and laws of the Indians of Peru, that it would be madness to question the truth of the accounts that have been given of them; for the ruins of these ancient works are still amazing.   On the other hand, I can hardly credit my own eyes, when I behold that nation involved as it were in Cimmerian darkness, rude, indocile, and living in a barbarism little better than those who have their dwelling among the wastes, precipices, and forests. But what is still more difficult to conceive is, how these people, whose former wisdom is conspicuous in the equity of their laws, and the establishment of a government so singular as that under which they live, should at present shew no traces of that genius and capacity which formed so excellent an œconomy, and so beautiful a system of social duties: though undoubtedly they are the same people, and still retain some of their ancient customs and manners.   Leaving therefore this intricate subject to be investigated by farther enquiries, I shall proceed to give an account of the present Indians, their genius, customs, and qualities, according to the best information I could obtain from a commerce with those people of all ranks, during ten years. Some particulars in this narrative will demonstrate that they still retain a few sparks of the industry and capacity of the ancient Indians of Peru; whilst others will shew that they are utterly destitute of the knowledge of certain sciences which were common among their ancestors; and that they are equally degenerated from their wisdom in making laws, and their regular observance of them.

It is no easy task to exhibit a true picture of the customs and inclinations of the Indians, and precisely display their genius and real turn of mind; for if considered as part of the human species, the narrow limits of their understanding seem to clash with the dignity of the soul; and such is their stupidity, that in certain particulars one can scarce forbear entertaining an idea that they are really beasts, and even destitute of that instinct we observe in the brute-creation. While in other respects, a more comprehensive judgement, better-digested schemes, and conducted with greater subtilty, are not to be found than among these people. This disparity may mislead the most discerning person : for should he form his judgement from their first actions, he must necessarily conclude them to be a people of the greatest penetration and vivacity. But when he reflects on their rudeness, the absurdity of their opinions, and their beastly manner of living, his ideas must take a different turn, and represent them in a degree little above brutes.

Such is the disposition of the Indians, that if their indifference to temporal things did not extend itself also to the eternal, they might be said to equal the happiness of the golden age, of which the ancient poets have given such inchanting descriptions. They possess a tranquillity immutable, either by fortunate or unfortunate events. In their mean apparel they are as contented as the monarch clothed with the most splendid inventions of luxury; and so far are they from entertaining a desire for better or more

com-

comfortable clothing, that they give themselyes no manner of concern about lengthening their own, though half their bodies continue naked. They shew the like disregard for riches; and even that authority or grandeur within their reach is so little the object of their ambition, that to all appearance it is the same thing to an Indian, whether he be created an alcalde, or forced·to perform the office of a common executioner.

And thus reciprocal esteem among them is neither heightened nor lessened by such circumstances. The same moderation appears in their food, never desiring more than what suffices; and they enjoy their coarse simple diet with the same complacency as others do their well-furnished tables. Nor do I indeed question but if they had their choice of either, they would prefer the latter; but at the same time they shew so little concern for the enjoyments of life, as nearly approaches to a total contempt of them : in short, the most simple, mean, and easiest preparation seems best adapted to their humour.

Nothing can move them, or alter their minds; even interest here loses all its power; it being common for them to decline doing some little act of service, though offered a very considerable reward. Fear cannot stimulate, respect induce, nor punishment compel them.   They are indeed of a very singular turn ; proof against every attempt to rouse them from their natural indolence, in which they seem to look down with contempt on the wisest of mortals: so firmly bigoted to their own gross ignorance, that the wisest measures to improve their understanding have been rendered abortive; so fond of their simplicity and indolence, that all the efforts and attention of the most vigilant have miscarried.   But in order to give a clearer idea of their tempers, we shall relate some particular instances of their genius and customs ; as

D d 2   other-

otherwise it will be impossible to draw their true character

THE Indians are in general remarkably slow, but very persevering; and this has given rise to a proverb, when any thing of little value in. itself requires a great deal of time and patience, that it is ' only fit to be done by an Indian.' In weaving carpets, curtains, quilts, and other stuffs, being unacquainted with any better method, at passing the woof they have the patience every time to count the threads one by one; so that two or three years is requisite to finish a single piece. This slowness undoubtedly is not entirely to be attributed to the genius of the nation ; it flows, in some measure, from the want of a method better adapted to dispatch. And perhaps with proper instructions they would make considerable progresses, as they readily comprehend whatever is shewn them relating to mechanicks: of this the antiquities still remaining, in the province of Quito, and over all Peru, are undeniable testimonies. But of these more will be said in the sequel. This indifference and dilatoriness of the Indians is blended with sloth, its natural companion ; and their sloth is of such a nature, that neither their own interest, nor their duty to their masters, can prevail on them to undertake any work. Whatever therefore is of absolute necessity to be done, the care of it is left to the Indian women. These spin, and make the half shirts and drawers, which constitute the whole apparel of their husbands. They cook the matalotage, or food, universally used among them ; they grind the barley, for machca, roast the maize for the camcha, and b ew the chicha; in the mean time, unless the master has been fortunate enough to get the better of the husband's sloth, and taken him to work, he sits squatting on his hams (being the usual posture of all the Indians), and looks on his wife while she is doing the necessary work of the family ; but, unless to drink,
he

he never moves from the fire-side, till obliged to come to table, or wait on his acquaintance. The only domestic service they do, is to plough their chacarita, or little spot of land, in order to its being sown; but the latter, together with the rest of the culture, makes another part, which is also done by the wife and children. When they are once settled in the above posture, no reward can make them stir; so that if a traveller has lost his way, and happens to come to any of these cottages, they hide themselves, and charge their wives to say that they are not at home; when the whole labour consists in accompanying the traveller a quarter of a league, or perhaps less, to put him in his way: and for this small service, he would get a rial, or half a rial at least. Should the passenger alight and enter the cottage, the Indian would still be safe; for, having no light but what comes through a hole in the door, he could not be discovered: and even if he should see the Indian, neither entreaties nor offers would prevail on the slothful wretch to stir a step with him. And it is the same if they are to be employed in any other business.

That the Indians may perform the works appointed by their masters, and for which they are properly paid, it will be of little signification to shew them their task; the master must have his eye continually upon them: for whenever he turns his back, the Indian immediately leaves off working. The only thing in which they shew a lively sensation and alacrity, is for parties of pleasure, rejoicings, entertainments, and especially dancings. But in all these the liquor must circulate briskly, which seems to be their supreme enjoyment. With this they begin the day, and continue drinking till they are entirely deprived both of sense and motion.

Such is their propensity to intemperance, that they are not restrained by any dignity of character; the cacique and the alcalde never fail to be of the com

pany

pany, at all entertainments; and drink like the rest, till the chicha has quite overcome them. It is worth notice, that the Indian women, whether maids or married, and also the young men before they are of an age to contract matrimony, entirely abstain from this vice; it being a maxim among them, that drunkenness is only the privilege of masters of families, as being persons who, when they are unable to take care of themselves, have others to take care of them.

THEIR manner of celebrating any solemnity is too singular to be omitted: the person who gives the entertainment invites all his acquaintance, and provides chicha sufficient for the number of his guests, at the rate of a jug for each; and this jug holds about two gallons. In the court of the house, if it be a large town, or before the cottage, if in a village, a table is placed and covered with a tucuyo carpet, only used on such festivities. The eatables consist wholly of camcha, and some wild herbs boiled. When the guests meet, one or two leaves of these herbs, with ten or twelve grains of camcha, finish the repast. Immediately the women present themselves with calabashes or round totumos, called pilches, full of chicha, for their husbands; and repeat it till their spirits are raised: then one of them plays on a pipe and tabor, whilst others dance, as they call it, though it is no more than moving confusedly from one side to the other, without measure or order. Some of the best voices among the Indian women sing in their own language. Thus their mirth continues while kept up by the liquor, which, as I have said before, is the soul of all their meetings. Another odd circumstance is, that those who do not dance, squat themselves down in their usual posture, till it comes to their turn. The table serves only for state, there being nothing on it to eat, nor do the guests sit down at it. When tired with intemperance, they all lie down together, without minding whether near the wife of another, or their

own

own sister, daughter, or a more distant relation; so shocking are the excesses to which they give themselves up on these solemnities, which are sometimes continued three or four days, till the priests find themselves obliged to go in person, throw away all the chicha, and disperse the Indians, lest they should buy more.

THE day after the festival is called concho, which signifies the day for drinking off the remains of the preceding: with these they begin; and if not sufficient to complete their revel, every one of the guests runs home to his house, and fetches a jug, or they club for more. This occasions a new concho for the next day: and thus, if left to themselves, from day to day, till either no more chicha is to be had, or they left without money or credit.

THEIR burials are likewise solemnized with excessive drinking. The house of mourning is filled with jugs of chicha; and not for the solace of the mourners and their visitors alone; the latter go out into the streets, and invite all of their nation who happen to pass by, whether married or single of both sexes, to come in and drink to the honour of the deceased; and to this invitation they will take no denial. The ceremony lasts four or five days, and sometimes more, strong liquor being their supreme wish, and the great object of all their labours.

IF the Indians are thus excessively addicted to intemperance, gaming is a fault with which they cannot be charged; though these two vices are generally seen together. They seem to have no manner of inclination for play; nor have they above one kind, and that of great antiquity among them; this they call pasa, i. e. a hundred, as he wins who first gets that number. They play at it with two instruments; one a spread eagle of wood with ten holes on each side, being tens; and are marked with pegs, to denote every man's gettings: the other is a bone in the man-

D d 4   ner

ner of a die, cut with seven faces; one of which has a particular mark, and is called guayro. The other five tell according to the number of them; and the last is a blank. The way of playing is only to toss up the bone; and the marks on the upper surface are so many got. But the guayro goes for ten.; and the like number is lost if the blank side appears. Though this game is peculiar to the Indians, it is very little used except at their revels.

THE common food of the Indians, as before observed, is maize made into camcha or mote, and machca; the manner of preparing the latter is, to roast the grain, and then reduce it to a flour; and this without any other apparatus or ingredient, they eat by spoonfuls; two or three of which, and a draught of chicha, qr, when that is wanting, of water, completes their repast. When they set out on a journey, their whole viaticum is a little bag, which they call gucrita, full of this meal, and a spoon. And this suffices for a journey of fifty or a hundred leagues. When hungry, or fatigued, they stop at some place where chicha is to be had, or at some water; where, after taking a spoonful of their meal into their mouth, they keep it some time, in order the more easily to swallow it; and with two or three such spoonfuls, well diluted with chicha, or, if that is not to be had, with water, they set forward as cheerfully as if risen from a feast.

THEIR habitations, as may be imagined, are very small; consisting of a little cottage, in the middle of which is their fire-place. Here both they and the animals they breed, live promiscuously. They have a particular fondness for dogs; and never are without three or four little curs in their hut: a hog or two, a little poultry, and cuyes, with some earthen ware, as pots, and jugs, and the cotton which their wives spin, constitute the whole inventory of an Indian's effects Their beds consist of two or three sheepskins,

skins, without pillows or any thing else; and on these they sleep in their usual squatting posture: and as they never undress, appear always in the same garb.

Though the Indian women breed fowl and other domestic animals in their cottages, they never eat them : and even conceive such a fondness for them that they will not even sell them, much less kill them with their own hands ; so that if a stranger, who is obliged to pass the night in one of their cottages, offers ever so much money for a fowl, they refuse to part with it, and he finds himself under the necessity of killing the fowl himself. At this his landlady shrieks, dissolves in tears, and wrings her hands, as if it had been an only son ; till, seeing the mischief past remedy, she wipes her eyes, and quietly takes what the traveller offers her.

Many of them in their journeys take their whole family with them ; the women carrying on their shoulders such children as are unable to walk. The cottages in the mean time are shut up; and there being no furniture to lose, a string, or thong of leather, serves for a lock : their animals, if the journey is to last for several days, they carry to the cottage of some neighbour or acquaintance : if otherwise, their curs are left guardians of the whole ; and these discharge their trust with such care, that they will fly at any one, except their masters, who offers to come near the cottage. And here it is worth observing, that dogs bred by Spaniards and Mestizos have such a hatred to the Indians, that, if one of them approaches a house where he is not very well known, they fall upon him, and, if not called off, tear him to pieces : on the other hand, the dogs of Indian breed are animated with the same rage against the Spaniards and Mestizos ; and, like the former, scent them at a distance.

The Indians, except those brought up in cities or towns, speak no language but their own, called
Quichua,

Quichua, which was established by the yncas, with an order for its being propagated all over the vast empire, that all their subjects might be able to understand each other; and therefore was distinguished by the name of the Yncas language. Some understand the Spanish, and speak it; yet very few have the good-nature to answer in it, though they know at the same time, that the person with whom they are conversing cannot understand them in Quichua. Nor is it of any consequence to desire and press them to explain themselves in Spanish, for this they absolutely refuse: whereas it is quite otherwise with the Indians born and bred in the towns; for, if spoken to in their own language, they are sure to answer in the Spanish.

Superstition is general among them; and they all, more or less, pretend to fortune-telling. This weakness is also of a long standing among them; and which neither the remonstrances of the priests, nor their own experience, can radically cure. Thus they employ artifices, supposed charms, and strange compositions, in order to obtain some visionary happiness for the success of a favourite scheme, or other weighty concern. In these prestiges their minds are so infatuated, that, to bring them to a sight of the folly and wickedness of such practices, and solidly to embrace the Christian religion, is a work of the greatest difficulty. And even when they have embraced it, are so superficial and fickle, that, if they attend divine service on Sundays and holidays, it is merely from fear of punishment; for otherwise there would be scarce one Indian, especially of the meaner sort, among the whole congregation. Pertinent to this, I shall relate, among many other instances, the following story, told me by a priest. An Indian had, for some time, absented himself from the service of the church; and the priest being informed that it was owing to his drinking early in the morning, on the
following

following Sunday, when he had been particularly ordered to make his appearance, charged him with his
fault, and directed that he should receive some lashes,
the usual punishment of such delinquents, be their
age or sex what it will, and perhaps the best adapted
to their stupidity. After undergoing the punishment,
he turned about to the priest, and thanked him for
having chastised him according to his deserts; to
which the priest replied with some words of exhor-
tation to him, and the audience in general, that they
would never omit any duty of Christianity. But he
had no sooner done, than the poor Indian stepped up
to him, and desired that he would order him a like
number of lashes for the next Sunday, having made
an appointment for a drinking match, so that he
should not be present. This may serve as a specimen
of the little impression made on them, notwithstanding
all the assiduity of the missionaries; and that though
continually instructed, from the first dawnings of
reason till the day of their death, they are found to
continue in a strange ignorance of the most essential
points of religion. Their indifference here is so very
deplorable, that they may be said to give themselves.
no more concern about their souls than about their
bodies: and though I with pleasure allow, that there
are many who, in the culture of their minds, sanctity
of manners, and delicacy of conscience, equal the
most wise and circumspect; yet the bulk of them, either by that gross ignorance which clouds their intellects, and renders them insensible of their eternal concerns, or their natural depravity, are hardened against
religious exhortations. For though they readily grant
every thing that is said to them, and never offer to
make the least objection; yet they secretly harbour
suspicions of some evil design, and leave room for
mental reservations, which spoil all. I am little inclined to lay any false charge to this or any nation,
and especially with regard to such an important sub-

2   ject:

ject: and in confirmation of what I have said, shall relate some further particulars.

Every Sunday in the year, the doctrinal priests instruct their parish in the articles of Christianity with indefatigable zeal: also, when any Indian is sick, they never fail to visit and exhort him to prepare for a comfortable passage into eternity, adding whatever they judge may conduce to the opening the eyes of his understanding; pathetically expatiating on the justice and mercy of God, the nature of death, the certainty of an approaching judgement, and his present danger. After speaking thus a considerable time, without a word from the patient, or the least sign of emotion in his countenance, the good man proceeds to remind him of his sins, and exhorts him to a sincere repentance, and to implore the mercy of his Creator; as, otherwise, his soul will be punished to all eternity. The Indian at length answers, with a serene faintness, " So it will be, father:" meaning, that things will happen as he has predicted; but does not understand in what these threatened sufferings consist. I have often heard priests of those towns, and men of parts and learning, talk with great concern on this subject. Hence it is that there are very few Indians to whom the holy eucharist is administered: nor would those of the house, where a sick person lies, ever give notice of it to the priest, were they not afraid of the punishment which the law in these cases inflicts: and even as it is, they often neglect this duty, and the patient dies without receiving the sacrament.

In their marriages, they run counter to the sentiments of all nations, esteeming what others detest; a virgin being never the object of their choice: for they look on it as a sure sign, that she who has not been known to others, can have nothing pleasing about her.

AFTER

AFTER a young man has asked the object of his affections of her father, and obtained his consent, they immediately begin to live together as man and wife, and assist the father-in-law in cultivating his chacara. At the end of three or four months, and often of a year, he leaves his bride, without ceremony, and perhaps for the wild reason above-mentioned: and even expostulates with the father-in-law, that he should endeavour to deceive him, by imposing upon him his daughter, whom nobody else had thought worthy of making his bedfellow. But if nothing of this happens, after passing three or four months in this commerce, which they call *Amanarse*, i. e. to habituate one's self, they then marry: and this custom is still very common, having hitherto proved too strong for the joint endeavours of the whole body of the clergy to extirpate. Accordingly, the first question at the ceremony of marriage is, whether they are *amannados*, in order to absolve them of that sin before they receive the nuptial benediction. They look upon no marriage to be legal which is not solemn, and according to them the whole consists in the nuptial benediction, which must be given them at the time they join their hands, as otherwise, on any caprice, they separate: and it is to no purpose to go about to persuade them that they were married; nor will any punishment have the least effect. For as it does not imply any infamy, the intention is lost. It is the same thing with them to be exposed to the public derision and insults, as to be ordered to shew their skill in dancing on a festival; the thing which, of all others, they most delight in. They are indeed sensible of corporal punishments during the time they are inflicting, but immediately afterwards are as placid and easy as if they had not been touched. This occasions many things to be connived at in them, and other means of prevention used.

IT

It is not uncommon among them to change their wives, without any other preliminary or agreement, than having been familiar with the wife of another. The former wife, together with the injured husband, concert a revenge ; and if reproached for such a proceeding, they cheerfully answer, that they had served them only as they deserved ; and it avails little to separate them, as they soon find means to return to the same manner. of living. Incests are very common among them, both as the consequence of their monstrous drunkenness, already mentioned, and from their making no distinction between honour and infamy, whereby their brutal appetites are under no restraint.

If the foregoing tempers or customs appear strange, their behaviour at confession is not less so : for, besides having but a slender acquaintance with the Spanish language, they have no form to direct them in it. On their coming to the confessor, which is always at his summons, he is obliged to instruct them in what they are going about, and with them repeat the *Confiteor* from one end to the other. For if he stops, the Indian also remains silent. Having gone through this, it is not enough for the priest to ask him, whether. he has committed this or that fault ; but if it be one of the common sort, he must affirm that he has committed it, otherwise the Indian would deny every thing. The priest further is obliged to tell him, that he well knows he has committed the sin, and he has proofs of it. Then the Indian, being thus pressed, answers, with great astonishment, that it is so ; and, imagining the priest really endued with some supernatural knowledge, adds circumstances which had not been asked him. It is not only difficult to bring them to declare their faults, but even to keep them from denying them, though publicly committed, and equally so to prevail on them to determine the number; this being only to be obtained by finesses; and then

little

little stress is to be laid on what they say. The natural dread, which more or less rises in all men at the approach of death, is what the Indians are less susceptible of than any other people. Their contempt of those evils which make the strongest impressions on the minds of men, is such, that they view the approach of death without peturbation: and the pain of the distemper affects them more than the danger of it. This I have often heard from several of the priests: and their words are confirmed by daily instances. For when the priests perform the last offices to dying persons, their answers are delivered with that composure and serenity, as leave no doubt but the inward state of their mind corresponds with these external appearances, being the principal and cause of them. The like is even seen in those whom their crimes have brought to die by the hands of justice; and among many other examples, I happened myself to be an eye-witness of one. Whilst I was at Quito, two malefactors were to be executed; one a Mestizo or Mulatto, and the other an Indian: both having been brought into the prison-chapel, I went to see them the night before the execution. The former was attended by several priests, who, in Spanish, exhorted him to die like a Christian, and shew a becoming fervour in his love to God, faith, and contrition, and a detestation for the crimes he had committed. On which, his aspect and whole deportment shewed a sense of his condition. The Indian had also ecclesiastics about him, performing, in his own language, the like kind offices. But to all appearance he was less concerned even than those about him, and seemed rather to be tilling a chacura, or tending a herd, than on the eve of eternity. His appetite was so far from leaving him, as was the case of his companion, that he was more eager, and, after dispatching his own, would have cleared his fellow-sufferer's plate; so that they were obliged to use some force to

prevent

prevent his eating to excess on such an exigency. He talked to the spectators with that ease and tranquillity, as if only going to take a short journey. He· answered to the exhortations without the least confusion: when he was ordered to kneel, he did so. The prayers and acts of devotion he also repeated word for word; but all the time rolling his eyes about, like a sportive child, whose weak age is diverted by trifling objects. Thus he behaved till brought to the gibbet, where his companion had been carried before him : nor did he shew the least alteration even in the awful moment. And this, to a civilized European so strange, is no more than what is common among the Indians of these parts.

THIS indifference with regard to death, or intrepidity, if we may term it so, shews itself upon many other occasions, particularly in the alacrity and resolution with which they face themselves before a bull, with no other view than for the bull to run full at him, and toss him so high in the air, that any other than an Indian would be killed by the fall. He however rises without receiving any hurt, and is· highly delighted with the victory, as he calls it, over the bull ; though the victory seems to lie on the bull's side. When they fight in a body against others, they fall on, without any regard to superiority of numbers, or who drops, or is wounded of their party. An action which in a civilized nation is counted the height of courage, is here merely the effect of barbarism and want of thought. They are very dextrous in haltering a bull at full speed; and, as·they fear no danger, attack him with what we should call great temerity. With the same dexterity they hunt bears : and a single Indian, with only a horse and his noose, never fails of getting the better of all the cunning and rage of this furious animal. This noose is made of cow-hide, so thin as not to be seized by the beast's paws, and yet so strong as not to be broken
by

by the struggles of the creature. On perceiving the bear, they immediately make towards him, whilst he sets up·in order to seize the horse. But the Indian being come within a proper distance, throws the noose about the creature's neck : then, with surprizing celerity having taken two or three turns with the other end about the saddle, claps spurs to his horse : in the mean time the bear, unable to keep pace with the horse, and struggling to clear himself of the noose, is choaked. This is considered as an achievement of admirable dexterity and bravery; and may be frequently seen in the province of Alausi, near the eastern Cordillera, where these animals abound.

A GREAT part of the rusticity in the minds of the Indians must be imputed to the want of culture; for they, who in some parts have enjoyed that advantage, are found to be no less rational than other men : and.if they do not attain to all the politeness of civilized nations, they at least think properly. The Indians of the mission of Paraguay are, among others, remarkable instances of this; where, by the zeal, address, and exemplary piety of the Jesuits, a regular well-governed republic of rational men has been established : and the people, from an ambulatory and savage manner of living, have been reduced to order, reason, and religion. One of the most effectual means for this was, the setting up schools for instructing the young Indians in Spanish, in which they also instruct their converts; and those who are observed to be of a. suitable genius, are taught Latin. In all the villages of the missions are schools for learning, not only to read and write, but also mechanic trades; and the artificers here are not inferior to those of Europe. These Indians, in their customs and intellects, are a different sort of people from those before-mentioned. They have a knowledge of things; a clear discernment of the turpitude of vice, and the amiableness of virtue; and act up to these sentiments : not that they have any natural advantage over the other: for I have observed throughout this whole kingdom, that the Indians of its several provinces through which I travelled are alike. And those of Quito are not more deficient in their understandings, than those of Valles or Lima; nor are these more acute or sagacious than the natives of Chili and Arauco.

WITHOUT going out of the province of Quito, we have a general instance in confirmation of what I have advanced. For all the Indians, brought up to the Spanish language, are far more acute and sensible than those who have spent their lives in little villages; and their behaviour more conformable to the dictates of a rational creature. They are men of abilities and skill, and have divested themselves of many of their errors. Whence they are called Ladinos, i. e. knowing men : and if they retain any of the culpable practices of the former, it is from the infection of intercourse, or from a mistaken notion that they should keep them up as transmitted to them from their ancestors. Among these are chiefly distinguished the barber-surgeons, who bleed with such dexterity, that, in the opinion of Mons. de Jussieu and Mons. Seniergues, surgeons to the French academists, they equal the most famous in Europe: and their intercourse with persons of a liberal education enlightens their understanding, so that they distinguish themselves to great advantage among their countrymen. It seems to me unquestionable, that if in villages care was taken to instruct the Indians in Spanish, conformable to the laws of the Indies, besides other acquirements, this people would have the benefit of conversing more frequently with the Spaniards, which would greatly improve their reason, and give them a knowledge of many things for which they have no word in their language. Accordingly it is observed that the Cholos (a name given to the Indian boys becoming acquainted with the Spanish language,

2   improve

improve so much in knowledge, that they look on their countrymen as savages, and take upon themselves the appellation of Ladinos.

I AM very far from imagining that the Spanish language itself has the virtue of improving the intellects of the Indians ; but only, that rational conversation with the Spaniards would lead them to a knowledge of many things : and consequently they might be brought to a greater purity of faith and practice. Whereas the conversation among themselves must be very low and confined : and what they have with the Spanish traders who understand their language, turns wholly on traffick.   But if they understood the Spanish, they would daily receive new lights by conversing with travellers whom they attend, as well as from the inhabitants of the cities, their masters, the priests, the corregidors, and others; and thus become more industrious and tractable, and acquainted with the nature of things of which before they had not so much as an idea.

ARE not the differences and advantages evident among ourselves, betwixt a young man whose stock of learning is his natural language, and him who is acquainted with others? What a superiority of knowledge, discernment, and facility in the latter! Hence we may form some idea of the abject state of the human mind, among rude country people, who cannot exchange a word with a stranger, and never stir out of their village : whereas, when any one happens to go to a neighbouring town, he returns home with enlarged knowledge, and entertains all the village with his narratives: but if he had not understood the language spoken in it, he would have been little the better; nor able to relate the strange things he saw and heard. This is the very case of the Indians ; and I am of opinion, that to teach them the Spanish tongue would be the best means of improving their reason, and consequently of making them better members of society : and that my superiors thought so, appears from the ordinances relating to America.

THE Indians in general are robust, and of a good constitution. And though the venereal distemper is so common in this country, it is seldom known among them : the principal cause of which unquestionably lies in the quality of the juices of their body, not being susceptible of the venom of this distemper. Many however attribute it to a quality in the chicha, their common drink. The disease which makes the greatest havock among them, is the small-pox : which is so fatal that few escape it. Accordingly it is looked upon in this country as a pestilence. This distemper is not continual as in other nations, seven or eight years, or more, passing without its being heard of; but when it prevails, towns and villages are soon thinned of their inhabitants. This desolation is owing partly to the malignity of the disease, and partly to the want of physicians and nurses. Accordingly, on being seized with this distemper, they immediately send for the priest to confess; and die for want of remedy and relief. The like happens in all other distempers; and were they frequent, would be equally fatal, these poor creatures dying for want of proper treatment and assistance; as is evident from the Creoles, who are also attacked by the distempers of the country. Some of the latter indeed die as well as of the former; but many more recover, having attendance and a proper diet: whereas the Indians are in want of every thing. What their houses and apparel are, has already been seen. Their bed is the same in health and sickness; and all the change in their food is in the manner of taking it, not in the species itself: for, however ill they may be, all they have is a small draught of machca dissolved in chicha ; so that, if any one does get the better of a distemper, it is more

owing

owing to the happiness of his constitution, than any relief he receives.

THEY are also subject to the bicho, or mal del valle; but this is soon cured. Sometimes, though seldom, they are also seized with tabardillos, or spotted fevers, for which they have an expeditious but singular cure. They lay the patient near the fire, on the two sheep-skins which compose his bed; and close by him place a jug of chicha. The heat of the fever, and that of the fire increasing the other, cause in him such a thirst, that he is incessantly drinking; whereby the eruptions are augmented, and the next morning he is either in a fair way of recovery, or so bad as to be carried off in a day or two.

THEY who either escape, or recover from, these distempers, reach to an advanced age; and both sexes afford many instances of remarkable longevity. I myself have known several, who, at the age of a hundred, were still robust and active; which unquestionably must, in some measure, be attributed to the constant sameness and simplicity of their food. But I must observe, that, besides the different kinds already mentioned, they also eat a great deal of salt with agi, gathering the pods of it; and having put some salt in the mouth, they bite the agi, and afterwards eat some machca or camcha: and thus they continue taking one after another, till they are satisfied. They are so fond of salt in this manner of eating it, that they prefer a pod or two of agi with some salt to any other food.

AFTER this account of the genius, customs, and qualities of the Indians, it will not be improper to speak a word or two of their diversions and occupations, premising, that this account does not extend to such Indians as live in cities and towns, or that occupy any public office or trade, they being looked upon as useful to the public, and live independently. Others in the kingdom of Quito are employed in the

manufac-

manufactories, the plantations, or in breeding of cattle.   In order to this, the villages are annually to furnish those places with a number of Indians, to whom their master pays wages as settled by the equity of the king: and at the end of the year they return to their villages, and are replaced by others.   This repartition is called mita.   And though these alterations should by order take place in the manufactories, yet it is not so: for being occupations of which none are capable but such as have been properly trained up, the Indian families, which are admitted, settle there, and the sons are instructed in weaving, from one generation to another.   The earnings of these are larger than those of the other Indians, as their trade requires greater skill and capacity.   Besides the yearly wages paid them by those whom they serve, they have also a quantity of land, and cattle given them to improve. They live in cottages built near the mansion-house, so that every one of these forms a kind of village; some of which consist of above an hundred and fifty families.

CHAP. VII.

An historical Account of the most remarkable Mountains and Paramos, or Deserts, in the Cordilleras of the Andes; the Rivers which have their Sources in these Mountains, and the Methods of passing them.

I NOW come to the most remarkable paramos, or deserts, of the kingdom of Quito, and the rivers flowing through that country, which, among many other natural curiosities, is peculiarly remarkable for the disposition of the ground, and its prodigious masses of snow, that exceed all comparison.

It

It has been before observed, that all the dependencies of the jurisdictions of this province are situated
betwixt the two Cordilleras of the Andes ; and that
the air is more or less cold according to the height of
the mountains, and the ground more or less arid.
These arid tracts are called Paramos, or deserts ; for
though all the Cordilleras are dry or arid, some of
them are much more so than others ; for the continual
snows and frost render them absolutely uninhabitable,
even by the beasts ; nor is there a single plant to be
found upon them.

Some of these mountains, seemingly as it were
founded on others, rise to a most astonishing height,
and are covered with snow even to their summits. The
latter we shall more particularly treat of, as they are
the most remarkable and curious objects.

The paramo of Asuay, formed by the junction of
the two Cordilleras, is not of this class ; for, though
remarkable for its excessive coldness and aridity, its
height does not exceed that of the Cordilleras in
general, and is much lower than that of Pichincha
and Corazon.   Its height is the degree of the climate,
where a continual congelation or freezing commences ;
and as the mountains exceed this height, so are they
perpetually covered with ice and snow ; that from a
determined point above Carabucu for instance, or the
surface of the sea, the congelation is found at the
same height in all the mountains. From barometrical
experiments made at Pucaguayco, on the mountain
Cotopaxi, the height of the mercury was 16 inches
5⅜ lines ; whence we determined the height of that
place to be 1023 toises above the plain of Carabucu,
and that of the latter above the superficies of the sea
about 1268.   Thus the height of Pucaguayco, above
the surface of the sea, is 2291 toises. The signal
which we placed on this mountain was thirty or forty
toises above the ice, or point of continual congelation ; and the perpendicular height from the com-

mencement of this point to the summit of the mountain, we found, from some geometrical observations
made for that purpose, to be about 880 toises. Thus
the summit of Cotopaxi is elevated 3126 toises above
the surface of the sea, or something above three geographical miles; and 639 toises higher than the top of
Pichincha. These are mountains I intend to speak
of; and the height of them all, considering the greatness of it, may be said to be nearly equal.

In these Cordilleras, the most southern mountain is
that of Mecas, more properly called Sanguay, though
in this country better known by the former, lying in
the jurisdiction of the same name. It is of a prodigious height, and the far greatest part of the whole
surface covered with snow. From its summit issues
a continual fire, attended with explosions, which are
plainly heard at Pintac, a village belonging to the
jurisdiction of Quito, and near forty leagues distant
from the mountain; and, when the wind is fair, the
noise is heard even at Quito itself. The country adjacent to this volcano is totally barren, being covered
with cinders ejected by it. In this Pacamo, the river
Sangay has its source. This river cannot be said to be
small, but after its junction with another, called the
Upano, forms the Payra, a large river which discharges itself into the Maranon.

In the same eastern Cordillera, about six leagues
west of the town of Riobamba, is a very high mountain, with two crests, and both of them covered with
snow; that on the north is called Collanes, and that
on the south Altar; but the space covered with snow
is much less than that of Sangay and others of this
class, its height being proportionally less.

North of the same town, and about seven leagues
distant, is the mountain of Tunguragua, of a conical
figure, and equally steep on all sides. The ground,
at its basis, is something lower than that of the Cordillera, especially on the north side, where it seems to
rise

rise from the plain on which the villages are situated. On this side, in a small plain betwixt its skirts and the Cordillera, has been built the village of Bannos, so called from its hot medicinal baths, to which there is a great resort from all parts of this jurisdiction. South of Cuença, and not far from another village, called Bannos also, belonging to this jurisdiction, are other hot waters on the summit of an eminence, gushing out through several apertures of four or five inches diameter, and of a heat which hardens eggs sooner than water boiling over the fire. These several streams unite and form a rivulet, the stones and banks of which are tinged with yellow, and the water is of a brackish taste. The upper part of this small eminence is full of crevices, through which issues a continual smoke: a sufficient indication of its containing great quantities of sulphureous and nitrous substances.

Noʀᴛʜ of Riobamba, inclining some degrees to the west, is the mountain of Chunborazo, by the side of which lies the road from Quito to Guayaquil. At first great numbers of the Spaniards perished in passing the vast and dangerous deserts on its declivity; but being at present better acquainted with them, and inured to the climate, such misfortunes are seldom heard of; especially as very few take this road, unless there is the greatest appearance of two or three days of calm and serene weather.

Noʀᴛʜ of this mountain stands that of Carguayraso, which has been already taken notice of.

Noʀᴛʜ of Latacunga, and about five leagues distant from it, is Cotopaxi, which, towards the northwest and south, extends itself beyond all the others; and which, as I have before observed, became a volcano at the time of the Spaniards' first arrival in this country. In 1743, a new eruption happened, having been some days preceded by a continual rumbling in its bowels. An aperture was made in its summit, and three about the same height near the middle of

its

its declivity, at that time buried under prodigious masses of snow. The ignited substances ejected on that occasion, mixed with a prodigious quantity of ice and snow, melting amidst the flames, were carried down with such astonishing rapidity, that in an instant the plain, from Callo to Latacunga, was overflowed ; and, besides its ravages in bearing down houses of the Indians and other poor inhabitants, great numbers of people lost their lives. The river of Latacunga was the channel of this terrible flood, till, being too small for receiving such a prodigious current, it overflowed the adjacent country like a vast lake near the town, and carried away all the buildings within its reach. The inhabitants retired to a spot of higher ground behind their town, of which those parts which stood within the limits of the current were totally destroyed. The dread of still greater devastations did not subside in three days, during which the volcano ejected cinders, while torrents of melted ice and snow poured down its sides. The fire lasted several days, and was accompanied with terrible roarings of the wind rushing through the volcano, and greatly exceeded the great rumblings before heard in its bowels. At last all was quiet, neither fire nor smoke were seen, nor was there any noise to be heard till the following year, 1744; when, in the month of May, the flames increased, and forced their passage through several other parts on the sides of the mountain ; so that in clear nights, the flames being reflected by the transparent ice, formed a very grand and beautiful illumination. November the 30th, it ejected such prodigious quantities of fire and ignited substances, that an inundation equal to the former soon ensued; so that the inhabitants of Latacunga gave themselves over for lost. And we ought to acknowledge the Divine protection, that it did not rage when we visited it, having occasion twice to continue some

time

time on its declivity, as we have already shewn in the third chapter of the fifth book.

FIVE leagues to the west of this mountain stands that of Illinisa, whose summit is also bifid, and constantly covered with snow. From it several rivulets derive their source; of which those flowing from the northern declivity continue that direction: as those from the southern side also run southward. The latter pay their tribute to the northern ocean, through the large river of the Amazons; while the former discharge themselves into the South-sea, by the river of Emeralds.

NORTH of Cotopaxi is another snowy mountain called Chinculagua, something less than the former, though even that is not to be compared to the others.

THE mountain of Cayamburo, which is one of the first magnitude, lies north, some degrees easterly, from Quito, at the distance of about eleven leagues from that city. There is neither appearance nor tradition of its having ever been a volcano. Several rivers issue from it, of which those from the W. and N. run either into the river of Emeralds or that of Mira, but all fall into the South-sea; while these from the E. discharge themselves into the river of the Amazons.

BESIDES the torrents which precipitate themselves from the snowy mountains, others have their source in the lower parts of the Cordilleras, and at their conflux form very large and noble rivers, which either pay the tribute to the north or south seas, as we shall hereafter observe.

ALL the springs issuing from the mountains in the neighbourhood of Cuença, on the west and south side as far as Talqui, with those of the eastern Cordillera, and northward. as far as the Parama de Burgay, unite at about half a league eastward of a chapel called Jadan, under the care of the curate of Paute, where forming a river, and passing near the village

from

from which it has its name, discharges itself into the river of the Amazons.   It is so deep at Paute as not to be fordable, though very wide there.

FROM the mountains of Assuay, Bueran, and the adjacent hills on the south, is formed a very considerable river, over which are several bridges.   It is called Cannar, from that town being the only one in its course; which it continues by Yocon to the bay of Guayaquil.

THE north parts of the Paramo of Asuay also gave rise to many streams, which, uniting with others coming from Mount Senegualap, and the western side of the eastern Cordillera, form the river Alausi, which discharges itself into the same bay.

ON the highest part of the Paramo de Tioloma, and near the signal one erected on this mountain for forming our series of triangles, are four lakes, the three nearest it being less than the other, which is about half a league in length, and called Coley; and the others, which are not greatly inferior, Pichabinnac, Pubillu, and Mactallan.   From these is formed the river Cebadas, which runs near the village of that name, and is joined by another arising from the springs on the Paramo of Lalanguso, and the streams from the Colta lake, after which, inclining a little from the north towards the east, passes by Pungala; and about a league from the village of Puni, is joined by the river Bamba, which has its source in the Parambo of Sisapongo.   Near the town of Cobigies is another, which flows from the mountain of Chimborazo, and which, after directing its course northward, till it is in an east and west direction with the mountain of Tunguragua, it winds to the east, and adds its water to those of the river of the Amazons.   At the town of Penipe, it is so deep and rapid as only to be crossed over a bridge made of bujucos.   Also before it reaches the town of los Bannos, it is increased by the rivers Latacunga and Bato, together with

with all the streams from both the Cordilleras, those from the southern summit of Elenisa, and the southern side of Ruminavi and Cotopaxi.

THE streams flowing from the north summit of Elenisa, I have already mentioned to run northward; and with these all from the same Cordillera unite, together with those issuing from the north and west sides of the mountain Ruminavi, those of Pasuchua; and from this junction rises the River Amaguanna. The two last mountains stand north and south from each other, in an intermediate space of the Cordilleras. From the north side of Cotopaxi the Paramo of Chinchulagua, which is also covered with snow, and the Cordillera de Guamani, other streams have their rise, and from their conflux is formed the river Ichubamba, which, running northward, joins the Amaguanna, a little to the north of Cono-Coto. Afterwards it receives the rivulets issuing from the eastern Cordillera, and changes its name to that of Guayllabamba. The waters which have their source in the western part of Cayamburo, and the southern part of Moxanda, form another river called Pisque, which first runs towards the west, and joining the Guayllabamba, takes the name of Alchipichi, which, a little to the north of St. Antonio, in the jurisdiction of Quito, is so broad and rapid, that there is no passing it but in a tarabita, which we shall presently describe. From hence it continues its course northwards, and at last falls into the river of Emeralds.

THE mountain of Majanda stands in the interval between the Cordilleras; and though it has only one side as it were, it is divided into two summits, one eastward and the other westward; and from both these, runs a small Cordillera, which afterwards joining, inclose this valley.

FROM the side of this mountain issue two large torrents, which meet in the lake of St. Pablo: from whence flows a river, which, being joined by others from

from the springs of the western Cordillera, form one stream, and after being increased by another brook from the heights of Oezillo, give rise to the river which washes the town of St. Miguel de Ibarra; after which it takes the name of Mira, and discharges itself into the South-sea, a little to the north of the river of Emeralds.

When the rivers are too deep to be forded, bridges are made at the most frequented places. Of these there are two kinds besides those of stone, which are very few: the former of wood, which are the most common; and the latter of bujucos. With regard to the first, they choose a place where the river is very narrow, and has on each side high rocks. They consist of only four long beams laid close together over the precipice, and form a path about a yard and a half in breadth, being just sufficient for a man to pass over on horseback; and custom has rendered these bridges so natural to them, that they pass them without any apprehension. The second, or those formed of bujucos, are only used where the breadth of the river will not admit of any beams to be laid across. In the construction of these, several bujucos are twisted together, so as to form a kind of large cable of the length required. Six of these are carried from one side of the river to the other, two of which are considerably higher than the other four. On the latter are laid sticks in a transverse direction, and, over these, branches of trees, as a flooring; the former are fastened to the four which form the bridge, and by that means serve as rails for the security of the passenger, who would otherwise be in no small danger from the continual oscillation. The bejuco bridges in this country are only for men, the mules swim over the rivers; in order to which, when their loading is taken off, they are drove into the water near half a league above the bridge, that they may reach the opposite shore near it, the rapidity of the stream carrying them

so

so great a distance. In the mean time, the Indians carry over the loading on their shoulders. On some rivers of Peru there are bejuco bridges so large, that droves of loaded mules pass over them; particularly the river Apurimac, which is the thoroughfare of all the commerce carried on between Lima, Cusco, La Plata, and other parts to the southward.

Some rivers, instead of a bejuco bridge, are passed by means of a tarabita; as is the case with regard to that of Alchipichi. This machine serves not only to carry over persons and loads, but also the beasts themselves; the rapidity of the stream, and the monstrous stones continually rolling along it, rendering it impracticable for them to swim over.

The tarabita is only a single rope made of bejuco, or thongs of an ox's hide, and consisting of several strands, and about six or eight inches in thickness. This rope is extended from one side of the river to the other, and fastened on each bank to strong posts. On one side is a kind of wheel, or winch, to straighten or slacken the tarabita to the degree required. From the tarabita hangs a kind of leathern hammock capable of holding a man; and is suspended by a clue at each end. A rope is also fastened to either clue, and extended to each side of the river, for drawing the hammock to the side intended. A push at its first setting off, sends it quickly to the other side.

For carrying over the mules, two tarabitas are necessary, one for each side of the river, and the ropes are much thicker and slacker. On this rope is only one clue, which is of wood, and by which the beast is suspended, being secured with girts round the belly, neck, and legs. When this is performed, the creature is shoved off, and immediately landed on the opposite side. Such as are accustomed to be carried over in this manner, never make the least motion, and even come of themselves to have the girts fastened round them; but it is with great difficulty they are first
brought

brought to suffer the girts to be put round their bodies, and when they find themselves suspended, kick
and fling, during their short passage, in a most terrible
manner. The river of Alchipichi may well excite
terror in a young traveller, being between'thirty and
forty fathoms from shore to shore; and its perpendicular height, above the surface of the water, twenty-
five fathoms. A representation of these bridges, and
the manner of conveying over the mules, was given
in the last plate, Nº V.

THE roads of this country are suitable to the bridges;
for though there are large plains between Quito and
the river Bambar, and the greatest part of the road
between the river Bamba and Alausi, and even to the
north of that city, lies along the mountains, yet these
are interrupted by fruitful breaches, the acclivities and
declivities of which are not only of a great length and
very troublesome, but also dangerous. In some places
there is a necessity for travelling along tracts on the
declivities of mountains, which are sometimes so narrow as hardly to allow room for the feet of the beast;
part of its body, and that of the rider, being perpendicular over a torrent fifty or sixty fathoms beneath
the road. So that certainly nothing but absolute
necessity, there being no other road, and long custom,
can get the better of that horror which must affect
the person at the sight of such imminent danger; and
there are too many instances of travellers losing their
effects, if not lives, their whole dependence being on
the sure foot of the mule. This danger is indeed, in
some measure, compensated by the security of the
roads; so that we see here. what none of the civilized nations can boast of, namely, single persons
travelling, unarmed, with a great charge of gold and
silver, but equally safe as if strongly guarded. If
the traveller happens to be fatigued in a desert, he
lays him down, and sleeps without the least apprehension of danger. Or if he takes up his lodgings

1   in

in a tambo, or inn, he sleeps with the same security, though the doors are always open: nor is he ever molested on the road. This is a convenience so favourable to commerce and intercourse, that it were greatly to be wished the same security could be established in the other parts of the world.

CHAP. VIII.

Continuation of the Account of the Paramos, or Deserts; with an Account of the Beasts, Birds, and other Particulars of this Province.

TO conclude my observations on the Paramos, which it was necessary to interrupt, in order to give a short account of the rivers, bridges and roads, I shall observe, that, these parts not being of a height sufficient to expose them to an eternal frost, they are covered with a kind of rush resembling the genista Hispanica, but much more soft and flexible. It is about half or three quarters of a yard in height, and, when of its full magnitude, its colour is like that of dried genista Hispanica. But where the snow remains some time on the ground without melting, none of these plants growing in habitable climates are found. There are indeed others, though few, and even these never exceed a certain height. Above this tract, nothing is seen but stones and sand all the way up to the beginning of the ice.

In these parts, where the above rush is the principal product, the soil is as little adapted to cultivation; but produces a tree, which the inhabitants call quinual, the nature of which very well suits the roughness of the climate. It is of middling height, tufted, and the timber strong; its leaf of a long, oval form, thick, and of a deep green colour. Though it bears the same name as the grain called quinua, of

Vol. I.   F f   which

which we have spoken elsewhere, and which grows in great plenty, the latter is not however the production of this tree; nor has the plant, on which it grows, any thing in common with it.

The climate proper for quinua is also adapted to the produce of a little plant, which the Indians call palo de luz. It is commonly about the height of two feet, consisting of stalks which grow out of the ground, and proceed from the same root. These stems are straight, and smooth up to the top, from which grow little branches with very small leaves. All of these nearly rise to the same height, except the outer ones, which are of a less size: it is cut close to the ground, where it is about three lines in diameter; and being kindled whilst green, gives a light equal to that of a torch, and, with care taken to snuff it, lasts till the whole plant is burnt.

In the same places grows also the achupalla, con-sisting of several stalks, something resembling those of the sabila; and as the new shoot up, the most out-ward grow old and dry, and form a kind of trunk, with a great number of horizontal leaves, hollow in the middle; and this, when not very large, is eatable like that of the palmitos.

Towards the extremity of the part where the rush grows, and the cold begins to increase, is found the vegetable called puchugchu, with round leaves grow-ing together so as to represent a very smooth bulb, having nothing in them but the roots: and as these increase, the outward case of leaves dilates into the form of a round loaf, usually a foot or two in height, and the same in diameter: on this account they are also called loaves or onions. When in their vigour, they are of so hardy a nature, that a stamp with a man's foot, or the tread of a mule, makes no im-pression on them; but when once fully ripe, they are easily broken. In the middle state, betwixt the full strength of their resistance and the decay of their

2   roots

roots by age, they have an elastic quality, yielding with a tremulous motion to the pressure of the foot, and on its being taken off recover their form.

In the places where the puchugchu thrives best, also grows the canchalagua, the virtues of which are well known in Europe. The form of this is like a very thin rush or straw; bears no leaves, but has a few small seeds at its extremity. It is medicinal, and particularly useful as a febrifuge; its taste is bitter, which it easily communicates either by infusion or decoction. In this country it is chiefly used as a sweetener of the blood, though thought to be of a hot quality. It grows in great quantities, and is found both among the puchugchu, and in other parts on the heath where the cold is less intense.

Another plant, not less valuable for its virtues, and growing chiefly in those dreadful deserts where, either from the severity of the cold or perpetual snows, or from the badness of the soil, nothing else is produced, is found the so celebrated calaguala; its height is about six or eight inches, and naturally spreads itself in thin stems along the sand, or climbs up the rocks. These branches in their form resemble the fibril of the roots of the other plants, being not above two or three lines in their greatest thickness, round, and full of little knots, where they bend round like the tendrils of a vine. They have a thin pellicle of a loose texture, which of itself separates when the plant dries. The most singular virtue of this plant is for all kind of imposthumes, internal or external, which it discusses and heals in a very little time. The manner of administering it is by decoction, of which a very little serves; or, after bruising it, to infuse it in wine, and take it fasting for three or four days, and no longer, its good effects in that time being usually conspicuous; and being extremely hot, it might prove pernicious, if taken in greater quantity than absolutely necessary; for which reason only three

F f 2   or

or four pieces, each about an inch and a half in length, are used for the infusion, and with such sort of wine as will best correct its bitterness. Though this excellent herb grows in most of those frozen deserts, yet the best is that in the southern province of Peru. The leaves are very small, and the few it bears grow contiguous to the stem.

THE paramos, or barren heaths, likewise yield the contrayerva, which makes a part of the materia medica in Europe, and is considered as an excellent alexipharmic. This is also a creeping plant, with a leaf of about three or four inches in length, and little more than one in breadth, thick, and the back part of it exceeding soft to the touch, and of a deep green. The other side is also smooth, but of a light green. On its stem grows a large blossom, consisting of many flowers inclining to a violet colour: but neither these nor the other flowers, which grow in great abundance in these countries, according to its several climates, are much esteemed; so that, when wanted, the readiest way is to send and have them cut from the plant.

THOUGH the severity of the air on the deserts is such, that all animals cannot live there, yet they afford many beasts of venery, which feed on the straw or rush peculiar to those parts; and some of these creatures are met with on the highest mountains, where the cold is intolerable to the human species. Among the rushes are bred great numbers of rabbits, and some foxes, both which in their appearance and qualities, resemble those of Carthagena and other parts of the Indies.

THE only birds known in those rigorous places are partridges, condors, and zumbadoies or hummers. The partridges differ something from those of Europe; they nearly resemble the quail, and are very scarce.

THE

The condor is the largest bird in these parts of the world; its colour and appearance resemble those of the galinazos, and sometimes it soars from the highest mountains so as to be almost out of sight: and by its being seldom seen in low places, a subtile air seems best to agree with it; though some, which have been tamed when young, live in the villages and plantations. Like the galinazos, they are extremely carnivorous, and are known frequently to seize and fly away with lambs that feed on the heaths: of this I happened to see an instance, in my way down from the signal of Lalanguso toward the plantation of Pul, lying near the bottom of those mountains. Observing, on a hill adjoining to that where I was, a flock of sheep in great confusion, I saw one of these condors flying upwards from it with a lamb betwixt its claws; and, when at some height, dropped it; then, following it, took it up, and let it fall a second time, when it winged its way out of sight, for fear of the Indians, who, at the cries of the boys and barkings of the dogs, were running towards the place.

In some deserts this bird is common; and as it preys on the flocks, the Indians are not wanting in their endeavours to catch them. One of the ways is, to kill a cow, or other beast, when of no further use, and to rub the flesh with the juice of some potent herbs, which they afterwards carry away: for otherwise the bird, sensible of them by natural instinct, would not touch the flesh. Further, to take off the smell, they bury the flesh till it becomes putrid, and then expose it; when the condors, allured by the smell of the carcase, hasten and greedily feed on it, till the herbs operate so as to render them quite senseless and incapable of motion: the Indians seize the opportunity, and destroy them. They likewise catch them with springes laid near some flesh: but such is the force of this bird, that, with a stroke of its wing, it sometimes knocks down the man who approaches it. Their wing also serves them as a shield, by which they ward off blows, without receiving any hurt.

The zumbador, or hummer, is a night bird, peculiar to the mountainous deserts; and they are seldom seen, though frequently heard, both by the singing and a strange humming made in the air by the rapidity of their flight, and which may be heard at the distance of fifty toises; and when near, is louder than that of a rocket. Their singing may indeed be called a kind of cry, resembling that of night-birds. In moonlight nights, when they more frequently make their appearance, we have often watched to see their size and the celerity of their motion; and though they passed very near us, we never were able to form any idea of their magnitude; all that we could see, was a white line which they formed in their flight through the air; and this was plainly perceivable, when at no great distance. We promised the Indians a reward if they would procure us one; but all they could do was to procure a young one, scarce fledged, though it was then of the size of a partridge, and all over speckled with dark and light brown; the bill was proportionate and strait; the aperture of the nostrils much larger than usual, the tail small, and the wings of a proper size for the body. According to our Indians, it is with the nostrils that it makes such a loud humming. This may, in some measure, contribute to it; but the effect seems much too great for such an instrument; especially as at the time of the humming it also uses its voice.

Among the valleys and plains formed by these mountains, are many marshy places, occasioned by the great variety of small streams of water; and in these breed great numbers of a bird called canclon, a name perfectly expressive of its manner of singing. It very much resembles the bandurria, though the species be different: it exceeds the bigness of a large goose,
has

has a long thick neck, and a head something resembling that bird. The bill is straight and thick, and its legs and feet thick and strong. The outward feathers of the wing are of a dark brown, those of the inside of a pure white; but the other parts of the body spotted. At the meeting of the wings they have two spurs, projecting to the length of an inch and a half, as their defence. The male and female are inseparable, whether flying, or on the ground, where they mostly keep themselves, never taking flight except across a valley, or when pursued. The flesh eats very well, after being kept three or four days to lessen its natural toughness. These birds are also found in places less cold than the mountainous deserts; but here, indeed, they are something different, having on the forehead a kind of cartilaginous horn; but both these and the other species have a crest on their head.

THE gardens of all kinds in the villages are much frequented by a bird very remarkable both for its smallness and the vivid colours of its feathers. It is generally called picaflores, or flower-peckers, from its hovering over them, and sucking their juices without lacerating or so much as disordering them. Its proper name is quinde, though it is also known by those of Rabilargo and Lisongero, and in England by that of humming bird. Its whole body, with its plumage, does not exceed the bigness of a middle-sized nutmeg; the tail is usually near three times the length of the whole body yet has but few feathers; its neck is short; the head proportioned, with a very brisk eye; the bill long and slender, white at the beginning and black at the end: the wings are also long and narrow. Most of the body is green, spotted with yellow and blue. Some are higher coloured than others; and all are variegated with streaks as it were of gold. Of this bird also there are various species, distinguished by their size and colours. This is thought to be the smallest of all known birds; the

female

female lays but two eggs at a time, and those no bigger than peas. They build in trees, and the coarsest materials of their nests are the finest straws they can pick up.

In the parts of this country, which are neither taken up by mountains nor forests, only tame animals are met with ; whence it is probable, that formerly its native species were but very few ; most of these having been introduced by the Spaniards, except the llama, to which the Indians added the name of runa, to de- note an Indian sheep, that beast being now understood by the runa-llama ; though properly llama is a general name importing beast, in opposition to the human species. This animal, in several particulars, resembles the camel : as in the shape of its neck, head, and some other parts ; but has no bunch, and is much smaller ; cloven-footed, and different in colour : for though most of them are brown, some are white, others black, and others of different colours : its pace resembles that of a camel, and its height equal to that of an ass betwixt a year and two old. The In- dians use them as beasts of carriage ; and they an- swer very well for any load under a hundred weight. They chiefly abound in the jurisdiction of Riobamba, there being scarce an Indian who has not one for car- rying on his little traffick from one village to another. Anciently the Indians used to eat the flesh of them, and still continue to make that use of those which are past labour. They say there is no difference betwixt it and mutton, except that the former is something sweeter ; it is a very docile creature, and easily kept. Its whole defence is, to eject from its nostrils some viscosities, which are said to give the itch to any on which they fall ; so that the Indians, who firmly be- lieve this, are very cautious of provoking the llama.

In the southern provinces of Peru, namely, in Cusco, La Paz, La Plata, and the adjacent parts, are two other animals, not very different from the llama :
                                              these

these are, the vicuna and the guanaco: the only difference between them being, that the vicuna is something smaller, its wool shorter and finer, and brown all
over the body, except the belly, which is whitish.
The guanàca on the contrary is much larger, its wool
long and harsh; but the shape of both is pretty near
alike. These last are of great service in the mines,
carrying metals in such rugged roads as would be im-
practible to any other beast.

In the houses is bred a creature called chucha; but
in the other southern provinces it is known by the Indian name of muca muca; it resembles a rat, but
considerably bigger. with a long snout, not unlike that
of a hog; the feet and tail are exactly the same as
those of a rat: but the hair is longer and black. In
the lower part of its belly, from the beginning of the
stomach to the natural orifice of the sex, runs a sort of
bag, formed of two membranous skins, which growing from the lower ribs, and joining in the middle,
follow the conformation of the belly, which they inclose: in the middle of it is an aperture extending
about two-thirds of its length, and which the creature
opens and shuts at pleasure by means of muscles,
doubtless formed by nature for this purpose. After
bringing forth her young, she deposits them in this
bag, and carries them as a second pregnancy till they
are fit for weaning; she then relaxes the muscles, and
the young come out as a second brood. Monsieun
de Jussieu and M. Seniergues, when at Quito, made
an experiment, at which Don George Juan and I
were both present. The dam had been dead three
days, and began to smell very disagreeably; the orifice
of the bag remained still shut, but the young ones we
found full of life within, each with a teat in its mouth;
from which, at the time we took them off, some small
drops of milk came out. The male I never saw:
but was told that it was of the same bigness and
shape as the female, except the bag; the testicles

3   of

of this creature are of an enormous disproportion, being of the size of a hen's egg. It is a very fierce enemy to all tame birds, and does a great deal of damage in the maize fields. The Indians eat the flesh, and say it is not at all disagreeable: but few Europeans have much veneration for their taste or cookery.

CHAP. IX.

Phænomena observed in the mountainous Deserts and other Parts of this Province. Hunting Matches. Dexterity of the American Horses.

TO the before-mentioned particulars of the mountainous deserts, I shall subjoin the phænomena seen there, as subjects equally meriting the curiosity of a rational reader. At first we were greatly surprized with two, on account of their novelty; but frequent observations rendered them familiar. One we saw in Pambamarca, on our first ascent thither; it was a triple circular iris. At break of day the whole mountain was encompassed with very thick clouds, which the rising of the sun dispersed so far as to leave only some vapours of a tenuity not cognizable by the sight: on the opposite side to that where the sun rose, and about ten toises distant from the place where we were standing, we saw, as in a looking-glass, the image of each of us, the head being as it were the centre of three concentric iris's: the last or most external colours of one touched the first of the following; and at some distance from them all, was a fourth arch entirely white. These were perpendicular to the horizon; and as the person moved, the phænomenon moved also in the same disposition and order. But what was most remarkable, though we were six or seven together, every one saw the phænomenon with
regard

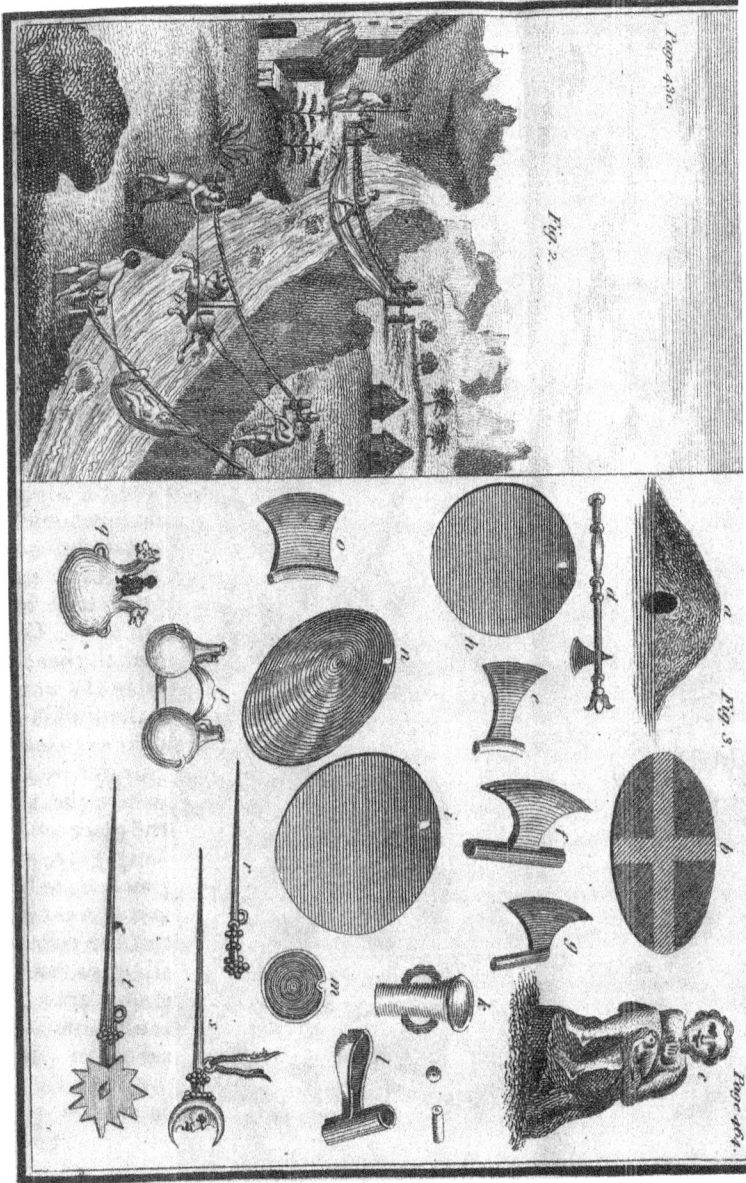

Fig. 2.

Fig. 3.

See Explanation of the Plates.

regard to himself, and not that relating to others. The diameter of the arches gradually altered with the ascent of the sun above the horizon; and the phænomenon itself, after continuing a long time, insensibly vanished. In the beginning, the diameter of the inward iris, taken from its last colour, was about five degrees and a half; and that of the white arch, which circumscribed the others, not less than sixty-seven degrees. At the beginning of the phænomenon, the arches seemed of an oval or elliptical figure, like the disk of the sun; and afterwards became perfectly circular   Each of the least was of a red colour, bordered with an orange; and the last followed by a bright yellow, which degenerated into a straw colour; and this turned to a green.   But in all, the external colour remained red.

On the mountains we also had frequently the pleasure of seeing arches formed by the light of the moon; particularly one on the 4th of April, 1738, about eight at night, on the plain of Turubamba.   But the most singular was one seen by Don George Juan, on the mountain of Quinoa-loma, on the 22d of May, 1739, at eight at night.   These arches were entirely white, without the mixture of any other colour; and formed along the slope or side of a mountain.   That which Don George Juan saw, consisted of three arches, touching in the same point: the diameter of the inner arch was sixty degrees; and the breadth of the white mark, or delineation, took up a space of five degrees; the two others were, in every respect, of the same dimensions.

The atmosphere, and the exhalations from the soil, seem more adapted than in any other place for kindling the vapours; meteors being here more frequent, and often very large; last longer, and are nearer the earth, than the like phænomena seen in other parts. One of these inflammations, of a very extraordinary largeness, was seen at Quito whilst we were there.
                                                                    I can-

I cannot exactly determine the date of its appearance, the paper on which I had wrote an account of it being lost, when I was taken by the English: but the particulars, which I remember, are as follow.

About nine at night, a globe of fire appeared to rise from the side of mount Pichincha; and so large, that it spread a light all over the part of the city facing that mountain. The house where I lodged looking that way, I was surprised with an extraordinary light darting through the crevices of the window shutters. On this appearance, and the bustle of the people in the streets, I hastened to the window, and came time enough to see it in the middle of its career, which continued from west to south, till I lost sight of it, being intercepted by the mountain of Panecillo, which lies in that quarter. It was round, and its apparent diameter about a foot. I said that it seemed to rise from the sides of Pichincha: for, to judge from its course, it was behind that mountain where this congeries of inflammable matter was kindled. In the first half of its visible course, it emitted a prodigious effulgency; then gradually began to grow dim, so that at its occultation behind the Panecillo, its light was very faint.

I shall conclude this chapter with an account of the manner of hunting, which is the only diversion in the country; and in which they passionately delight. Indeed the most remarkable circumstance in it is the ardour and intrepidity of the hunters; and which a stranger, at first, will naturally consider as mere rashness, till he sees persons of the greatest prudence, after having made one single trial, join in these parties; trusting entirely to their horses; so that it is rather to be termed a dextrous and manly exercise, and proves the superiority both of the riders and horses to the most celebrated in Europe; and that the boasted fleetness of the latter is dulness, when compared to the

the celerity with which those of America run over mountains and precipices.

The hunting is performed by a great number of people, who are divided into two classes; one on horseback, the other on foot, who are generally Indians. The business of the latter is, to rouse the beast; and that of the others, to hunt it. They all, at break of day, repair to the place appointed, which is generally on the summit of the paramos. Every one brings his greyhound; and the horsemen place themselves on the highest peaks, whilst those on foot range about the breaches, making a hideous noise in order to start the deer. Thus the company extend themselves three or four leagues, or more, according to their numbers. On the starting of any game, the horse which first perceives it sets off; and the rider, being unable to guide or stop him, pursues the chace sometimes down such a steep slope, that a man on foot, with the greatest care, could hardly keep his legs; from thence up a dangerous ascent, or a long side of a mountain, that a person, not used to this exercise, would think it much safer to throw himself out of the saddle, than commit his life to the precipitate ardour of the horse. Thus they continue till they come up with the game, or till, after following it four or five leagues, the horses tire. Those in the other stations, on perceiving one horse on its speed, immediately start; and thus the whole company are soon in motion; some hastening to meet the beast, and others following the chace; so that in such multitudes it is very seldom his good fortune to escape. The horses here do not wait for the riders to animate them; they set forward immediately on seeing another on full speed on a different mountain, or at the shouts of the huntsmen or cries of the dogs, though at ever so great a distance, or even by observing in a dog the least motion that he scents the game. One such circumstance is sufficient for these horses: and it then becomes prudence in the

rider

rider to give him his way, and at the same time to let him feel the spur to carry him over the precipices. But, at the same time, let him be very attentive to keep the saddle; for on such declivities the least neglect throws the rider over the horse's head : the consequence of which, either by the fall or by being trampled upon, is generally fatal. These horses are called paramos, being backed and exercised in running over such dangerous places. Their usual pace is trotting. There is indeed another species called aguilillas, equally remarkable for their swiftness and security. Though the aguilillas only pace, they equal the longest trot of the others: and some of them are so fleet, that no other horse can match them even at full gallop. I once was master of one of this kind; and which, though none of the racers, often carried me in twenty-nine minutes from Callao to Lima, which is two measured leagues and a half, though notwithstanding great part of the road was very bad and stony; and in twenty-eight or twenty-nine minutes brought me back again, without ever taking off the bridle. This I can assert from my own experience. These horses are very seldom known to gallop or trot; and it is a very difficult matter even to bring them to it by teaching, though the trotting horses soon come into pacing. The pace of the aguilillas is by lifting up the fore and hind leg of the same side at once; but instead of putting the hinder foot in the place where the fore foot was, as is the usual way of other pacing horses, they advance it farther, equal to that on the contrary side, or something beyond it; that thus, in each motion, they advance twice the space of the common horses. Besides, they are very quick in their motions, and remarkably easy to the rider.

OTHER horses, not of this breed, are taught the same manner of pacing, and perform it with ease and expedition, as those in whom it is a natural quality:

neither

neither species are handsome, but very gentle and docile; full of spirit and intrepidity.

CHAP. X.

*A short Account of the many Silver and Gold Mines
in the Province of Quito ; and the Method of extracting the Metal.*

THE chief riches of the kingdom of Peru, and
the greatest part of the Spanish possessions on
the continent, being the mines, which spread their
ramifications through the whole extent of these countries ; that province is justly accounted the most valuable where the mines are most numerous, or at
least where the greatest quantity of metal is procured.   The fertility of the soil, the exuberant harvests with which the labourer's toil is rewarded, would
lose much of their advantage, had not the precious
contents in the bowels of the earth exercised the
ingenuity of the miner.   The fertile pastures which
so richly cover the country, are disregarded, if the
stones upon trial are not found to answer the avidity of the artists: and the plentiful productions of
the earth, which are in reality the most excellent
gifts of nature, for the support and comfort of human life, are undervalued and slighted, unless the
mountains contain rich veins of a fine silver.   Thus,
contrary to the nature of things, the name of rich
is bestowed on that province where most mines are
worked, though so entirely destitute of the other more
necessary products, that the great number of people,
employed in the mines, are under a necessity of being
supplied from other parts : and those provinces, whose
pastures are covered with flocks, and herds, whose
fields yield plentiful harvests, and their trees bend
beneath rich fruits, under the fertilizing influence of
a be-

a benign climate, but destitute of mines, or forgotten through neglect, are looked upon as poor: and indeed, except in the plentiful surface of the earth, make no wealthy appearance. This is the case here; and the reason of it is evident: those countries are as staples for silver and gold, which are taken from the bowels of the earth only to be sent into distant nations with all possible diligence, their native country being that where they make the least stay: and the same practice is observed to be carried on, no less eagerly, throughout every town and village in the Indies: for, as they cannot well do without European goods, the gold and silver of America must be paid in exchange for them.

In a province where no mines are worked, the fertility of the soil, and goodness of its products are neglected; for the scarcity of money reduces them to such a low price, that the husbandman, for want of an incentive to any assiduous industry, instead of sowing and planting all he could, consults only what he may vend according to the common consumption, besides what is necessary for the support of his family. And as the whole return of what he receives for his fruits and grain, even when he is so fortunate as to export any, goes away again in exchange for European goods, the scarcity of money still continues, and he is so poor as sometimes possibly to want even necessaries. It is otherwise in provinces abounding with mines; for these being the objects of the attention and labours of its inhabitants, there is a continual circulation of money. What is carried out, is replaced by that drawn from the mines. Nor are they even in want of European goods, or the produce of the more fertile countries, plenty of traders from all parts resorting to places near the mines, as the original seats of gold and silver. But that province where the richness of the mines and of the soil concenter, is doubtless preferable to those where nature has given
only

only one of these advantages. Quito may justly be classed among the former, being that province which of all Peru is the most fertile in grain and fruits ; the most populous, and especially in Spaniards; abounds most in cattle; has the most manufactures, and excels in them ; and in mines, if not the richest, yet equal to any of the others, on which nature has poured out these her choicest favours. But it seems as if nature, unwilling to distinguish this by an absolute happiness, has denied it a suitable concourse of people, that it might not at once have a full enjoyment of all the benefits lavished on it, there being no reason which can disculpate the inhabitants of Quito in the neglect of the mines. For though the number of them discovered be very great, and afford a very probable conjecture that the Cordilleras must contain many more; yet very few are worked, particularly within these jurisdictions. Thus the riches of the country lie buried, and without them the fertility of the soil cannot supply their want; so as to spread through the province an opulence like that observable in the other provinces of Peru, where, by the circulation of silver, there is an universal appearance of affluence, gaiety, and splendour.

Of the great number of mines within the province of Quito, some were formerly worked, which at present are abandoned. The country then was sensible of its advantage; and the remembrance of the general opulence of those times, resulting from the riches taken out of the mines, still subsists. Not only the capital, but the towns and villages were then very populous: and many of its inhabitants were famous all over Peru for their prodigious wealth. The rich mines within the jurisdiction of Mecas, were irrecoverably lost by a revolt of the Indians; and in process of time the very remembrance of their situation was obliterated. The mines of Zaruma have been abandoned, the art of working the ore being lost, for want of a sufficient number of people to apply themselves to it: and the same decline is now seen all over the province. The fertility, as natural to the climate, still continues in all its plenty: but scarce the shadow of its former lustre and magnificence remains; and that enormous wealth, in which it gloried, is now no more. For if its products and manufactures bring in considerable quantities of silver from Lima and Valles, all is expended on European goods; so that, as I observed, little of that gold and silver, so common in the more southern provinces, is to be seen here.

The only part of the province of Quito, which, under this unhappy change, preserves its ancient opulence, is the department within the government of Popayan, which throughout abounds in gold mines, and great numbers of them are still worked. To gratify the curious, I shall give an account of the principal, and the manner of working the gold ore; as it is different from that used in the mines of Caxa. After which, I shall mention the other mines known within that province.

Every part of the jurisdiction of Popayan abounds in mines of gold; and though in some departments more are worked than in others, yet they all yield gold: and new mines are daily discovered and worked; which, under all the inclemencies of the air, in some parts fills its towns with inhabitants. Among the departments belonging to the province of Quito, the richest in gold are those of Cali, Buga, Almaguar, and Barbacoas, some of its mines being always more or less worked; and with this singular advantage in its gold, of never being mixed with any heterogeneous body; consequently no mercury is requisite in extracting it.

THE

THE gold mines in these parts are not Caxa mines, as those of silver and many of gold are; that is, they are not contained and confined as it were betwixt two natural walls; but the gold is found dispersed and mixed with the earth and gravel; as sands are found mingled with earths of different species. Thus the whole difficulty consists in separating the grains of gold from the earth; and this is very easily done, though otherwise it would be impracticable, by running conduits of water. This method is also equally necessary in the Caxa mines, where the silver and gold are intimately united with other bodies, as, after having gone through the operation of the quicksilver, which their quality renders indispensable, it is washed in order to separate the remaining filth. After the last operation the amalgama is pure, consisting entirely of quicksilver, and gold or silver, according to the species which has been worked.

THE manner, throughout the whole jurisdiction of Popayan, for extracting the gold, is, to dig the ore out of the earth, and lay it in a large cocha, or reservoir made for that purpose; and when this is filled, water is conveyed into it through a conduit: they then vigorously stir the whole, which soon turns to a mud, and the lightest parts are conveyed away through another conduit, which serves as a drain; and this work is continued till only the most ponderous parts, as little stones, sand, and the gold, remain at the bottom.   The next part of the progress is, to go into the cocha with wooden buckets made for this purpose, in which they take up the sediment; then moving them circularly and uniformly, at the same time changing the waters, the less ponderous parts are separated; and at last the gold remains at the bottom of the bucket, clear from all mixture.   It is generally found in grains as small as those of sand; and for that reason called oro en polvo; though sometimes

times pepitas, or seeds, are found among it, of different sizes, but generally they run small. The water issuing from this cocha is stopped in another contrived a little beneath it, and there undergoes a like operation; in order to secure any small particles of gold, which, from their extreme smallness, might be carried off by the current of the water being mixed with earth and other substances: and lastly, this water is passed into a third cocha. But the savings here are generally inconsiderable.

THIS is the method practised in all the mines belonging to the jurisdiction of Popayan. The labourers are Negro slaves, purchased by the owners: and whilst some are employed in washing, others bring earth; so that the washers are kept in continual employment. The fineness of this gold is generally of twenty-two carats; sometimes more, even to twenty-three: sometimes indeed it is under, though very seldom below twenty-one.

IN the district of Choco are many mines of Lavadero, or wash gold, like those we have just described. There are also some, where mercury must be used, the gold being enveloped in other metallic bodies, stones, and bitumens. Several of the mines have been abandoned on account of the platina; a substance of such resistance, that, when struck on an anvil of steel, it is not easy to be separated; nor is it calcinable; so that the metal, inclosed within this obdurate body, could not be extracted without infinite labour and charge. In some of these mines the gold is found mixed with the metal called tumbaga, or copper, and equal to that of the East; but its most remarkable quality is, that it produces no verdigrease, nor is corroded by any acids, as common copper is well known to be.

THE gold taken out of all these lavaderos, or mines, in the province of Quito, is partly circulated
in

in it: but after no long stay, like the other gold of these countries, goes away to Lima; yet these circulations, however temporary, preserve it from that decay which other parts have felt. A large quantity of this gold is carried to Santa Fé or Carthagena, so that Quito sees very little of it.

In the district of the town of Zaruma, within the jurisdiction of Loxa, are several gold mines worked; and though of no great fineness, being only betwixt sixteen and eighteen carats, they are so rich, that, when refined to twenty carats, they prove more advantageous to the miners than those where the gold is naturally of that fineness, but less abundant. Anciently it was usual to work veins; but the inhabitants are now so indolent, that most of them are neglected. These ores are worked with quicksilver; and all the mines here are Caxa mines. Of the same kind also are other gold mines within the jurisdiction of the government of Jaen Bracamoros, which, about eighty or a hundred years ago, yielded great quantities of metal. But the Indians of those parts, encouraged by the success of their brethren of Macas, having revolted, the situation of them was entirely forgotten; and no care has since been taken to search after them. The gold extracted from these mines, though not so fine as that of Popayan, far exceeded the Zaruma gold. The Indians still extract some small quantities, when absolute necessity drives them to this resource for paying the tribute. In order to this, they go to some brook or river, and there wait till it overflows its bank, then wash the sands till they have procured a sufficient quantity to answer their present necessity; then they immediately leave off, not thinking it worth while to fatigue themselves any longer about it. Several mines discovered all over this province, have undergone the same fate. One of these was in the jurisdiction of the town of Latacunga,

G g 3   near

near the village of Angamarca; the owner of which was an inhabitant of the village called Sanabria. The quantity of metal he procured from it was so great, that in order to lose no time, he caused it to be worked day and night, and had for that purpose a great number of Negro slaves, who laboured in the night; and the Indians continued the work in the day time. But in the height of his prosperity, the mine in a violent storm gave way, and sunk so low, that, though frequent searches have been made after it, the vein could not be found. At last, in the year 1743, a person discovered it by an accident of the same nature that had destroyed it; a violent tempest happened, during which, a torrent of water gushed out through the former entrance of the mine. The person, interpreting this accident as a providential indication, immediately undertook the working of it; and it has fully answered his expectations.

Within the jurisdiction of this province are many other mines, which appear to have been worked at different times, and to have yielded a great quantity of metal. The nature of the country seems best adapted to gold mines; though there are several silver veins, which appear to be very rich: and accordingly an account of them is entered in the several revenue offices, and in the records of the audience of Quito. Some have been lately worked, though with little encouragement: of this number may be said to be that of Guacaya, in the jurisdiction of Zicchos, on the frontiers of Latacunga; and another likewise of silver, about two leagues from the former. Both were worked some time; but never beyond the surface of the earth, the undertakers not having a sufficient stock of their own to work them in form; and the assistance they solicited was denied. The most celebrated silver mine in all this district is that called Sarapullo, about eighteen leagues from the same town

of

of Zicchos. This also was opened, but discontinued through the instability of the undertaker, and the want of proper assistance.

In the other jurisdiction, as well as in that of Latacunga, are all the indications of rich mines, though the number of them discovered is much less. The mountain of Pichincha is, by the inhabitants of Quito, thought to contain immense treasures: and the grains of gold, found in the sands of the waters which issue from it, greatly countenance the opinion; though there is not the least vestige all over the mountain, that formerly any mine was discovered or worked there. But the latter is no great objection : as the disruptions caused by storms, or process of time, are such as sometimes might entirely choak them up, and cover them so as to leave no traces of their existence; and a suitable diligence and care have not been used for the discovery of any. Besides this mountain, its whole Cordillera, together with the eastern chain of Guamani, and many other parts, equally abound with the like appearances of rich mines.

In the districts of Otabalo, and the town of San Miguel de Ibarra, in the territories of the village of Cayambe, along the sides and eminences of the vast mountain Cayamburo, are still remaining some monuments in confirmation of the tradition, that, before the Conquest, mines were worked there, which yielded a vast quantity of metal. Among several mountains near the village of Mira, famed for their ancient riches, is one called Pachon, from which an inhabitant of that village is certainly known to have collected, a few years ago, a vast fortune. None of these are worked; a particular nothing strange to him who sees how the mines lately discovered are neglected, though their quality is sufficiently known.

THE whole country of Pallactanga, in the jurisdiction of the town of Rio Bamba, is full of mines of
gold and silver; and the whole jurisdiction abounds
with them to such a degree, that one person with
whom I was acquainted in that town, and who, by
his civilities to us and the French academicians,
seemed to have a soul suitable to his opulence, had
entered, on his own account, at the mine-office of
Quito, eighteen veins of gold and silver, and all of
a good quality. The ore of one of these veins, by
the miners called negrillos, being assayed at Lima, in
1728, it appeared, from a certificate of Don Juan
Antonio de la Mota Torres, that it produced eighty
marks of silver per chest; a very astonishing circumstance, the usual produce in rich mines being only
eight or ten marks per chest, each chest containing
fifty quintals of ore. This is the case of Potosi and
Lipes, which, after the expence of carrying the ore
to other places, in order to its being refined, and other
charges, not only answers them all at ten marks per
chest, but the surplus is then very considerable. There
are likewise other mines where, after being refined, a
chest yields only five or six marks of silver, and in some
only three; which yet will bear the expence of refi-
ning, being in a cheap country, where great numbers
of people are willing to work for low wages. Besides the riches contained in the mountains belonging to the jurisdiction of Cuença, though this rests
only on an old Indian tradition, several mines have
lately been discovered and worked, but not with the
care requisite to reap all the advantages they offer.
One of these was in the district of Alausi, at about
six leagues from a plantation called Susna; the
owner of which, during the intervals of rural labour, used to employ his Indians and Negroes in
taking out the ore, which he found to be very rich:
but, for want of a sufficient fund to prosecute this

1   work,

work, and at the same time not neglect his plantation, he never was able to get from the mine that immense quantity of silver which its richness seemed to promise, if worked in form. All that country is indeed so full of mines, that, with an industrious turn in the minds of the inhabitants, they would be found in number and richness to equal those which have proved the sources of such infinite wealth to the southern provinces of Peru: but it is far otherwise. This supineness is thought to be owing to the great plenty; and consequently a low rate of all kinds of provisions: for the inhabitants, having all they desire for little or nothing, cannot be prevailed on to slave in digging the earth for gold: whence the inhabitants of the cities and towns are hindered from acquiring large fortunes, and consequently increasing them, by undertaking to work more mines. Add to this the prejudice, or rather apprehension of the difficulties; which are thought so great, that when a person expresses his intention of working in some mine, others look upon him as a man running headlong to his destruction, and who risks certain ruin for remote and uncertain hopes. They endeavour therefore to divert him from his purpose; and if they cannot succeed in this, they fly from him as if they were afraid lest he should communicate the infection to them. It is not therefore strange that these mines, so rich in all appearance, should be neglected, and no person found desirous of reaping the great advantages which would doubtless result from working them. This occupation, for want of being sufficiently acquainted with it, is universally dreaded: whereas in the southern provinces of Quito it is quite otherwise; the celebrated miners being men of great power, vast fortunes, and the most eminent families in the country. Besides which, are

great

great numbers of other miners of more limited circumstances, all eagerly embracing any opportunity of employing their substance in undertaking mines.

THE governments of Quijos and Majos are no less abundant in mines than the jurisdictions of Quito; those in Jaen are of infinite richness; and those of Maynas and Atacames not inferior to them. With regard to the first, it is very well known, that the Indians on the banks of the Maranon, by washing the sands of some of the rivers running into it, procure what gold they want, though their desires in this point are as moderate as the avidity of other nations are insatiable. This gold is an evident sign that the adjacent country abounds in mines. As to the second, experience has shewn that the borders of the rivers of Santiago and Mira are full of veins of gold, the Mulattos and Mestizos supplying themselves with that metal by washing the sands. But neither of them have applied themselves to discover the original veins. Besides gold and silver mines, the province of Quito has also those of other metals, and quarries of fine stone : but these are utterly disregarded by the inhabitants. Yet this province could not attain the complete possession of its riches, if to the mines of gold and silver, nature had not added those materials which are necessary in extracting the treasures they contain, and in the other services of life : nor could this country be properly said to be rich in mines, if it afforded only those of gold and silver; but nature, that there might be no deficiency in her gift, hath also furnished it with mines of azogue or quicksilver, which are found in the southern extremity of the province, near a village of the same name belonging to the jurisdiction of Cuenca. Formerly the quicksilver for the gold and silver mines was furnished from hence; but this has been suppressed;

so

so that at present only those of Guança Velica are allowed to be worked: by which means a stop has been put to those frauds discovered in the payments of the fifths; the miners; instead of applying to the mine-offices within their department, or the principal staple, supplying themselves with contraband mercury. And the end has been fully answered with regard to the revenue; frauds being now much more difficult, and consequently less frequent, since the quicksilver can be had only from one mine, than when several were open. But at the same time it is certain, that this prohibition was the principal cause of the decay of the silver mines in the province of Quito: and had the case been duly examined, many other remedies might have been found to prevent these clandestine practices, besides an absolute obstruction to so great a part of the riches of that country.

It is the opinion of some naturalists, and the marks of it are indeed very evident, that the ground on which the city of Cuenca stands, is entirely an iron mine, its veins shewing themselves in the chasms of some breaches; and the pieces taken out of the sloughs prove it beyond dispute, not only by their colour and weight, but by being attracted by the magnet, when reduced to small pieces; and many intelligent persons in these species of mines affirm, that it not only is an iron-mine, but also of extreme richness; though this has not been ascertained by experiment.

It is also equally unquestionable, that, were it possible to turn the industry of the inhabitants into this channel, mines of copper, tin, and lead, might also be discovered, though no such thing is at present known. But it is natural to suppose, that, where there are so many mines of the most precious metals, those of copper and lead are not wanting. In the next chapter I shall give some account of other mines; together with the quarries of curious stone, and several

veral ancient monuments of antiquity, that nothing
may be wanting towards the complete knowledge of
this province, from which Spain derives such great
advantages.

CHAP. XI.

*Monuments of the ancient Indians, in the Juris-
diction of Quito. Account of the several Gems
and Quarries found near that City.*

THE ancient inhabitants of Peru were far enough
from carrying the sciences to any perfection,
before the conquest of the country by the Spaniards.
They were not destitute of all knowledge of them;
but it was so faint and languid, that it was far from
being sufficient for cultivating their minds. They
had also some glimmerings of the mechanic arts;
but their simplicity, or want of taste, was so remarkable, that, unless forced by absolute necessity, they
never departed from the models before them. The
progress and improvements they made were owing
to industry, the common directress of mankind. A
close application supplied the want of science. Hence,
after a long series of time, and excessive labour, they
raised works, not so totally void of art and beauty,
but that some particulars raise the admiration of an
attentive spectator. Such, for instance, were some
of those structures of which we have still superb ruins,
in which, considering the magnitude of the works,
and the few tools they were masters of, their con-
trivance and ingenuity are really admirable. And
the work itself, though destitute of European symmetry, elegance, and disposition, is surprizing, even
in the very performance of it.

THESE

THESE Indians raised works both for the convenience and veneration of posterity. With these the plains, eminences, or lesser mountains, are covered; like the Egyptians, they had an extreme passion for rendering their burial-places remarkable. If the latter erected astonishing pyramids, in the centre of which their embalmed bodies were deposited; the Indians, having laid a body without burial in the place it was to rest in, environed it with stones and bricks as a tomb; and the dependents, relations, and intimate acquaintance of the deceased, threw so much earth on it as to form a tumulus or eminence which they called guaca. The figure of these is not precisely pyramidical; the Indians seeming rather to have affected the imitation of nature in mountains and eminences. Their usual height is about eight or ten toises, and their length betwixt twenty and twenty-five, and the breadth something less; though there are others much larger. I have already observed, that these monuments are very common all over this country; but they are most numerous within the jurisdiction of the town of Cayambe, its plains being as it were covered with them. The reason of this is, that formerly here was one of their principal temples, which they imagined must communicate a sacred quality to all the circumjacent country, and thence it was chosen for the burial-place of the kings and caciques of Quito; and in imitation of them the caciques of all these villages were also interred there.

THE remarkable difference in the magnitude of these monuments seems to indicate that the guacas were always suitable to the character, dignity, or riches of the person interred; as indeed the great number of vassals under some of the most potent caciques, concurring to raise a guaca over his body, it must certainly be considerably larger than that of
a pri-

a private Indian, whose guaca was raised only by his family and a few acquaintance: with them also were buried their furniture, and many of their instruments both of gold, copper, stone, and earth: and these now are the objects of the curiosity or avarice of the Spaniards inhabiting the country; that many of them make it a great part of their business to break up those guacas, in expectation of finding something valuable: and, misled by finding some pieces of gold here and there, they so devote themselves to this search, as to spend in it both their substance and time: though it must be owned, that many, after a long perseverance under disappointments, have at length met with rich returns for all their labour and expence. Two instances of this kind happened while we were in the country; the first guaca had been opened near the village of Cayambe, in the plain of Pesillo, a little before our arrival at Quito; and out of it were taken a considerable quantity of gold utensils; some of which we saw in the revenue-office, having been brought there as equivalents for the fifths. The second was more recently discovered in the jurisdiction of Pastos, by a Dominican friar, who, from a turn of genius for antiquities, had laid out very large sums in this amusement; and at last met with a guaca in which he is said to have found great riches. This is certain, that he sent some valuable pieces to the provincial of his order, and other persons at Quito. The contents of most of them consist only of the skeleton of the person interred; the earthen vessels in which he used to drink chica, now called guaqueros; some copper axes, looking-glasses of the ynca-stone, and things of that kind, being of little or no value, except for their great antiquity, and their being the works of a rude illiterate people.

THE

THE manner of opening the guacas is, to cut the lower part at right angles, the vertical and horizontal line meeting in the centre, where the corpse and its furniture are found.

THE stone mirrors taken out of the guacas are of two sorts; one of the ynca-stone, and the other of the gallinazo-stone : the former is not transparent, of a lead colour, but soft ; they are generally of a circular form, and one of the surfaces flat, with all the smoothness of a crystal looking-glass ; the other oval and something spherical, and the polish not so fine. They are of various sizes, but generally of three or four inches diameter, though I saw one of a foot and a half ; its principal surface was concave, and greatly enlarged objects ; nor could its polish be exceeded by the best workmen among us. The great fault of this stone is, its having several veins and flaws, which, besides the disadvantage to the surface of the mirror, render it liable to be broken by any little accident. Many are inclined to think that it is not natural, but artificial. There are, it must indeed be owned, some appearances of this, but not sufficient for conviction. Among the breaches in this country, some quarries of them are found ; and quantities continue to be taken out, though no longer worked for the use the Indians made of them. This does not however, absolutely contradict the fusion of them, in order to heighten their quality, or cast them into a regular form.

THE gallinazo-stone is extremely hard, but as brittle as flint : it is so called from its black colour, in allusion to the colour of the bird of that name ; and is in some measure diaphanous. This the Indians worked equally on both sides ; and reduced it into a circular figure. On the upper part they drilled a hole for a string to hang it by ; the surfaces were as smooth as those of the former, and very exactly reflect objects. The mirrors made of this stone were of different kinds, some plain, some concave, and others convex. I have seen them of all kinds: and from the delicacy of the workmanship one would have thought these people had been furnished with all kinds of instruments, and completely skilled in opticks. Some quarries of this stone are likewise met with; but they are entirely neglected, though its transparency, colour, and hardness, besides its having no flaws or veins, render it very beautiful.

The copper axes of the Indians differ very little in their shape from ours: and it appears that these were the instruments with which they performed most of their works: for if not the only, they are the most common edge-tools found among them; and the whole apparent difference betwixt those they use, consists only in size and shape: for though they all resemble an axe, the edge in some is more circular than in others. Some have a concave edge, others a point on the opposite side, and a fluted handle These instruments were not all of copper, some having been found of gallinazo, and of another stone something resembling the flint, but less hard and pure. Of this stone, and that of the gallinazo, are several points, supposed to have been heads of spears, as these were their two chief instruments or weapons: for, had they used any other, some would doubtless have been found among the infinite number of guacas which have been opened.

The guaqueros, or drinking-vessels, are of a very fine black earth: but the place where they were made is utterly unknown. They are round, and with a handle in the middle, the mouth on one side, and on the other the head of an Indian, whose features are so naturally expressed, that very few of our workmen could equal it. Others, though of the same form, are of a red earth. Besides which there are found
larger

larger and smaller vessels of both kinds of earth used in making and keeping the chicha.

Aᴍᴏɴɢ the gold pieces are the nose-jewels, which in form resemble the foot of a chalice, and very little less : these were appended to the septum, which divides the two nostrils. There are also found collars, bracelets, and ear-pendants, resembling the nose-jewels : but all these are no thicker than paper: the idols, which are at full length, are every where hollow within; and as they are all of one piece, without any mark of soldering, the method they used in making them is not easily conceived. If it be said that they were cast; still the difficulty remains, how the mould could be of such a fragility as to be taken away without damaging works, which, in all their parts, are so extremely thin.

Tʜᴇ maize has ever been the delight of the Indians; for, besides being their food, their favourite liquor chicha was made of it; the Indian artists therefore used to shew their skill in making ears of it in a kind of very hard stone; and so perfect was the resemblance, that they could hardly be distinguished by the eye from nature; especially as the colour was imitated to the greatest perfection; some represented the yellow maize, some the white; and in others the grains seemed as if smoke-dried by the length of time they had been kept in their houses. The most surprizing circumstance of the whole is, the manner of their working, which, when we consider their want of instruments and wretched form of those they had, appears an inexplicable mystery: for either they worked with copper tools, a metal little able to resist the hardness of stones; or, to give the nice polish conspicuous on their works, other stones must have been used for tools. But the labour, time, and patience, requisite to make

Vᴏʟ. I.   H h   only

only a hole in the gallinazos, as was made in the mirrours; and much more to give their surfaces such a smoothness and polish, that they are not to be distinguished from the finest glass, must have been prodigious.   These are works which the most ingenious of our artists would be extremely at a loss to produce, if they were allowed only pieces of copper and stones without any other tools or materials.   It is the greatest proof of the ingenuity of these people, that by mere dint of genius, and unassisted by information, they should attain to such contrivances and such a delicacy of workmanship.

YET all that we have said is surpassed by the ingenuity of the Indians in working emeralds, with which they were supplied from the coast of Manta, and the countries dependent on the government of Atacames, Coaquis or Quaques.   But these mines are now entirely lost, very probably through negligence.   These curious emeralds are found in the tombs of the Indians of Manta and Acatames: and are, in beauty, size, and hardness, superior to those found in the jurisdiction of Santa Fé; but what chiefly raises the admiration of the connoisseur is, to find them worked, some in spherical, some cylindrical, some conical, and of various other figures; and all with a perfect accuracy.   But the unsurmountable difficulty here is, to explain how they could work a stone of such hardness; it being evident, that steel and iron were utterly unknown to them.   They pierced emeralds, and other gems, with all the delicacy of the present times furnished with so many tools: and the direction of the hole is also very observable; in some it passes through the diameter; in others, only to the centre of the stone, and coming out at its circumference they formed triangles at a small distance from one another: and thus the figure of the stone to give it relief was varied with the direction of the holes.

AFTER

AFTER this account of the guacas of these idolatrous nations, the custom which equally prevailed among the southern nations of Peru, I proceed to their superb edifices, whether temples, palaces, or fortresses : and though those in the kingdom of Quito are not the most stately and magnificent, the court and residence of the yncas having been in the province of Cusco; yet some of the former sufficiently denote the grandeur of the Indians who then inhabited it, and their fondness for such edifices; intending as it were to hide the rusticity of their architecture under richness and magnificence which they profusely bestowed on their edifices, whether of brick or stone.

THE greatest part of one of these works is still existing, near the town of Cayambe, being a temple built of unbaked bricks. It stands on an eminence of some height; its figure is perfectly circular, and its diameter eight toises. Of this structure nothing now remains but the walls, which are in good condition; and about two toises and a half in height, and four or five feet in thickness. The cement of the bricks is of the same earth with that of which they are made : and the hardness of them may be conceived, from remaining so long in a good condition exposed to the injuries of weather, having no cover.

BESIDES the ancient tradition that this structure was one of the temples of those times, the manner of its construction countenances such a conjecture : for its circular form, without any separation in the inside, shews it to have been a place of public resort, and not any habitation. The smallness of the door renders it probable, that, though the yncas entered into their palaces in the chairs in which they were carried, as will be seen hereafter, this place they entered on foot, in token of veneration; the dimensions of the

door

door not admitting of any other manner. And, as I have before observed, that one of the principal temples was not far from hence, this was probably the very structure.

At the extremity of the plain which runs northward from Latacunga, are still seen the walls of a palace of the yncas of Quito; and is still called by its ancient name Callo. At present it serves for the mansion-house of a plantation belonging to the Augustines at Quito. If it wants the beauty and grandeur which characterise the works of the ancient Egytians, Greeks, Romans, and other nations versed in the fine arts; yet, if we make proper allowance for the rusticity of the Indians, and compare this with their other buildings, the dignity of the prince will be abundantly conspicuous, in the prodigious magnitude of the materials, and the magnificence of the structure. You enter it through a passage five or six toises in length, leading into a court, round which are three spacious saloons, filling the three other sides of its squares. Each of these saloons has several compartments; and behind that which faces the entrance, are several small buildings, which seem to have been offices, except one; and this, from the many divisions in it, was, in all probability, a menagerie. Though the principal parts still continue, the ancient work is something disfigured, dwellings having been lately built among them, and alterations made in the chief apartments.

This palace is entirely of stone, equal in hardness to flint; and the colour almost black. They are exceedingly well cut, and joined so curiously, that the point of a knife, or even so much as a piece of the finest paper cannot be put betwixt them; so that they only shew the walls to be of different stones; and not one entire composition; but no cement is perceivable. The stones without are all of a convex figure;

The Balza described Vol. 1. Pag. 182.

Fig. 3

Fig. 4

Scale v. Strand.

Fig. 2.

Fig. 2.

See Explan. of the Plates.

Fig. 1.

figure; but at the entrance of the door are plane. But there is a visible inequality, both in the stones and in their courses; which gives a more singular air to the work; for a small stone is immediately followed by one large and ill-squared; and that above is made to fit the inequalities of the other two, and at the same time fill up all the interstices between the projections and irregularity of their faces; and this in such perfection, that, whatsoever way they are viewed, all parts appear joined with the same exactness. The height of these walls is about two toises and a half, and about three or four feet in thickness. The doors are about two toises high, and their breadth at the bottom about three or four feet; but runs narrowing upwards, where the aperture is only two feet and a half. The doors of the palaces, where the yncas resided, were made of such a height, to allow room for the chairs in which the monarch was carried on men's shoulders into his apartment, the only place in which his feet touched the ground. It is not known whether this or the other palaces of the yncas had any stories, nor how they were roofed: for those we examined were either open, or had been roofed by the Spaniards: But it is highly probable that they covered them with boards, in the form of a terrace, that is, supported by beams laid across: for in the walls there is nothing near the ground that affords room for a conjecture, that they ever supported any roofs: on this horizontal roof they contrived some slope for carrying off the waters. The reason of contracting their doors at the top was, that the lintel might be of one stone; for they had no idea either of arches or of key-stones, as may be concluded from no such works occurring among all their edifices.

About fifty toises north of this palace, fronting its entrance, is a mountain, the more singular as

being

being in the midst of a plain : its height is betwixt twenty-five and thirty toises, and so exactly, on every side, formed with the conical roundness of a sugar-loaf, that it seems to owe its form to industry; especially as the end of its slope on all sides forms exactly with the ground the same angle in every part. And what seems to confirm this opinion is, that guacas, or mausoleums, of prodigious magnitude, were greatly affected by the Indians in those times. Hence the common opinion, that it is artificial, and that the earth was taken out of the breach north of it, where a little river runs, does not seem improbable. But this is no more than conjecture, not being founded on any evident proof. In all appearance this eminence, now called Panecillo de Callo, served as a watch-tower, commanding an uninterrupted view of the country, in order to provide for the safety of the prince on any sudden alarm of an invasion, of which they were under continual apprehensions, as will appear from the account of their fortresses.

About two leagues north-east of the town of Atun-Canar, or great Canar, is a fortress or palace of the yncas. It is the most entire, the largest, and best built in all the kingdom. Close by its entrance runs a little river, and the back part of it terminates in a high and thick wall at the slope of a mountain. In the middle of it is a kind of oval tower; about two toises high from the ground within the fort, but without it rises six or eight above that of the hill. In the middle of the tower is a square of four walls; which, on the side facing the country, leave no passage; and all its angles touch the circumference of the oval. On the opposite side only, is a very narrow pass, answering to the inward part of the tower. In the middle of this square is an apartment of two small rooms, without

any

any communication; and the doors of them opposite to the space which separates them. In the
sides towards the country are loop-holes; and in
critical times it was made a court of guard. From
the outside of this oval tower, a wall is extended on
the left side about forty toises, and about twenty-
five on the right; this wall was continued in a great
number of irregular angles, and inclosed a large spot
of ground. It had only one entrance, which was
in the side opposite to the tower; and facing the
last angle on the right near the rivulet. From
this gate or entrance was a passage, just broad
enough for two persons to walk abreast; and at the
wall turned short off towards the tower; but always
of the same breadth. After this it winded towards
the breach, and widened so as to form a parade before the tower. In these passages, at the distance of
every two or three paces, one sees niches formed
within the wall, like sentry-boxes: and on the other
side two doors, which were entrances to the same
number of soldiers de logis, and seem to have served
the corps of the garrison for barracks. In the inner square, to the left of the tower, were several
apartments, of which the height, disposition, and
doors, are a sufficient proof that this was once the
prince's palace. All the walls being full of hollows,
resembling cupboards, in which, as likewise in the
two chambers of the tower, the niches, and along
the passages, were stone pegs, with a head betwixt
six and eight inches long, and three or four in diameter: the use of these probably was for hanging
up their arms.

THE whole main wall on the slope of the
mountain, and descending laterally from the oval
tower, is very thick, and the outside perpendicular. Within is a large rampart, and on it a parapet of an unusual height; and though the ram-

part

part reached quite round the wall, there was only one ascent to it, which was adjoining to the oval tower. The outward and inward walls are all of the same kind of stone, very hard and well-polished: and disposed like those of Callo. The apartments also were without ceiling or flooring, like those of the above-mentioned palace.

At Pomallacta, within the jurisdiction of the town of Guasuntos, are some rudera of another fortress like the former: and it is a common opinion here, that there was a subterraneous communication between these two fortifications; but this does not seem at all probable. For besides the distance of six leagues, the ground is very uneven, and interrupted by some of the smaller branches of the cordilleras, breaches, and brooks. The inhabitants are, however, very tenacious of their opinion: and some affirm, that a few years before our arrival in the country, a person entered this subterraneous passage at the fort of Canar, but, his light going out, he was obliged to return. They farther say, that the entrance is within the fort at the foot of the tower, where indeed there is a small low door, but now choaked up with earth; and was doubtless for some use. But this does not imply that it led to the other fortress, as, besides a great quantity of lights, there must also have been here and there vent-holes or spiracles, which, considering the mountains, is utterly impracticable.

Many other walls and ruins are seen all over the country, both in the plains, on the sides of the hills, and on their summits; but most in desert places, and without any vestige of a town or village near them; and except these three, they are either of adoves or unknown stone, without any arrangement. The more irregular are thought to be the works of Indians before they were reduced by the

yncas:

yncas: but those of Callo, and the other two fortress-es, by their superior symmetry, shew that they are of a later date, and built under the direction of the yncas, who applied themselves with exemplary at-tention to promote necessary arts throughout all their conquests; possibly from this political view, that the people, sensible of the happy change, might be the better subjects.   All these remains of an-tique edifices the Indians call Inca perca, the Yncas walls.

ANOTHER Indian method of fortification, and of which there are still some remains, was, to dig three or four ranges of moats quite round the tops of such mountains, as, though high and steep, were not subject to frosts: and every one on the inside strengthened by a parapet, whence they could safely annoy the enemy.   These they called Pucuras; and within the last range of moats they built barracks for the garrison.   These kinds of forts were so com-mon, that one scarce meets with a mountain with-out them.   On the peaks of Pambamarca, are three or four; and one of them on the place where we fixed our signal for the meridian triangles.   In like manner we found them on almost all the other mountains; and the outward moat of circumvalla-tion was above a league in extent.   The breadth and depth of each was alike; but in respect of one another, there was not the same uniformity, some of them having a breadth of two toises and even more, and others not one; and the like difference is observable in their depth.   It was, however, their constant care to make the inward bank at least three or four feet higher than the outward, to have the greater advan-tage over the assailants.

THE junction and polish so much admired in all the remaining stone-works of the Indians, plainly shew, that they made use of some stones
to

to polish others, by rubbing them together; it being highly improbable that they could bring them to such perfection with the few and awkward tools they used: as for the working of iron, they were undoubtedly strangers to it, there being many mines of that metal in this country, and not one of them with any marks of having ever been touched. And no iron was found among them at the arrival of the Spaniards. But, on the contrary, they shewed an extreme fondness for any thing made of that metal.

I HAVE already mentioned the quarries, or mines, producing the two kinds of stone of which the Indians made their mirrors; and which were those most esteemed. There are likewise quarries of other stones, which, in a country where gold and silver mines do not abound, would be thought valuable. Of these one is in the plain of Talqui, south of Cuença; out of which are taken very large and beautiful blocks of white and very clear alabaster. Its only fault is its softness: yet that is not such as to hinder all kinds of works from being made of it; or rather, its easiness contributes to their pefection: nor is there any danger of large flakes flying off, which often spoil an entire piece. The only quarries of this stone are near Cuença; but those of rock crystal I have seen in many parts, from whence I have had some very large, clear, and transparent pieces, and of a remarkable hardness: but, as it is not esteemed here, no use is made of it; so that what is found is purely by accident. In the same jurisdiction of Cuença, and about two leagues north-west of the city, not far from the villages of Racan and Saansay, is a small mountain, entirely covered with flints; mostly black, some of a reddish cast, and others whitish. But, being strangers to the manner
of

of cutting and filing them for fire-arms, the people make no use of them: and on some occasions, flints, either for muskets or pistols, have been sold at Cuença, Quito, and all over the country, for two rials each; but one is the common price of them, being brought from Europe. Consequently, as there is here a whole quarry of them, their exorbitant price is wholly owing to a want of industry, as this would in a short time render them as expert at cutting flints as the Europeans.

AFTER the mines of metals, and the quarries of large stones, it would be improper to omit the gems found in this province. I have already observed, that the jurisdiction of Atacames and Manta formerly abounded in emeralds of a fineness surpassing those of the mines of Santa Fé. Not a small number of them was destroyed by an error of the first Spaniards, who came hither, imagining that, if they were real gems, they would stand the stroke of a hammer on an anvil. The loss of the mines of Atacames, and the neglect of many others of gold and silver, was in some measure compensated by the discovery of several in the jurisdiction of Cuença; but which have been but little improved, though they exhibit the most inviting signs of their great riches, namely, fragments of rubies; and which, intelligent persons say, are very fine. These are usually found among the sands of a rapid river, not far from the village of Azogues. The Indians, and others, frequently make it their business to go and wash those sands, where they find small sparks, about the bigness of a lentil, and sometimes larger; and it is not to be questioned but these are washed away by the continual allision of the water in its passage along the mine. But the inhabitants, content with this piddling work, do not trouble themselves to trace the origin of the mine; though there is all
the

the appearance in the world that it would turn to very good account. I myself, when I was at that village, saw some of these sparks in their natural state; and both their colour and hardness sufficiently shewed that they were of a very fine sort.

ANOTHER kind of stone is found in great plenty all over this country. It is of a fine green, and harder than alabaster, though not pellucid: but no more valued than any of the former; except that a few toys or utensils are made of it.

HERE are also some mines of sulphur, and some parts afford vitriol; but no farther known than as nature has placed them in view; not only the improvement of them being entirely neglected, but scarcely any notice taken of those which lie on the surface of the ground; either because the inhabitants stand in no need of those minerals, or from their strong aversion to any thing that requires labour.

NORTH of Quito, betwixt two plantations, at the foot of mount Anlagua, one of which bears the same name, and the other that of Courogal, runs a very large river, which petrifies any wood, leaves, &c. thrown into it. I have had whole branches thus petrified; and the porosity of the stem, the fibres of the rind, even the smallest veins of the leaves, and the meander of its fibrille, equally discernible as when fresh cut from the tree. I have also had large pieces of timber petrified, which at first sight appeared to be wood thoroughly dried; no visible alteration having been made in them, except in colour.

WITH all these appearances, I cannot think that the wood, leaves, and the like, which are put into the river, are really turned into stone of such a hardness as that I experienced: but as the appearance is undeniable, I shall offer an explanation of this supposed transmutation.   IT

It must be observed, that the rocks and all the parts which this river washes, are covered with a crust of hardness little inferior to that of the main rock; and this increases its volume, and distinguishes itself from the original rock, which is something yellowish. The inference I would draw from hence is, that the water of the river is mixed with petrifying, viscid, and glutinous particles, which adhere to the body they surround : and as by their extreme subtility they insinuate themselves through its pores, they fill the place of the fibres, which the water insensibly rots off and separates, till at length all that was leaf or wood gives way to that petrifying matter; which still retains the impression of the parts of the original, with its several veins, fibres, and ramifications. For at the time of its insinuation, the ducts of the wood, or leaves, serve for a kind of mould, by which it naturally takes the entire figure of the body into which it has obtruded itself.

An observation I made with some branches confirms me in this opinion : for, having opened them, I found some leaves and bits of wood, which snapped on breaking; and the inside was as large as real stone, the texture only remaining of its first substance. But in others, the parts consolidated by the stony matter snapped; and the fibres, not having yet undergone a total corruption, retained the appearance of wood, though some were more rotten and decayed than others. I had also some leaves, the surface of which was only covered with a very fine lapideous tegument, but within were entire leaves, except here and there a little mark of decay.

It is to be observed, that this matter much more easily fastens on any corruptible substance, than on the more compact and solid, as stones, and the like: the reason of which is, that in one it meets with pores, in which it fixes itself; but

having

1

having no such hold on the harder bodies, it is soon washed off by the agitation of the water; that if now and then such crusts are seen on stones, they never make any sensible addition to their volume, though some excrement is now conspicuous from the difference of the colour; that of the petrified leaves, both within and without, is of a pale yellow; and the same prevails in the stems: though in these always with a mixture of that of the wood itself when dry.

Though all the jurisdictions of the kingdom of Quito, from N. to S. are not molested by the vicinity of wild Indians, yet it is the misfortune of the governments of Quixos and Macas, Jean and Maynas, to be surrounded and intermixed with those barbarians; so that by only passing the eastern Cordillera of the Andes, towards that part you usually meet with them: and from some parts of those eminences the smoke of their cottages may be seen. This sight is most frequently beheld from the mountain on the back of the town of Cayambe; and all along to the northward, from the village of Mira within the jurisdiction of the town of San Miguel di Ibarra. The sportsmen, when hunting on those hills, often see the smoke both on this side and likewise on the same Cordillera, from the jurisdiction of Riobamba, to that of Cuenca. The village of Mira has often been surprized with the sudden appearance of some of these Indians; but they have as suddenly turned back, and with the same haste they came. It is not uncommon for Indians of these jurisdictions, from a fondness for sloth and licentiousness, to leave their houses and go over to the savages; as among them they may, without controul, follow their natural idolatry, and give themselves up to drunkenness and all manner of vice; and, what they think a supreme happiness, be served and attended by women,

2

whose

whose office it is to take care of and support them: all their occupation being hunting, whenever compelled by necessity, or induced by a sudden fit of industry. Thus they live in a debasement of human nature; without laws or religion; in the most infamous brutality; strangers to moderation; and without the least controul or restraint on their excesses.

END OF THE FIRST VOLUME.

J. Bʀᴀᴛᴛᴇʟʟ, Printer, Marshall-Street, Golden-Square.

Printed by Printforce, United Kingdom